MATHEMATICS
Its Content, Methods, and Meaning

VOLUME THREE

MATHEMATICS
Its Content, Methods, and Meaning

EDITED BY

A. D. Aleksandrov, A. N. Kolmogorov, M. A. Lavrent'ev

TRANSLATED BY

K. Hirsch

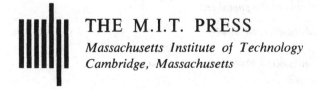

THE M.I.T. PRESS
*Massachusetts Institute of Technology
Cambridge, Massachusetts*

МАТЕМАТИКА
ЕЕ СОДЕРЖАНИЕ, МЕТОДЫ И ЗНАЧЕНИЕ
Издательство Академии Наук СССР
Москва 1956

First MIT Press paperback edition, 1969

*Translation aided by grant NSF–G 16422 from the
National Science Foundation*

Copyright © 1963 by the American Mathematical Society

*All rights reserved. No portion of this book may be reproduced
without the written permission of the publisher.*

Library of Congress Card Number: 64-7547

ISBN 978-0-262-01012-2 (hc. : alk. paper) — 978-0-262-51003-5 (pb. : alk. paper)

PREFACE TO THE RUSSIAN EDITION

Mathematics, which originated in antiquity in the needs of daily life, has developed into an immense system of widely varied disciplines. Like the other sciences, it reflects the laws of the material world around us and serves as a powerful instrument for our knowledge and mastery of nature. But the high level of abstraction peculiar to mathematics means that its newer branches are relatively inaccessible to nonspecialists. This abstract character of mathematics gave birth even in antiquity to idealistic notions about its independence of the material world.

In preparing the present volume, the authors have kept in mind the goal of acquainting a sufficiently wide circle of the Soviet intelligentsia with the various mathematical disciplines, their content and methods, the foundations on which they are based, and the paths along which they have developed.

As a minimum of necessary mathematical knowledge on the part of the reader, we have assumed only secondary-school mathematics, but the volumes differ from one another with respect to the accessibility of the material contained in them. Readers wishing to acquaint themselves for the first time with the elements of higher mathematics may profitably read the first few chapters, but for a complete understanding of the subsequent parts it will be necessary to have made some study of corresponding textbooks. The book as a whole will be understood in a fundamental way only by readers who already have some acquaintance with the applications of mathematical analysis; that is to say, with the differential and integral calculus. For such readers, namely teachers of mathematics and instructors in engineering and the natural sciences, it will be particularly important to read those chapters which introduce the newer branches of mathematics.

PREFACE TO THE RUSSIAN EDITION

Naturally it has not been possible, within the limits of one book, to exhaust all the riches of even the most fundamental results of mathematical research; a certain freedom in the choice of material has been inevitable here. But along general lines, the present book will give an idea of the present state of mathematics, its origins, and its probable future development. For this reason the book is also intended to some extent for persons already acquainted with most of the factual material in it. It may perhaps help to remove a certain narrowness of outlook occasionally to be found in some of our younger mathematicians.

The separate chapters of the book are written by various authors, whose names are given in the Contents. But as a whole the book is the result of collaboration. Its general plan, the choice of material, the successive versions of individual chapters, were all submitted to general discussion, and improvements were made on the basis of a lively exchange of opinions. Mathematicians from several cities in the Soviet Union were given an opportunity, in the form of organized discussion, to make many valuable remarks concerning the original version of the text. Their opinions and suggestions were taken into account by the authors.

The authors of some of the chapters also took a direct share in preparing the final version of other chapters: The introductory part of Chapter II was written essentially by B. N. Delone, while D. K. Faddeev played an active role in the preparation of Chapter IV and Chapter XX.

A share in the work was also taken by several persons other than the authors of the individual chapters: §4 of Chapter XIV was written by L. V. Kantorovič, §6 of Chapter VI by O. A. Ladyženskaja, §5 of Chapter 10 by A. G. Postnikov; work was done on the text of Chapter V by O. A. Oleĭnik and on Chapter XI by Ju. V. Prohorov.

Certain sections of Chapters I, II, VII, and XVII were written by V. A. Zalgaller. The editing of the final text was done by V. A. Zalgaller and V. S. Videnskiĭ with the cooperation of T. V. Rogozkinaja and A. P. Leonovaja.

The greater part of the illustrations were prepared by E. P. Sen'kin.

Moscow
1956   EDITORIAL BOARD

FOREWORD BY THE EDITOR OF THE TRANSLATION

Mathematics, in view of its abstractness, offers greater difficulty to the expositor than any other science. Yet its rapidly increasing role in modern life creates both a need and a desire for good exposition.

In recent years many popular books about mathematics have appeared in the English language, and some of them have enjoyed an immense sale. But for the most part they have contained little serious mathematical instruction, and many of them have neglected the twentieth century, the undisputed "golden age" of mathematics. Although they are admirable in many other ways, they have not yet undertaken the ultimate task of mathematical exposition, namely the large-scale organization of modern mathematics in such a way that the reader is constantly delighted by the obvious economizing of his own time and effort. Anyone who reads through some of the chapters in the present book will realize how well this task has been carried out by the Soviet authors, in the systematic collaboration they have described in their preface.

Such a book, written for "a wide circle of the intelligentsia," must also discuss the general cultural importance of mathematics and its continuous development from the earliest beginnings of history down to the present day. To form an opinion of the book from this point of view the reader need only glance through the first chapter in Part 1 and the introduction to certain other chapters; for example, Analysis, or Analytic Geometry.

In translating the passages on the history and cultural significance of mathematical ideas, the translators have naturally been aware of even greater difficulties than are usually associated with the translation of scientific texts. As organizer of the group, I express my profound gratitude to the other two translators, Tamas Bartha and Kurt Hirsch, for their skillful cooperation.

The present translation, which was originally published by the American Mathematical Society, will now enjoy a more general distribution in its new format. In thus making the book more widely available the Society has been influenced by various expressions of opinion from American mathematicians. For example, ". . . the book will contribute materially to a better understanding by the public of what mathematicians are up to. . . . It will be useful to many mathematicians, physicists and chemists, as well as to laymen. . . . Whether a physicist wishes to know what a Lie algebra is and how it is related to a Lie group, or an undergraduate would like to begin the study of homology, or a crystallographer is interested in Fedorov groups, or an engineer in probability, or any scientist in computing machines, he will find here a connected, lucid account."

In its first edition this translation has been widely read by mathematicians and students of mathematics. We now look forward to its wider usefulness in the general English-speaking world.

August, 1964

S. H. GOULD
Editor of Translations
American Mathematical Society
Providence, Rhode Island

CONTENTS

PART 5

CHAPTER XV   THEORY OF FUNCTIONS OF A REAL VARIABLE   3
S. B. Stečkin

§1. Introduction   3
§2. Sets   5
§3. Real Numbers   12
§4. Point Sets   18
§5. Measure of Sets   25
§6. The Lebesgue Integral   30
Suggested Reading   36

CHAPTER XVI   LINEAR ALGEBRA   37
D. K. Faddeev

§1. The Scope of Linear Algebra and Its Apparatus   37
§2. Linear Spaces   48
§3. Systems of Linear Equations   61
§4. Linear Transformations   74
§5. Quadratic Forms   84
§6. Functions of Matrices and Some of Their Applications   91
Suggested Reading   95

CHAPTER XVII   NON-EUCLIDEAN GEOMETRY   97
A. D. Aleksandrov

§1. History of Euclid's Postulate   98
§2. The Solution of Lobačevskiĭ   101
§3. Lobačevskiĭ Geometry   105

§4. The Real Meaning of Lobačevskiĭ Geometry 114
§5. The Axioms of Geometry; Their Verification in the Present Case 122
§6. Separation of Independent Geometric Theories from Euclidean Geometry 129
§7. Many-Dimensional Spaces 136
§8. Generalization of the Scope of Geometry 151
§9. Riemannian Geometry 164
§10. Abstract Geometry and the Real Space 178
Suggested Reading 189

PART 6

CHAPTER XVIII TOPOLOGY 193
P. S. Aleksandrov

§1. The Object of Topology 193
§2. Surfaces 197
§3. Manifolds 202
§4. The Combinatorial Method 204
§5. Vector Fields 212
§6. The Development of Topology 218
§7. Metric and Topological Spaces 221
Suggested Reading 224

CHAPTER XIX FUNCTIONAL ANALYSIS 227
I. M. Gel'fand

§1. n-Dimensional Space 228
§2. Hilbert Space (Infinite-Dimensional Space) 232
§3. Expansion by Orthogonal Systems of Functions 237
§4. Integral Equations 245
§5. Linear Operators and Further Developments of Functional Analysis 252
Suggested Reading 261

CHAPTER XX GROUPS AND OTHER ALGEBRAIC SYSTEMS 263
A. I. Mal'cev

§1. Introduction 263
§2. Symmetry and Transformations 264
§3. Groups of Transformations 273
§4. Fedorov Groups (Crystallographic Groups) 285
§5. Galois Groups 293

§6. *Fundamental Concepts of the General Theory of Groups*   297
§7. *Continuous Groups*   305
§8. *Fundamental Groups*   308
§9. *Representations and Characters of Groups*   314
§10. *The General Theory of Groups*   319
§11. *Hypercomplex Numbers*   320
§12. *Associative Algebras*   330
§13. *Lie Algebras*   339
§14. *Rings*   342
§15. *Lattices*   347
§16. *Other Algebraic Systems*   349
     Suggested Reading   351

INDEX   353

CONTENTS OF THE SERIES   373

PART 5

CHAPTER XV

THEORY OF FUNCTIONS OF A REAL VARIABLE

§1. Introduction

At the end of the 18th and the beginning of the 19th century, the differential and integral calculus was essentially worked out. Up to that time (in fact, throughout the 19th century) mathematicians were engaged in constructing its several branches, in discovering more and more new facts, and in developing more and more new domains of application of the differential and integral calculus to various problems of mechanics astronomy, and technology. Now it became possible to survey the results obtained, to study them systematically, and to delve into the meaning of the basic concepts of analysis. And here it became apparent that all was not well with the foundations of analysis.

Already in the 18th century there was no consensus among the greatest mathematicians of that time as to what a function is. This came out in prolonged controversies whether this or that solution of a problem, this or that concrete mathematical result were correct or incorrect. Gradually it became clear that also other basic concepts of analysis had to be made more precise. An inadequate understanding of the meaning of continuity and of the properties of continuous functions led to a number of erroneous statements, for example that a continuous function is always differentiable. Mathematics came to operate with such complicated functions that it became impossible to rely on intuition and guesswork. So there arose a real need to bring order into the fundamental concept of analysis.

The first serious attempt in this direction was made by Lagrange, and then Cauchy followed on the same path. Cauchy sharpened the definitions of limit, continuity, and integral and brought them into common use, as they survive to our days. Approximately at the same time, the Czech mathematician Bolzano made a rigorous study of the basic properties of continuous functions.

Let us consider these properties of continuous functions in more detail. Suppose that a continuous function $f(x)$ is given on some interval $[a, b]$, i.e., for all numbers satisfying the inequalities $a \leqslant x \leqslant b$. Previously it was regarded as obvious that if the function assumes values of opposite signs at the end points of the interval, then it must be zero at some intermediate point. Now this fact received a rigorous foundation. In the same way it was proved rigorously that a continuous foundation given on an interval assumes at certain points its greatest and its least value.

The study of these properties of continuous functions made it necessary to go deeper into the nature of the real numbers. As a result the theory of real numbers appeared; the basic properties of the numerical line were clearly formulated.

Further developments of mathematical analysis necessitated the study of more and more "bad," in particular discontinuous, functions. Discontinuous functions appear, for example, as limits of continuous functions, where it is not known *a priori* whether the limit function is continuous or not, and also in schematizing processes with sudden sharp variations. Here was a new task, namely to generalize the apparatus of analysis to discontinuous functions.

Riemann investigated the problem to what classes of discontinuous functions the concept of integral could be extended. As a result of this work on the foundation of analysis, there arose a new mathematical discipline: the theory of functions of a real variable.

If the classical mathematical analysis operates essentially with "good" (for example, continuous or differentiable) functions, the theory of functions of a real variable investigates considerably wider classes of functions. If in mathematical analysis the definition of some operation (for example integration) is given for continuous functions, then it is characteristic of the theory of functions of a real variable to find out to what classes of functions this definition is applicable, how the definition has to be modified so as to become wider. In particular, only the theory of functions of a real variable could give a satisfactory answer to the question what the length of curve is and for what curves it makes sense to talk of length.

The foundation on which this theory of functions of a real variable is built is the *theory of sets*.

Accordingly, we begin our exposition with an account of the elements of the theory of sets, next we turn to the study of point sets, and we conclude the chapter with an explanation of one of the fundamental concepts of the theory of functions of a real variable, namely the Lebesgue integral.

§2. Sets

People have constantly to deal with various collections of objects. As was already explained in Chapter 1, this entailed the development of the concept of *number* and later that of a *set*, which is one of the basic primitive mathematical concepts and does not lend itself to an accurate definition. The following remarks are meant to illustrate what a set is but do not pretend to serve as a definition.

Set is the name for an aggregate, ensemble, or collection of things that are combined under a certain criterion or according to a certain rule. The concept of a set arises by an abstraction. By considering a certain collection of objects as a set, we disregard all the connections and relations between the various objects that make up the set, but we preserve the individual features of the objects. Thus, the set consisting of five coins and the set consisting of five apples are different sets. But the set of five coins arranged in a circle and the set of the same coins arranged one next to the other is one and the same set.

Let us give some examples of sets. We can talk of the grains forming a heap of sand, of the set of all planets of our solar system, of the set of all people that are in a certain house at a given moment, or of the set of all pages of this book. In mathematics we constantly come across various sets such as the set of all roots of a given equation, the set of all natural numbers, the set of all points on a line, etc.

The mathematical discipline that studies general properties of sets, i.e., properties that do not depend on the nature of the constituent objects, is called the *theory of sets*. This discipline began to be developed rigorously at the end of the 19th and the beginning of the 20th century. The founder of the scientific theory of sets is the German mathematician G. Cantor.

Cantor's work on the theory of sets grew from studying questions of convergence of trigonometric series. This is a very common phenomenon: Very often the occupation with concrete mathematical problems leads to the construction of very abstract and general theories. The value of such abstract constructions lies in the fact that they turn out to be connected not only with the concrete problem from which they have sprung but have also applications to a number of other problems. In particular, this is the case in the theory of sets. The ideas and concepts of the theory of sets penetrated literally into all branches of mathematics and changed its face entirely. Therefore it is impossible to form a proper picture of contemporary mathematics without being acquainted with the elements of the theory of sets. For the theory of functions of a real variable the theory of sets is of particularly great significance.

A set is considered as given when one can tell of every object whether it belongs to the set or not. In other words, a set is completely determined by all the objects that belong to it. If a set M consists of the objects $a, b, c, \cdots$ and of no others, then we write

$$M = \{a, b, c, \cdots\}.$$

The objects that form a certain set are usually called its *elements*. The fact that an object m is an element of a set M is written in the form

$$m \in M$$

and is read: "m belongs to M" or "m is an element of M". If an object n does not belong to a set M, then one writes: $n \bar{\in} M$. Every object can only be one element of a given set; in other words, all the elements of one and the same set are distinct from one another.

The elements of a set M can themselves be sets; however, to avoid contradiction it is convenient to postulate that a set M cannot be one of its own elements, $M \bar{\in} M$.

The set that contains no elements is called the *empty set*. For example, the set of all real roots of the equation

$$x^2 + 1 = 0$$

is empty. Henceforth the empty set will be denoted by ϕ.

If for two sets M and N every element x of M is also an element of N, then we say that M enters into N, that M is part of N, that M is a subset of N, or that M is contained in N; this is written in the form

$$M \subseteq N \quad \text{or} \quad N \supseteq M.$$

For example, the set $M = \{1, 2\}$ is part of the set $N = \{1, 2, 3\}$.

Clearly we always have $M \subseteq M$. It is convenient to regard the empty set as part of any set.

Two sets are *equal* if they consist of the same elements. For example, the set of roots of the equation $x^2 - 3x + 2 = 0$ and the set $M = \{1, 2\}$ are equal.

Now we define rules of *operations* on sets.

Union or sum. Suppose that $M, N, P, \cdots$ are sets. The union or sum of these sets is the set X consisting of all elements that belong to at least one of the "summands" $M, N, P, \cdots$

$$X = M + N + P + \cdots.$$

§2. SETS

Here, even if an element x belongs to several summands, it occurs in the sum X only once. Clearly

$$M + M = M,$$

and if $M \subseteq N$, then

$$M + N = N.$$

Intersection. The intersection or common part of the sets $M, N, P, \cdots$ is the set Y consisting of all those elements that belong to all the sets $M, N, P, \cdots$.

Clearly $M \cdot M = M$ and if $M \subseteq N$, then $M \cdot N = M$.

If the intersection of the sets M and N is empty, $M \cdot N = \phi$, then we say that these sets are *disjoint*.

As a notation for the operations of sum and intersection of sets, we also use the symbols Σ and Π. Thus,

$$E = \Sigma E_i$$

is the sum of the sets E_i, and

$$F = \Pi E_i$$

their intersection.

We recommend that the reader prove that sum and intersection of sets are connected by the usual distributive law

$$M(N + P) = MN + MP,$$

and also by the law

$$M + NP = (M + N)(M + P).$$

Difference. The difference of two sets M and N is the set Z of all those elements of M that do not belong to N,

$$Z = M - N.$$

If $N \subseteq M$, then the difference $Z = M - N$ is also called the *complement* of N in M.

It is not hard to show that always

$$M(N - P) = MN - MP$$

and

$$(M - N) + MN = M.$$

Thus, the rules for operations on sets differ considerably from the usual rules of arithmetic.

Finite and infinite sets. Sets consisting of a finite number of elements are called finite sets. If the number of elements of a set is unbounded, then the set is called infinite. For example, the set of all natural numbers is infinite.

Let us consider two arbitrary sets M and N and ask whether the number of elements in those sets is the same or not.

If the set M is finite, then the collection of its elements is characterized by a certain natural number, namely, the number of its elements. In this case, in order to compare the numbers of elements of M and N, it is sufficient to count the number of elements in M and the number of elements in N and to compare the numbers so obtained. Also it is natural to reckon that if one of the sets M and N is finite and the other infinite, then the infinite set contains more elements than the finite.

However, if both sets M and N are infinite, then a simple count of the elements yields nothing. Therefore the following problem arises at once: Do all infinite sets have the same number of elements or do there exist infinite sets with larger or smaller numbers of elements? If the latter is true, then how can the numbers of elements in infinite sets be compared? We shall now turn our attention to these problems.

One-to-one correspondences. Again let M and N be two finite sets. How can we find out which of these sets contains more elements without counting the number of elements in each set? To this end let us form *pairs* by combining in a pair one element of M and one element of N. Then, if for some element of M there is no longer an element of N to be paired with it, M has more elements than N. Let us illustrate this argument by an example.

Suppose that in a room there are a certain number of people and a certain number of chairs. In order to find out of which there are more, it is sufficient to ask the people to sit down. If somebody is left without a place, it means that there are more people and if, say, all are placed and all places are taken, then there are as many people as chairs. This method of comparing the number of elements in sets has the advantage over a direct count of the elements because it is applicable without essential modifications not only to finite but also to infinite sets.

Let us consider the set of all natural numbers

$$M = \{1, 2, 3, 4, \cdots\}$$

and the set of all even numbers

$$N = \{2, 4, 6, 8, \cdots\}.$$

Which set contains more elements? At first sight it seems to be the

§2. SETS

former. However, we can form pairs from elements of these sets, as set out in Table 1:

Table 1.

M	1	2	3	4	⋯
N	2	4	6	8	⋯

No element of M nor of N remains without a partner. True, we could also have formed pairs as in Table 2:

Table 2.

M	1	2	3	4	5	⋯
N	—	2	—	4	—	⋯

Then many elements of M remain without a partner. On the other hand, we could have formed pairs as in Table 3:

Table 3.

M	—	1	—	2	—	3	—	⋯
N	2	4	6	8	10	12	14	⋯

Now many elements of N remain without a partner.

Thus, if the sets A and B are infinite, then distinct methods of forming pairs lead to different results. If there is one method of forming pairs in which every element of A and every element of B is paired off with some element, then we say that a one-to-one correspondence can be set up between A and B. For example, we can establish a one-to-one correspondence between the sets M and N as is clear from Table 1.

If between the sets A and B a one-to-one correspondence can be set up, then we say that they have the *same* number of elements. If for *every* method of pairing there are always some elements of A without a partner, then we say that the set A contains more elements than B or that A has a greater *cardinality* than B.

Thus we have obtained an answer to one of the questions raised earlier: how to compare the number of elements in infinite sets. However, this has by no means brought us nearer to an answer to the other question: Do there exist infinite sets at all having distinct cardinalities? In order to get an answer to this question let us study some simple types of infinite sets.

Countable sets. If we can set up a one-to-one correspondence between the elements of a set A and the elements of the set of all natural numbers

$$Z = \{1, 2, 3, \cdots\},$$

then we say that the set A is *countable*. In others words, a set is countable if all its elements can be enumerated by means of the natural numbers i.e., written down in the form of a sequence

$$a_1, a_2, \cdots, a_n, \cdots.$$

Table 1 shows that the set of all even numbers is countable (the upper numbers can now be regarded as the suffix of the corresponding lower number).

Countable sets are, so to speak, the very smallest infinite sets: Every infinite set contains a countable subset.

If two nonempty finite sets do not intersect, then their sum contains more elements than either summand. For infinite sets this cannot hold. For example, let E be the set of all even numbers, O the set of all odd numbers, and Z the set of all natural numbers. As Table 4 shows, the sets E and O are countable. However, the set $Z = E + O$ is again countable.

Table 4.

E	2	4	6	8	$\cdots$
O	1	3	5	7	$\cdots$
Z	1	2	3	4	$\cdots$

The violation of the rule "the whole is larger than the parts" in infinite sets shows that the properties of infinite sets differ *qualitatively* from those of finite sets. The transition from the finite to the infinite proceeds in complete agreement with the well-known principle of dialectics, qualitative variation of properties.

Let us show that *the set of all rational numbers is countable*. For this purpose we arrange the rational numbers in Table 5.

Here all the natural numbers are placed in the first row in ascending order, in the second row zero and all the negative numbers in decreasing order, in the third row the positive reduced fractions with denominator 2 in ascending order, in the fourth row the negative reduced fractions with denominator 2 in descending order, etc. It is clear that every rational number occurs once and only once in this table. Let us now enumerate all

§2. SETS

Table 5.

(1)	(2)	(3)	(4)	(5)	(6)	
1	2	3	4	5	6	...
0	−1	−2	−3	−4	−5	...
$\frac{1}{2}$	$\frac{3}{2}$	$\frac{5}{2}$	$\frac{7}{2}$	$\frac{9}{2}$	$\frac{11}{2}$	...
$-\frac{1}{2}$	$-\frac{3}{2}$	$-\frac{5}{2}$	$-\frac{7}{2}$	$-\frac{9}{2}$	$-\frac{11}{2}$	...
$\frac{1}{3}$	$\frac{2}{3}$	$\frac{4}{3}$	$\frac{5}{3}$	$\frac{7}{3}$	$\frac{8}{3}$	...
$-\frac{1}{3}$	$-\frac{2}{3}$	$-\frac{4}{3}$	$-\frac{5}{3}$	$-\frac{7}{3}$	$-\frac{8}{3}$	...
...	...	...	...	...	...	...

the numbers of the table in the order indicated by the arrows. Then all the rational numbers are arranged in a single sequence:

Number of the place occupied by rational number	1	2	3	4	5	6	7	8	9	...
Rational number	1,	2,	0,	3,	−1,	$\frac{1}{2}$,	4,	−2,	$\frac{3}{2}$,	...

So we have established a one-to-one correspondence between all the rational numbers and all the natural numbers. Therefore the set of all rational numbers is countable.

Sets with the cardinal number of the continuum. If we can set up a one-to-one correspondence between the elements of a set M and the points of the interval $0 \leqslant x \leqslant 1$, then we say that the set M has the *cardinal number of the continuum*. In particular, by this definition the set of points of the segment $0 \leqslant x \leqslant 1$ has itself the cardinal number of the continuum.

From figure 1 it is clear that the set of points of any interval AB has the cardinal number of the continuum. Here the one-to-one correspondence is established geometrically by means of a projection.

It is not hard to show that the sets of points of any open interval $a < x < b$ and of the whole numerical line have the cardinal number of the continuum.

Of considerably greater interest is the following fact: The set of points of the square $0 \leqslant x \leqslant 1, 0 \leqslant y \leqslant 1$ has the cardinal number of the continuum. Thus, roughly speaking, there are "as many" points in the square as on the segment.

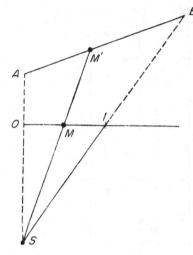

Fig. 1.

§3. Real Numbers*

The development of the concept of number has been described in detail in Chapter I. Here we shall give the reader a brief account of the theories of the real numbers that have arisen in the 19th century in connection with the foundation of the basic concepts of analysis.

Rational numbers. We assume that the reader is familiar with the main properties of rational numbers. Without going into details we recall these properties. Rational numbers, i.e., numbers of the form m/n, where m and n are integers and $n \neq 0$, form a set of numbers in which two operations (addition and multiplication) are defined. These operations are subject to a number of laws (axioms). In what follows $a, b, c, \cdots$ denote rational numbers

1. *Axioms of addition.*

1. $a + b = b + a$ (commutativity)
2. $a + (b + c) = (a + b) + c$ (associativity)
3. the equation
$$a + x = b$$
has a unique solution (existence of the inverse operation).

From these axioms it follows immediately that the expression $a + b + c$ has a unique meaning, that there exists a rational number 0 (null) for which $a + 0 = a$ and that addition has an inverse operation (subtraction) so that the expression $b - a$ has a meaning.

---

* During the writing of this section, we have had valuable consultations with A. N. Kolmogorov.

§3. REAL NUMBERS

Thus, from the algebraic point of view, all the rational numbers form a commutative group under the operation of addition.

II. Axioms of multiplication.

1. $ab = ba$ (commutativity)
2. $a(bc) = (ab)c$ (associativity)
3. the equation
$$ay = b,$$
where $a \neq 0$ has a unique solution (existence of the inverse operation).

From these axioms it follows that the expression abc has a meaning, that there exists a rational number 1 for which $a \cdot 1 = a$ and that for rational numbers other than 0 the inverse operation (division) exists. All the rational numbers except 0 form a commutative group under the operation of multiplication.

III. Axiom of distributivity.

1. $(a + b)c = ac + bc$.

The axioms I-III together indicate that under the operations of addition and multiplication the rational numbers form a so-called *algebraic field*.

IV. Axioms of order.

1. For any two rational numbers a and b one and only one of the following three relations holds: either $a < b$, or $a > b$, or $a = b$.
2. If $a < b$, and $b < c$, then $a < c$.
3. If $a < b$, then $a + c < b + c$ (monotonicity of addition).
4. If $a < b$ and $c > 0$, then $ac < bc$ (monotonicity of multiplication by $c > 0$).

All these axioms together allow us to call the set of rational numbers an *ordered field*.

Apart from the rational numbers there exist also other systems of objects that satisfy these axioms and are therefore ordered fields.

We mention two important properties of rational numbers.

Density: For arbitrary a and b, $a < b$, there is a c such that $a < c < b$.

Countability: The set of all rational numbers is countable (see §2).

On measuring quantities. The insufficiency of the rational numbers alone in mathematics already becomes apparent in dealing with such an important task as that of measuring quantities. As one of the simplest examples, we shall here consider the problem of measuring the length of intervals.

Let us take a line on which a definite direction, an origin (the point 0), and a unit of scale are marked. Then it is clear what a segment OA with its end point at $1/2$, $1/3$, $2/3$, $-1/3$, etc. is. More generally, with every rational number a we can associate a point a on the line, namely the point with the coordinate $x = a$. In this case the number a determines the length of the directed segment OA. However, not every segment gets a certain (rational) number as the measure of its length in this construction. For example, it was already known to the ancient Greeks that the length of the diagonal of a square with unit side cannot be measured by any rational number. The natural outcome of this situation is the setting up of a one-to-one correspondence between numbers and lengths, i.e., a further extension of the concept of number.

Real numbers. We were led to the conclusion that the rational numbers alone are insufficient to measure quantities and that the concept of number must be extended so that there exists a one-to-one correspondence between numbers and points on a line. With this in mind let us try to find out whether it is not possible to determine the position of an *arbitrary* point on a line by means of the rational points only. A similar construction within the domain of rational numbers will then lead us to the concept of a real number.

Let α be an arbitrary point on the line. Then all the rational points a can be divided into two parts: We put into one part all the points a that are to the left of α, and into the other all those to the right of α. As regards the point α itself (if it happens to be rational), it can be put into either part. Such a division of the rational points is usually called a *cut*. Cuts are taken to be identical if the collection of rational points in the left and the right parts of the cuts coincide (to within a single point). Now it is not difficult to see that distinct points α and β determine different cuts. For since the rational points are everywhere dense on the line, we can find rational points r_1 and r_2 lying strictly between α and β. Then in one of the cuts they come into the right part and in the other into the left part.

Thus every point on the line determines a cut in the domain of rational points, and different cuts correspond to different points. It is very important that a cut can also be defined in another way and without specific reference to the number α. For let us define a cut in the domain of rational points as a division of all rational points into two nonempty disjoint sets A and B

§3. REAL NUMBERS

such that $a < b$ for every $a \in A$, $b \in B$. Under this definition we can assign to a cut in a unique fashion that point (boundary) which produces it. In other words, by means of cuts in the domain of rational points, we can determine *every* point on the line. The construction we have just explained was proposed by the German mathematician R. Dedekind and is known under the name of *Dedekind cut*.

A cut is not the only possible method of determining the position of an arbitrary point by means of the rational points. Nearer to the usual practice of measuring is the following method of G. Cantor. Again let α be an arbitrary point on the line. Then we can find two arbitrarily close rational points a and b such that α lies between a and b. The points a and b determine the position of α approximately. Let us imagine this process of approximate determination of α continued indefinitely and in such a way that at each successive step the accuracy is increased more and more. Then we obtain a system of intervals $[a_n, b_n]$ with their ends at rational points such that $[a_{n+1}, b_{n+1}] \subseteq [a_n, b_n]$ and $b_n - a_n \to 0$ $(n \to \infty)$. A system of intervals satisfying these conditions is called a *nest of intervals*. It is clear that such a nest of intervals determines the position of α uniquely.

By means of similar constructions in the domain of rational numbers, we can define the real numbers. Next we define the operations among real numbers and ascertain that they satisfy the same axioms as the operations on rational numbers. Now every point on the line corresponds to a real number and vice versa. On account of this, the set of all real numbers is often called the numerical line.

Principles of continuity. There are essential differences between the set of all rational numbers and the set of all real numbers. In fact, the set of all real numbers has a number of properties that characterise the *continuity* of this set, whereas the set of all rational numbers does not have these properties. They are usually called principles of continuity. We shall enumerate the most important of them.

Dedekind's principle. If the set of all real numbers is divided into two nonempty sets X and Y without elements in common, so that for arbitrary $x \in X$, $y \in Y$ the inequality $x < y$ holds, then there exists a unique number ξ (the boundary) for which $x \leq \xi \leq y$ for arbitrary $x \in X$, $y \in Y$.

The set of all real numbers x satisfying the inequalities $a \leq x \leq b$ is called an interval of the numerical line and is denoted by $[a, b]$. A system of intervals $[a_n, b_n]$ is called *nested* if $[a_{n+1}, b_{n+1}] \subseteq [a_n, b_n]$ and $b_n - a_n \to 0$ $(n \to \infty)$.

Cantor's principle. For every nested system of intervals $[a_n, b_n]$ there exists one and only one real number ξ that belongs to each of these intervals.

Weierstrass' principle. Every nondecreasing sequence of real numbers that is bounded above converges.

Let us say that a sequence of real numbers $\{x_n\}$ is a *fundamental* sequence if for every $\epsilon > 0$ we can find a natural number N such that for all $n > N$ and all natural p

$$|x_{n+p} - x_n| < \epsilon.$$

Cauchy's principle. Every fundamental sequence of real numbers converges.

Since we have not given an accurate construction of the real numbers we are not in a position to establish that these principles hold for the set of real numbers. Our next object is to investigate how these principles are interrelated. Let us then assume that one of the principles of continuity holds for the real numbers and examine which of the remaining principles of continuity follows from it.

The over-all result that we shall arrive at is that all the principles of continuity are equivalent

We say that a number b is the (least) upper bound of a set E

$$b = \sup E,$$

if (1) $x \leqslant b$ for every $x \in E$ and (2) there exists no number $b' < b$ with the same property.

Let us show that the following proposition follows from Dedekind's principle: Every nonempty set E of numbers that is bounded above has a least upper bound. We divide all the real numbers into two classes X and Y according to the following criterion: We put $x \in X$ if there exists an $a \in E$ such that $a \geqslant x$, and we put $y \in Y$ if for every $a \in E$ we have $a < y$. It is easy to verify that this is a cut. By Dedekind's principle it has a boundary ξ; this boundary is the least upper bound of E.

We shall now show that Weierstrass' principle follows from Dedekind's. Let $\{x_n\}$ be a nondecreasing sequence of real numbers, bounded above. By what we have just proved it has a least upper bound ξ. By definition of an upper bound $x_n \leqslant \xi$ ($n = 1, 2, \cdots$); for every $\epsilon > 0$ we can find an index n_0 such that $x_{n_0} > \xi - \epsilon$. Since the sequence $\{x_n\}$ is monotonic, this implies that $\xi - \epsilon < x_n \leqslant \xi$ for all $n > n_0$, i.e., the sequence $\{x_n\}$ converges to the limit ξ.

To prove the converse relation between the principles of Dedekind and Weierstrass we note that Weierstrass' principle implies:

Archimedes' principle. No matter what the real numbers $a > 0$ and b are, we can always find a natural number n such that $na > b$.

This principle means that for every real number b the sequence $\{b/n\}$ ($n = 1, 2, \cdots$) converges to zero.

§3. REAL NUMBERS

Suppose that Weierstrass' principle holds, but Archimedes' does not hold. The latter means that there exists an $a > 0$ such that the sequence $x_n = na$ is bounded. Moreover, it is increasing. By Weierstrass' principle it has a limit ξ. Hence it follows that the interval $[\xi - a/2, \xi]$ contains some point $x_n = na$ of our sequence. But then $x_{n+1} = (n+1)a > \xi$, and this contradicts the fact that ξ is the least upper bound of $\{x_n\}$.

Weierstrass' principle implies Dedekind's principle. Let the set of all real numbers be divided into two disjoint sets X and Y such that $x < y$ for all $x \in X$, $y \in Y$. We shall show that this cut has a unique boundary ξ. Let m be an integer and n a natural number. We denote by x_n the largest element of the form $m/2^n \in X$ such that $x_n + 1/2^n \in Y$. Since the set of elements of the form $m/2^n$ is contained in the set of elements of the form $m/2^{n+1}$, we have $x_n \leqslant x_{n+1}$. Moreover, the sequence $\{x_n\}$ is bounded (for example, by the number $x_1 + 1/2$). Hence by Weierstrass' principle it has a limit ξ. We shall show that ξ is the boundary of our cut. For if $x < \xi$ then $x \in X$. And if $y > \xi$, then $y \in Y$, because it follows from Archimedes' principle that we can find a number n such that $1/2^n < y - \xi = a$. But $x_n < \xi$, $x_n + 1/2^n \in Y$, and then $y = \xi + a > x_n + 1/2^n$, therefore $y \in Y$.

One can also show that Cantor's and Cauchy's principles are equivalent. However, when Cauchy's principle holds it does not follow that Dedekind's principle holds. This statement has to be understood in the following sense: There exists an ordered field for which Cauchy's principle holds, but Dedekind's does not hold. If it is assumed beforehand that Archimedes' principle holds, then all four principles are equivalent.

Uncountability of the continuum. Let us show that the set of all points of the segment $0 \leqslant x \leqslant 1$ is uncountable. We shall give an indirect proof. Suppose that the set of all points of the segment $0 \leqslant x \leqslant 1$ is countable. Then all the points x of this segment can be indexed by means of the natural numbers

$$x_1, x_2, \cdots, x_n, \cdots. \tag{1}$$

In [0, 1] we choose an interval σ_1 so that its length is less than 1 and that it does not contain the point x_1. Such an interval can readily be found. Next, within σ_1 we choose an interval σ_2 so that its length is less than $1/2$ and that σ_2 does not contain the points x_1, x_2. Generally, when an interval σ_{n-1} has already been chosen we choose in it an interval σ_n so that its length is less than $1/n$ and that it does not contain the points $x_1, x_2, \cdots, x_n$. In this way we construct an infinite sequence of intervals

$$\sigma_1, \sigma_2, \cdots, \sigma_n, \cdots,$$

such that each is contained in the preceding one and their lengths tend to zero with increasing n. Then by Cantor's principle there exists a unique point x in the interval [0, 1] that belongs to all the intervals σ_n. Since by our hypothesis all the points of [0, 1] are accounted for in (1), the point x which is common to all σ_n coincides with some point x_m of that sequence. But by our construction σ_m does not contain x_m so that $x \neq x_m$. Thus we have arrived at a contradiction. Therefore the initial hypothesis, that the set of all points of [0, 1] is countable, is false and so this set is uncountable. This is what we set out to prove.

This theorem shows that there exist distinct infinite cardinalities and therefore gives a positive answer to the first question raised.

§4. Point Sets

In the preceding section we have already come across sets whose elements are *points*. In particular, we have considered the set of all points of an arbitrary interval and the set of all points (x, y) of the square $0 \leqslant x \leqslant 1, 0 \leqslant y \leqslant 1$. We shall now turn to a more detailed study of properties of such sets.

A set whose elements are points is called a *point set*. Thus, we can speak of point sets on a line, in a plane, or in an arbitrary space. For simplicity's sake we shall here confine ourselves to the study of points sets *on a line*.

There is a close connection between the real numbers and the points on a line: With every real number we can associate a point on the line and vice versa. Therefore, in speaking of point sets we may include with them sets consisting of real numbers, sets on the numerical line. Conversely, in order to define a point set on a line we shall, as a rule, give the coordinates of all the points of the set.

Point sets (and, in particular, point sets on a line) have a number of special properties that distinguish them from arbitrary sets and make the theory of point sets into a self-contained mathematical discipline. First of all, it makes sense to speak of the *distance* between two points. Furthermore, we can establish a relation of *order* (left, right) between the points on a line; accordingly, one says that a point set on a line is an *ordered* set. Finally, as we have already mentioned earlier, Cantor's principle holds for the line; this property of the line is usually characterized as *completeness* of the line.

We introduce a notation for the simplest sets on a line.

An *interval* $[a, b]$ is the set of points whose coordinates satisfy the inequality $a \leqslant x \leqslant b$.

An open *interval* (a, b) is the set of points whose coordinates satisfy the conditions $a < x < b$.

The *semi-intervals* $[a, b)$ and $(a, b]$ are defined by the conditions $a \leqslant x < b$ and $a < x \leqslant b$, respectively.

§4. POINT SETS

Open intervals and semi-intervals can be *improper*. Thus, $(-\infty, \infty)$ denotes the whole line and, for example, $(-\infty, b]$ the set of all points for which $x \leq b$.

We begin with an account of the various possibilities for the position of a set *as a whole* on a line.

Bounded and unbounded sets. A set E of points on a line can either consist of points whose distances from the origin of coordinates do not exceed a certain positive number or it has points arbitrarily far from the origin of coordinates. In the first case E is called *bounded*, in the latter *unbounded*. An example of a bounded set is the set of all points of the interval $[0, 1]$, and an example of an unbounded set is the set of all points with integral coordinates

It is easy to see that, when a is a fixed point on the line, a set E is bounded if and only if the distances from a of arbitrary points $x \in E$ do not exceed a certain positive number.

Sets bounded above and below. Let E be a set of points on a line. If there is a point A on the line such that every point $x \in E$ lies to the left of A, then we say that E is *bounded above*. Similarly, if there is a point a on the line such that every $x \in E$ lies to the right of a, then E is called *bounded below*. Thus, the set of all points on the line with positive coordinates is bounded below, and the set of all points with negative coordinates bounded above.

It is clear that the definition of a bounded set is equivalent to the following: A set E of points on a line is called bounded if it is bounded above and below. Notwithstanding that these two definitions are very similar, there is an essential difference between them: The first is based on the fact that a distance is defined between the points on a line, and the second that these points form an ordered set.

We can also say that a set is bounded if it lies entirely in some interval $[a, b]$.

The least upper and greatest lower bound of a set. Suppose that a set E is bounded above. Then there exist points A on the line such that there are no points of E to their right. Using Cantor's principle we can show that among all the points A having these properties there is a leftmost. This point is called the *least upper bound* of E. The *greatest lower bound* of a point set is defined similarly.

If there is a rightmost point in E, then it is obviously the least upper bound of E. However, it can happen that E has no rightmost point.

For example, the set of points with the coordinates

$$\frac{0}{1}, \frac{1}{2}, \frac{2}{3}, \frac{3}{4}, \frac{4}{5}, \ldots$$

is bounded above and has no rightmost point. In this case the least upper bound a does not belong to E, but there are points of E arbitrarily near to a. In the example above $a = 1$.

Distribution of a point set near an arbitrary point on the line. Let E be a point set and x an arbitrary point on the line. We consider the various possibilities for the distribution of the set E near x. The following cases are possible:

1. Neither the point x nor the points sufficiently near to it belong to E.
2. The point x does not belong to E, but there are points of E arbitrarily near it.
3. The point x belongs to E, but all points sufficiently near to it do not belong to E.
4. The point x belongs to E and there are other points of E arbitrarily near it.

In the case 1, x is called *exterior* to E, in the case 3, an *isolated* point of E, and in the cases 2 and 4, a *limit* point of E.

Thus, if $x \bar{\in} E$, then x can be either exterior to E or a limit point, and if $x \in E$, it can be either an isolated point of E or a limit point.

A limit point may or may not belong to E and is characterized by the condition that there are points of E arbitrarily near to it. In other words, a point x is a limit point of E if every open interval δ containing x contains infinitely many points of E. The concept of a limit point is one of the most important in the theory of points sets.

If x and all points sufficiently near to it belong to E, then x is called an *interior* point of E. Every point x that is neither an exterior nor an interior point of E is called a *boundary* point of E.

Let us give some examples to illustrate all these concepts.

Example 1. Let E_1 consist of the points with the coordinates

$$1, \frac{1}{2}, \frac{1}{3}, \ldots, \frac{1}{n}, \ldots.$$

Then every point of this set is an isolated point of it, the point 0 is a limit point of E_1 (and does not belong to it), and all the remaining points on the line are exterior to E_1.

§4. POINT SETS

Example 2. Let E_2 consist of all the *rational* points of the interval [0, 1]. This set has no isolated points, every point of the interval [0, 1] is a limit point of E_2, and all the remaining points on the line are exterior to E_2. Clearly among the limit points of E_2 there are some that belong and others that do not belong to the set.

Example 3. Let E_3 consist of *all* points of the interval [0, 1]. As in the preceding example, E_3 has no isolated points and every point of [0, 1] is a limit point. However, in contrast to the preceding example, all the limit points of E_3 belong to the set.

Example 4. Let E_4 consist of all the points on the line with integral coordinates. Every point of E_4 is isolated; E_4 has no limit points.

We also point out that in Example 3 every point of the open interval (0, 1) is an interior point of E_3, and in Example 2 every point of the interval [0, 1] is a boundary point of E_2.

From the preceding examples it is clear that an infinite point set on a line may have isolated points (E_1, E_4) or not (E_2, E_3), that it may have interior points (E_3) or not (E_1, E_2, E_4). As regards limit points, only E_4 in Example 4 does not have any. As the following important theorem shows, this is connected with the fact that E_4 is unbounded.

The Theorem of Bolzano-Weierstrass: *Every bounded infinite point set on a line has at least one limit point.*

Let us prove this theorem. Suppose that E is a bounded infinite point set on a line. Since E is bounded, it lies entirely in some interval $[a, b]$. We divide this interval in half. Since E is infinite, at least one of the intervals so obtained contains infinitely many points of E. We denote that interval by σ_1 (if both halves of $[a, b]$ contain infinitely many points of E, then σ_1 shall denote, say, the left half). Next we divide σ_1 into two equal halves. Since the part of E that lies in σ_1 is infinite, at least one of the intervals so obtained contains infinitely many points of E. We denote it by σ_2. We continue the process of dividing an interval in half indefinitely and each time select that half which contains infinitely many points of of E. So we obtain a sequence of intervals $\sigma_1, \sigma_2, \cdots, \sigma_n, \cdots$. This sequence of intervals has the following properties: Every interval σ_{n+1} is contained in the preceding one σ_n; every interval σ_n contains infinitely many points of E; and the lengths of the intervals tend to zero. The first two properties of the sequence follow immediately from its construction, and to prove the last property it is sufficient to note that if the length of $[a, b]$ is l, then the length of σ_n is $l/2^n$. By Cantor's principle there exists a unique point x that belongs to all σ_n. We shall show that this x is a limit point of E.

For this it is sufficient to make sure that if δ is some open interval containing x, then it contains infinitely many points of E. Since every interval σ_n contains x and the lengths of the σ_n tend to zero, for a sufficiently large n the interval σ_n is entirely contained in δ. But by hypothesis σ_n contains infinitely many points of E. Therefore δ too contains infinitely many points of E. Thus, x is in fact a limit point of E and the theorem is proved.

Exercise. Show that if a set E is bounded above and has no rightmost point, then its least upper bound is a limit point of E (and does not belong to E).

Closed and open sets. One of the fundamental tasks of the theory of point sets is the study of properties of various types of points sets. We shall acquaint the reader with this theory for two examples. We shall now study properties of the so-called closed and open sets.

A set is called *closed* if it contains all its limit points. If a set has no limit points, then it is usually also taken to be closed. Apart from its limit points a closed set can also contain isolated points. A set is called *open* if every point of it is interior.

Let us give some examples of closed and open sets. Every interval $[a, b]$ is a closed set, and every open interval (a, b) an open set. The improper semi-intervals $(-\infty, b]$ and $[a, \infty)$ are closed, and the improper intervals $(-\infty, b)$ and (a, ∞) are open. The whole line is at the same time closed and open. It is convenient to regard the empty set also as both closed and open. Every finite point set on the line is closed, since it has no limit points. The set consisting of the points

$$0, 1, \frac{1}{2}, \frac{1}{3}, \frac{1}{4}, \cdots \frac{1}{n}, \cdots$$

is closed; this set has the single limit point $x = 0$ which belongs to the set.

Our task is to examine the structure of an arbitrary closed or open set. For this purpose we require a number of auxiliary facts which we assume without proof.

1. The intersection of any number of closed sets is closed.

2. The union of any number of open sets is open.

3. If a closed set is bounded above, then it contains its least upper bound. Similarly, if a closed set is bounded below, then it contains its greatest lower bound.

§4. POINT SETS

Let E be an arbitrary point set on a line. The *complement* of E, denoted by CE, is defined as the set of all points on the line that do not belong to E. Clearly, if x is an exterior point of E, then it is an interior point of CE and vice versa.

4. If a set F is closed, then its complement CF is open and vice versa.

Proposition 4 shows that there is a very close link between closed and open sets: They are complements of each other. Because of this it is sufficient to study either closed or open sets only. A knowledge of the properties of sets of one type enables us at once to read off properties of sets of the other type. For example, every open set is obtained by deleting some closed set from the line.

Let us now proceed to study properties of closed sets. We make one definition. Let F be a closed set. An open interval (a, b) having the property that none of its points belong to F, while a and b belong to F, is called an *adjacent interval* of F. Among the adjacent intervals we also count the improper intervals (a, ∞) or $(-\infty, b)$, provided a or b belong to F but the intervals do not intersect F. We shall show that if a point x does not belong to the closed set F, then it belongs to one of its adjacent intervals.

We denote by F_x that part of F that lies to the right of x. Since x itself does not belong to F, we can represent F_x as an intersection

$$F_x = F \cdot [x, \infty).$$

Both F and $[x, \infty)$ are closed. Therefore, by proposition 1, F_x is closed. If F_x is empty, then the whole semi-interval $[x, \infty)$ does not belong to F. Let us assume then that F_x is not empty. Since this set lies entirely on the semi-interval $[x, \infty)$, it is bounded below. We denote by b its greatest lower bound. By proposition 3, $b \in F_x$ so that $b \in F$. Furthermore, since b is the greatest lower bound of F_x, the semi-interval $[x, b)$ lying to the left of b does not contain points of F_x, consequently not of F either. Thus we have constructed a semi-interval $[x, b)$ containing no points of F, and either $b = \infty$ or b belongs to F. Similarly, we can construct a semi-interval $(a, x]$ not containing points of F with either $a = -\infty$ or $a \in F$. Now it is clear that the open interval (a, b) contains x and is an adjacent interval of F. It is easy to see that if (a_1, b_1) and (a_2, b_2) are two adjacent intervals of F, then they either coincide or are disjoint.

From the preceding it follows that every closed set on the line is obtained by deleting from the line a certain number of open intervals, namely the adjacent intervals of F. Since every open interval contains at least one rational point and all the rational points on the line form a countable set, we see that the number of adjacent intervals cannot be more than countable. Hence we reach a remarkable conclusion. Every closed set on the line is obtained by deleting from the line at the most a countable set of disjoint open intervals.

By proposition 4 it follows at once that every open set on the line is the sum of not more than a countable number of open intervals. By propositions 1 and 2 it is also clear that every set of the structure we have indicated is in fact closed (open).

It will be seen from the following example that closed sets can have a very complicated structure.

Cantor's perfect set. We shall construct a particular closed set that has a number of remarkable properties. First of all we delete from the line the improper intervals $(-\infty, 0)$ and $(1, \infty)$. After this operation we are left with the interval $[0, 1]$. Next we delete from this the open interval $(1/3, 2/3)$ which forms its middle third. From each of the remaining intervals $[0, 1/3]$ and $[2/3, 1]$, we delete its middle third. This process of deleting the middle thirds of the remaining intervals can be continued indefinitely. The point set on the line that remains after all these open intervals have been deleted is called Cantor's perfect set; we shall denote it by the letter P.

Let us investigate some properties of this set: P is closed, because it is formed by deleting from the line a certain set of disjoint open intervals; P is not empty; in any case it contains the end points of all the removed intervals.

A closed set F is called *perfect* if it has no isolated points, i.e., if every point of it is a limit point. We shall show that P is perfect. For if x were an isolated point of P, then it would have to be a common end point of two adjacent intervals of the set. But by our construction the adjacent intervals of P do not have common end points.

The set P does not contain any open interval. For suppose that a certain open interval δ entirely belongs to P. Then it belongs entirely to one of the intervals obtained at the nth step of the construction of P. But this is impossible, because for $n \to \infty$ the length of these intervals tends to zero.

One can show that P has the cardinality of the continuum. From this it follows, in particular, that Cantor's perfect set contains, apart from the end points of the adjacent intervals, other points. For the end points of the adjacent intervals form only a countable set.

Various types of point sets occur constantly in the most diverse branches of mathematics, and a knowledge of their properties is absolutely indispensable in studying many mathematical problems. Of particularly great importance is the theory of point sets in mathematical analysis and topology.

We shall now give a few examples of point sets that appear in the classical parts of analysis. Let $f(x)$ be a continuous function given on the interval $[a, b]$. We fix a number α and consider the set of all points x for which $f(x) \geqslant \alpha$. It is easy to show that this can be an arbitrary closed set on the interval $[a, b]$. Similarly, the set of those points x for which $f(x) > \alpha$ can be any open set $G \subset [a, b]$. If $f_1(x), f_2(x), \cdots, f_n(x), \cdots$ is a sequence of continuous functions given on $[a, b]$, then the set of all points x where this sequence converges cannot be arbitrary and belongs to a certain well-defined type.

The mathematical discipline whose object is to study the structure of point sets is called the *descriptive theory of sets*.

In the development of the descriptive theory of sets Soviet mathematicians have made great contributions, N. N. Luzin and his pupils, P. S. Aleksandrov, M. Ja. Suslin, A. N. Kolmogorov, M. A. Lavrent'ev, P. S. Novikov, L. V. Keldyš, A. A. Ljapunov, and others.

The investigations of N. N. Luzin and his pupils have shown that there are strong ties between descriptive set theory and mathematical logic. Difficulties arising in the study of a number of problems of descriptive set theory (in particular, problems of determining the cardinality of certain sets) are difficulties of a logical nature. On the other hand, the methods of mathematical logic enable us to penetrate more deeply into certain problems of descriptive set theory.

§5. Measure of Sets

The concept of the measure of a set is a far-reaching generalization of the concept of the length of an interval. In the simplest case (the only one we shall consider here) the task is to give a definition of length not only for intervals but also for more complicated point sets on a line.

Let us agree that the unit of measurement is the interval $[0, 1]$. Then the length of an arbitrary interval $[a, b]$ is obviously $b - a$. Similarly, if we have two disjoint intervals $[a_1, b_1]$ and $[a_2, b_2]$, it is natural to interpret the length of the set E consisting of these two intervals as the number $(b_1 - a_1) + (b_2 - a_2)$. However, it is by no means clear what we have to understand by the length of a set on the line of a more complicated nature; for example, what is the length of the Cantor set of §4 of this chapter? Hence the conclusion, the concept of length of a set on the line requires a rigorous mathematical definition.

The problem of defining the lengths of sets or, as we now say, of measuring sets is very important, because it is of vital significance in generalizing the concept of an integral. The concept of measure of a set also has applications to other problems in the theory of functions, in probability, topology, functional analysis, etc.

We shall give an account of the definition of measure of sets which is due to the French mathematician H. Lebesgue and is the foundation of the definition of integral given by him.

Measure of an open and a closed set. We begin with the definition of measure of an arbitrary open or closed set. As we have mentioned in §4, every open set on the line is a finite or countable sum of pairwise disjoint open intervals.

The measure of an open set is defined as the sum of the lengths of its constituent open intervals.

Thus, if

$$G = \sum (a_i, b_i)$$

and the intervals (a_i, b_i) are pairwise disjoint, then the measure of G is equal to $\sum (b_i - a_i)$. Generally, the measure of a set E being denoted by μE, we can write

$$\mu G = \sum (b_i - a_i).$$

In particular, the measure of a single open interval is equal to its length

$$\mu(a, b) = b - a.$$

Every closed set F contained in $[a, b]$ and such that the end points of $[a, b]$ belong to F is obtained from $[a, b]$ by deleting from it a certain open set G. The *measure of the closed set* $F \subseteq [a, b]$, where $a \in F$, $b \in F$ is defined as the difference between the length of $[a, b]$ and the measure of the open set G complementary to F (relative to $[a, b]$).

Thus

$$\mu F = (b - a) - \mu G. \tag{2}$$

It is not difficult to verify that according to this definition the measure of an arbitrary interval is equal to its length

$$\mu[a, b] = b - a,$$

and the measure of a set consisting of a finite number of points is zero.

General definition of measure. In order to give a definition of measure of sets of more general nature than open and closed sets, we have to make use of an auxiliary concept. Let E be a set lying on the interval $[a, b]$. We consider all possible *coverings* of E, i.e., all open sets $V(E)$ containing

§5. MEASURE OF SETS

E. The measure of each of these sets $V(E)$ is already defined. The aggregate of measures of all sets $V(E)$ is a certain set of positive numbers. This set is bounded below (for example by 0) and therefore has a greatest lower bound which we denote by $\mu_e E$. The number $\mu_e E$ is called the *outer measure* of E.

Let $\mu_e E$ be the outer measure of a set E and $\mu_e CE$ the outer measure of its complement relative to $[a, b]$.

If the relation

$$\mu_e E + \mu_e CE = b - a \tag{3}$$

holds, then the set E is called *measurable* and the number $\mu_e E$ its *measure*: $\mu E = \mu_e E$; if the relation (3) does not hold, then we say that E is not measurable; a nonmeasurable set has no measure.

We note that always

$$\mu_e E + \mu_e CE \geqslant b - a. \tag{4}$$

Let us give a few clarifications. The length of the simplest sets (for example, open or closed intervals) has a number of remarkable properties. We mention the most important of these.

1. If the sets E_1 and E are measurable and $E_1 \subseteq E$, then

$$\mu E_1 \leqslant \mu E;$$

i.e., the measure of a part of E does not exceed the measure of the whole set E.

2. If E_1 and E_2 are measurable, then the set $E = E_1 + E_2$ is measurable and

$$\mu(E_1 + E_2) \leqslant \mu E_1 + \mu E_2;$$

i.e., the measure of a sum does not exceed the sum of the measures of the summands.

3. If sets E_i ($i = 1, 2, \cdots$) are measurable and pairwise disjoint, $E_i E_j = \phi$ ($i \neq j$), then their sum $E = \sum E_i$ is measurable and

$$\mu(\sum E_i) = \sum \mu E_i ;$$

i.e., the measure of a finite or countable sum of pairwise disjoint sets is equal to the sum of the measures of the summands.

This property of the measure is called its full additivity.

4. The measure of a set E does not change if it is displaced as a rigid body.

It is desirable that the fundamental properties of length are preserved for the more general concept of measure of sets. But it can be proved quite rigorously that this turns out to be *impossible* if a measure is to be ascribed to an arbitrary point set on the line. Consequently, in the sense of this definition there are sets that have a measure or are measurable and others that have no measure or are nonmeasurable. Besides, the class of measurable sets is so wide that this circumstance does not lead to any essential disadvantages. In fact, the construction of an example of a nonmeasurable set is rather difficult.

We shall now present some examples of measurable sets.

Example 1. *The measure of Cantor's perfect set P* (see §4). In constructing the set P from the interval $[0, 1]$, we have first thrown out an adjacent interval of length $1/3$, then two adjacent intervals of length $1/9$, then four adjacent intervals of length $1/27$, etc. Generally, at the nth step we have thrown out 2^{n-1} adjacent intervals of length $1/3^n$. Thus, the sum of the lengths of the intervals removed is equal to

$$S = \frac{1}{3} + \frac{2}{9} + \frac{4}{27} + \cdots + \frac{2^{n-1}}{3^n} + \cdots.$$

The terms of this series form a geometric progression with the first term $1/3$ and the common ratio $2/3$. Therefore the sum S of the series is

$$\frac{\frac{1}{3}}{1 - \frac{2}{3}} = 1.$$

Thus, the sum of the length of the intervals adjacent to the Cantor set is 1. In other words, the measure of the open set G complementary to P is 1. Therefore, the set itself has the measure

$$\mu P = 1 - \mu G = 1 - 1 = 0.$$

This example shows that a set may have the cardinality of the continuum and yet have measure zero.

Example 2. *Measure of the set R of all rational points of the interval* $[0, 1]$. First of all we shall show that $\mu_e R = 0$. In §2 we had found that R is countable. We arrange the points of R in a sequence

$$r_1, r_2, \cdots, r_n, \cdots.$$

Next, for given $\epsilon > 0$ we enclose the point r_n by an open interval δ_n of

§5. MEASURE OF SETS

length $\epsilon/2^n$. The sum $\delta = \Sigma \delta_n$ is an open set covering R. The open intervals δ_n may intersect so that

$$\mu(\delta) = \mu\left(\sum \delta_n\right) < \sum \mu \delta_n = \sum \frac{\epsilon}{2^n} = \epsilon.$$

Since ϵ can be chosen arbitrarily small, we have $\mu_e R = 0$.

Further, by (2)

$$\mu_e R + \mu_e CR \geqslant 1,$$

i.e., $\mu_e CR \geqslant 1$. But since CR is contained in $[0, 1]$, we have $\mu_e CR \leqslant 1$. Hence

$$\mu_e R + \mu_e CR = 1,$$

and*

$$\mu R = 0, \quad \mu CR = 1. \tag{5}$$

This example shows that a set may be everywhere dense on an interval and yet have measure zero.

Sets of measure zero play no role in many problems of the theory of functions and can be neglected. For example, a function $f(x)$ is Riemann integrable if and only if it is bounded and the set of its points of discontinuity has measure zero. We could add a considerable number of such examples.

Measurable functions. We now proceed to one of the most brilliant applications of the concept of measure of sets, namely, a description of that class of functions with which mathematical analysis and the theory of functions actually operate. The precise statement of the problem is as follows. If a sequence $\{f_n(x)\}$ of functions given on a certain set E converges at every point of E except at the points of a set N of measure zero, then we shall say that the sequence $\{f_n(x)\}$ converges *almost everywhere*.

What functions can be obtained from continuous functions by repeated application of the operation of forming the limit of an almost everywhere convergent sequence of functions and of algebraic operations?

To answer this question we require some new concepts.

Let $f(x)$ be a function defined on a set E and α an arbitrary real number. We denote by

$$E[f(x) > \alpha]$$

the set of all points of E for which $f(x) > \alpha$. For example, if $f(x)$ is defined on $[0, 1]$ and $f(x) = x$ on this interval, then set $E[f(x) > \alpha]$ is equal to $[0, 1]$ for $\alpha < 0$, to $(\alpha, 1]$ for $0 \leqslant \alpha < 1$, and empty for $\alpha \geqslant 1$.

---

* The same argument shows that every countable point set on the line has measure zero.

A function $f(x)$ defined on a set E is called *measurable* if E itself is measurable and the set $E[f(x) > \alpha]$ is measurable for every real number α.

One can show that every continuous function given on an interval is measurable. However, among the measurable functions there are also many discontinuous functions, for example the Dirichlet function which is equal to 1 at the irrational points of [0, 1] and 0 elsewhere.

We mention without proof that measurable functions have the following properties.

1. If $f(x)$ and $\phi(x)$ are measurable functions defined on one and the same set E, then the functions

$$f + \phi, f - \phi, f \cdot \phi \text{ and } \frac{f}{\phi}$$

are also measurable (the latter if $\phi \neq 0$).

This property shows that algebraic operations on measurable functions lead again to measurable functions.

2. If a sequence $\{f_n(x)\}$ of measurable functions defined on a set E converges almost everywhere to a function $f(x)$, then this function is also measurable.

Thus, the operation of forming the limit of an almost everywhere convergent sequence of measurable functions again leads to measurable functions.

These properties of measurable functions were established by Lebesgue. A deeper study of measurable functions was carried out by the Soviet mathematicians D. F. Egorov and N. N. Luzin. In particular, N. N. Luzin has proved that every measurable function on an interval can be made continuous by changing its values on a certain set of arbitrarily small measure.

This classical result of N. N. Luzin and the properties of measurable functions listed enable us to prove that measurable functions form that class of function of which we talked at the beginning of this subsection. Measurable functions are also of great importance in the theory of integration, namely, the concept of an integral can be generalized in such a way that every bounded measurable function turns out to be integrable. A detailed account of this will be given in the next section.

§6. The Lebesgue Integral

We shall now proceed to the central theme of this chapter, the definition of the Lebesgue integral and an account of its properties.

To understand the underlying principle of this integral, let us consider

§6. THE LEBESGUE INTEGRAL

the following example. Suppose that there is a large collection of coins of various denominations and we have to add up the total sum of money involved. This can be done in two ways. We can arrange the coins in a row and add the value of each new coin to the total value of the preceding ones. However, we can also proceed differently: We put the coins in heaps such that the coins in each heap are of equal value; then we count the number of coins in each heap, multiply this number by the value of the corresponding coin, and finally add up the numbers so obtained. The first method of counting money corresponds to the process of Riemann integration, and the second to the process of Lebesgue integration.

Going over from coins to functions, we can say that for the computation of the Riemann integral the domain of definition of the function (the axis of abscissas; figure 2a) is divided into small parts, while for the computa-

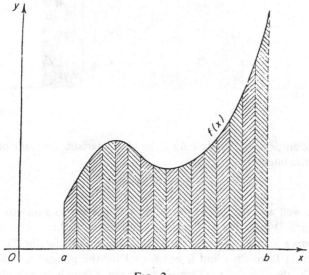

Fig. 2a.

tion of the Lebesgue integral it is the domain of values of the function (the axis of ordinates; figure 2b) that is so divided. The latter principle was applied in practice long before Lebesgue for the computation of integrals of functions of oscillating character; however, Lebesgue was the first to develop it in all generality and to give it a rigorous foundation by means of the theory of measure.

Let us examine how the measure of sets and Lebesgue integral are connected. Let E be an arbitrary measurable set on an interval $[a, b]$.

We construct a function $\phi(x)$ which is equal to 1 when x belongs to E and zero when x does not belong to E. In other words, we set

$$\phi(x) = \begin{cases} 1 & x \in E. \\ 0 & x \bar{\in} E. \end{cases}$$

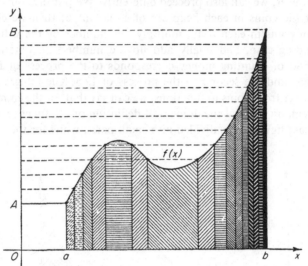

FIG. 2b.

The function $\phi(x)$ is usually called the characteristic function of E. We consider the integral

$$I = \int_a^b \phi(x)\, dx.$$

We are well accustomed to regarding the integral as equal to the area of the figure D bounded by the axis of abscissas, the lines $x = a, x = b$, and the curve $y = \phi(x)$ (see Chapter II). Since in our case the "height" of D is different from zero and is equal to 1 for the points $x \in E$ and these points only, the area (by the formula that area is length times width) must be equal numerically to the length (measure) of E. Thus, I must be equal to the measure of E

$$I = \mu E. \tag{6}$$

The Lebesgue integral of the function $\phi(x)$ is defined just so.

The reader should realize clearly that the equation (6) is the *definition* of the integral $\int_a^b \phi(x)\, dx$ as a Lebesgue integral. It can happen that the integral I does not exist in the sense in which it was understood in Chapter II, i.e., as a limit of integral sums. But if this is the case, then the integral I exists as a Lebesgue integral and is equal to μE.

§6. THE LEBESGUE INTEGRAL

As an example let us calculate the integral of the Dirichlet function $\Phi(x)$ equal to 0 at the rational points and to 1 at the irrational points of [0, 1]. Since the measure of the set of irrational points of [0, 1] is 1, by (5), the Lebesgue integral

$$\int_0^1 \Phi(x)\, dx$$

is equal to 1. It is easy to verify that the Riemann integral of this function does not exist.

An auxiliary proposition. Suppose now that $f(x)$ is an arbitrary bounded measurable function defined on $[a, b]$. We shall show that every such function can be represented with arbitrarily prescribed accuracy as a linear combination of characteristic functions of sets. In order to see this we divide up the interval of the ordinate axis between the greatest lower bound A and the least upper bound B of the values of the function by points $y_0 = A, y_1, \cdots, y_n = B$ into intervals of length less than ϵ, where ϵ is an arbitrary fixed positive number. Next, if at the point $x \in [a, b]$,

$$y_i \leqslant f(x) < y_{i+1} \quad (i = 0, 1, \cdots, n-1),$$

then we set at that point

$$\phi(x) = y_i,$$

and if at x

$$f(x) = y_n = B,$$

then we set

$$\phi(x) = y_n.$$

The construction of the function $\phi(x)$ is shown in figure 3.

By the construction of $\phi(x)$ we have for every point of $[a, b]$

$$|f(x) - \phi(x)| < \epsilon.$$

Moreover, since the function $\phi(x)$ assumes only a finite number of values $y_0, y_1, \cdots, y_n$, it can be written in the form

$$\phi(x) = y_0 \cdot \phi_0(x) + y_1 \cdot \phi_1(x) + \cdots + y_n \cdot \phi_n(x), \tag{7}$$

where $\phi_i(x)$ is the characteristic function of the set of points for which $\phi(x) = y_i$, i.e., $y_i \leqslant f(x) < y_{i+1}$ (at every point $x \in [a, b]$ only one summand on the right-hand side of (7) is different from zero)! Thus, our proposition is proved.

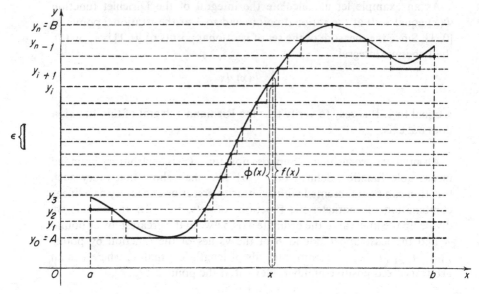

Fig. 3.

Definition of the Lebesgue integral. We now proceed to the definition of the Lebesgue integral of an arbitrary measurable function. Since $\phi(x)$ differs by little from $f(x)$, we can take, as an approximation of the value of the integral of $f(x)$, that of $\phi(x)$. But when we bear in mind that the functions $\phi_i(x)$ are characteristic functions of sets and formally use the ordinary rules of computation for integrals, we find

$$\int_a^b \phi(x)\,dx = \int_a^b \{y_0\phi_0(x) + y_1\phi_1(x) + \cdots + y_n\phi_n(x)\}\,dx$$
$$= y_0\int_a^b \phi_0(x)\,dx + y_1\int_a^b \phi_1(x)\,dx + \cdots + y_n\int_a^b \phi_n(x)\,dx$$
$$= y_0\mu e_0 + y_1\mu e_1 + \cdots + y_n\mu e_n,$$

where μe_i is the measure of the set e_i of those x for which

$$y_i \leqslant f(x) < y_{i+1}.$$

Thus, an approximate value of the Lebesgue integral of $f(x)$ is the "Lebesgue integral sum"

$$S = y_0\mu e_0 + y_1\mu e_1 + \cdots + y_n\mu e.$$

§6. THE LEBESGUE INTEGRAL

In accordance with this the Lebesgue integral is defined as the limit of the Lebesgue integral sum S when

$$\max | y_{i+1} - y_i | \to 0,$$

which corresponds to uniform convergence of $\phi(x)$ to $f(x)$.

It can be shown that the Lebesgue integral sums have a limit for every bounded measurable function, i.e., every bounded measurable function is Lebesgue integrable. The Lebesgue integral can also be extended to certain classes of unbounded measurable functions, but we shall not deal with this here.

Properties of the Lebesgue integral. The Lebesgue integral has all the desirable properties of the ordinary integral, namely, the integral of a sum is equal to the sum of the integrals, a positive factor can be taken before the integral sign, etc. However, the Lebesgue integral has one remarkable property that the ordinary integral does not have: If the functions $f_n(x)$ are measurable and uniformly bounded:

$$| f_n(x) | < K$$

for every n and every x in $[a, b]$ and if the sequence $\{f_n(x)\}$ converges almost everywhere to $f(x)$, then

$$\int_a^b f_n(x) \, dx \to \int_a^b f(x) \, dx.$$

In other words, the Lebesgue integral permits unrestricted passage to the limit. In fact, this property of the Lebesgue integral makes it a very convenient and often an indispensable tool in many investigations. In particular, the Lebesgue integral is absolutely necessary in the theory of trigonometric series, the theory of function spaces (see Chapter XIX), and other branches of mathematics.

Let us give an example. Let $f(x)$ be a periodic function with period 2π and

$$\frac{a_0}{2} + \sum_{n=1}^{\infty} (a_n \cos nx + b_n \sin nx)$$

its Fourier series. If, for example, $f(x)$ is continuous, then it is easy to show that

$$\frac{1}{\pi} \int_0^{2\pi} f^2(x) \, dx = \frac{a_0^2}{2} + \sum_{n=1}^{\infty} (a_n^2 + b_n^2). \tag{8}$$

This identity is known as Parseval's equality. Now we raise the question: For what class of periodic functions is Parseval's equality (8) valid? This is the answer: Parseval's equality (8) is valid if and only if $f(x)$ is measurable on $[0, 2\pi]$ and $f^2(x)$ is Lebesgue integrable on the same interval.

Suggested Reading

F. Hausdorff, *Set theory*, 2nd ed., Chelsea, New York, 1962.
J. M. Hyslop, *Real variable*, Interscience, New York, 1960.
E. Kamke, *Theory of sets*, Dover, New York, 1950.
E. Landau, *Foundations of analysis: the arithmetic of whole, rational, irrational and complex numbers*, Chelsea, New York, 1951.
E. J. McShane and A. T. Betts, *Real analysis*, Van Nostrand, New York, 1959.

CHAPTER **XVI**

LINEAR ALGEBRA

§1. The Scope of Linear Algebra and Its Apparatus

Linear functions and matrices. Among the functions of a single variable, by far the simplest is the so-called *linear function* $l(x) = ax + b$. Its graph is, of course, the simplest of curves, namely the straight line.

All the same, the linear function is one of the most important. This is due to the fact that every "smooth" curve on a small segment is like a straight line, and the less curved the segment is, the nearer it comes to a straight line. In the language of the theory of the functions, this means that every "smooth" (continuously differentiable) function is, for a small change of the independent variable, close to a linear function. The linear function can be characterized by the fact that its increment is proportional to the increment of the independent variable. Indeed: $\Delta l(x) = l(x_0 + \Delta x) - l(x_0) = a(x_0 + \Delta x) + b - (ax_0 + b) = a\, \Delta x$. Conversely, if $\Delta l(x) = a\, \Delta x$, then $l(x) - l(x_0) = a(x - x_0)$ and $l(x) = ax + l(x_0) - ax_0 = ax + b$, where $b = l(x_0) - ax_0$. But from the differential calculus, we know that in the increment of an arbitrary differentiable function we can single out in a natural way the principal part, the so-called differential of the function, which is proportional to the increment of the independent variable, and that the increment of the function differs from its differential by an infinitesimal of higher order than the increment of the independent variable. Thus, a differentiable function is, for an infinitely small change of the independent variable, really close to a linear function to within an infinitesimal of higher order.

The situation is similar with functions of several variables.

A *linear function of several variables* is a function of the form $a_1 x_1 + a_2 x_2 + \cdots + a_n x_n + b$. If $b = 0$, the linear function is said to be *homogeneous.* A linear function of several variables is characterized by the following two properties:

1. The increment of a linear function, computed under the assumption that only one of the independent variables receives some increment while the values of the remaining variables are unchanged, is proportional to the increment of this independent variable.

2. The increment of a linear function, computed under the assumption that all the independent variables obtain increments, is equal to the algebraic sum of the increments obtained by changing each variable separately.

The linear function of several variables plays the same role among all the functions of these variables as the linear function of one variable among all the functions of one variable. For every "smooth" function (i.e., a function having continuous partial derivatives with respect to all variables) is close to some linear function for small changes of the independent variables. In fact, the increment of such a function $w = f(x_1, x_2, \cdots, x_n)$ is equal, to within infinitesimals of higher order, to the total differential $(\partial f/\partial x_1)\, dx_1 + \cdots + (\partial f/\partial x_n)\, dx_n$, which is a linear homogeneous function of the increments $dx_1, \cdots, dx_n$ of the independent variables. Hence it follows that the function w itself, which is equal to the sum of its initial value and its increment, can be expressed in terms of its independent variables for small changes of these in the form of a linear inhomogeneous function to within infinitesimals of higher orders.

Problems whose solution requires the investigation of functions of several variables arise in connection with the study of the dependence of one quantity on several factors. A problem is called *linear* if the dependence under consideration turns out to be linear. By the properties of linear functions that we have indicated earlier, a linear problem can be characterized by the following properties.

1. *The property of proportionality.* The result of the action of each separate factor is proportional to its value.

2. *The property of independence.* The total result of an action is equal to the sum of the results of the actions of the separate factors.

The fact that every "smooth" function can be replaced in a first approximation by a linear one, for small changes of the variables, is a reflection of a general principle, namely that every problem on the change of some quantity under the action of several factors can be regarded in a first approximation, for small actions, as a linear problem, i.e., as having the properties of independence and proportionality. It often turns out that this attitude gives an adequate result for practical purposes (the classical theory of elasticity, the theory of small oscillations, etc.)

The physical quantities to be studied are often characterized by certain

§1. THE SCOPE AND APPARATUS

numbers (a force by the three projections on the coordinate axes, the tension at a given point of an elastic body by the six components of the so-called stress tensor, etc.). Hence there arises the necessity of considering simultaneously several functions of several variables, and, in a first approximation, of several linear functions.

A linear function of one variable is so simple in its properties that it does not require any special study. Things are different with linear functions of several variables, where the presence of many variables introduces some special features. The situation is still more complicated when we go from a single function of several variables $x_1, x_2, \cdots, x_n$ to a set of several functions $y_1, y_2, \cdots, y_m$ of the same variables. As a "first approximation" there appears here a set of linear functions:

$$y_1 = a_{11}x_1 + \cdots + a_{1n}x_n + b_1,$$
$$y_2 = a_{21}x_1 + \cdots + a_{2n}x_n + b_2,$$
$$\cdots\cdots\cdots\cdots\cdots\cdots\cdots\cdots\cdots\cdots\cdots\cdots$$
$$y_m = a_{m1}x_1 + \cdots + a_{mn}x_n + b_m.$$

A set of linear functions is already a rather complicated mathematical object, and the study of its full of interesting and nontrivial results.

The study of linear functions and their systems also constitutes the initial object of that branch of algebra that is called linear algebra.

Historically, the first task of linear algebra is that of solving a system of linear equations:

$$a_{11}x_1 + \cdots + a_{1n}x_n = b_1,$$
$$a_{21}x_1 + \cdots + a_{2n}x_n = b_2,$$
$$\cdots\cdots\cdots\cdots\cdots\cdots\cdots\cdots\cdots\cdots$$
$$a_{m1}x_1 + \cdots + a_{mn}x_n = b_m.$$

The simplest case of this problem is treated in a school course on elementary algebra. The problem of finding methods for the simplest possible and least laborious numerical solution of systems for large n still attracts the close attention of many researchers, because the numerical solution of such systems enters as an important constituent part into many calculations and investigations.

Linear homogeneous functions are also known as *linear forms*. A given system of linear forms

$$y_1 = a_{11}x_1 + \cdots + a_{1n}x_n,$$
$$\cdots\cdots\cdots\cdots\cdots\cdots\cdots\cdots\cdots\cdots$$
$$y_m = a_{m1}x_1 + \cdots + a_{mn}x_n$$

is completely described by its system of coefficients, since the properties of such a system of forms depend only on the numerical values of the coefficients and the names of the variables are inessential.

For example, the system of forms

$$3x_1 + x_2 - x_3, \qquad 3t_1 + t_2 - t_3,$$
$$2x_1 + x_2 + 3x_3, \quad \text{and} \quad 2t_1 + t_2 + 3t_3,$$
$$x_1 - x_2 - x_3 \qquad t_1 - t_2 - t_3$$

obviously have identical properties and need not be regarded as essentially distinct.

The set of coefficients of a system of linear forms can be given in a natural way in the form of a rectangular array

$$\begin{bmatrix} a_{11} & \cdots & a_{1n} \\ \cdots & \cdots & \cdots \\ a_{m1} & \cdots & a_{mn} \end{bmatrix}$$

Such arrays bear the name of *matrices*. The numbers a_{ij} are called the elements of the matrix. The need of considering matrices arises necessarily from the very scope of linear algebra.

Important special cases of matrices are the matrices that consist of a single column, which are simply called columns, those that consist of a single row, called rows, and finally the square matrices, i.e., those in which the number of rows is equal to the number of columns. The number of rows (or columns) of a square matrix is called its *order*. The "matrix" (a) consisting of a single number is identified with that number.

In connection with the simplest operations on a set of linear forms, it is natural to define operations on matrices.

Suppose that two systems of linear forms are given,

$$y_1 = a_{11}x_1 + \cdots + a_{1n}x_n,$$
$$\cdots\cdots\cdots\cdots\cdots\cdots\cdots\cdots$$
$$y_m = a_{m1}x_1 + \cdots + a_{mn}x_n,$$

and

$$z_1 = b_{11}x_1 + \cdots + b_{1n}x_n,$$
$$\cdots\cdots\cdots\cdots\cdots\cdots\cdots\cdots$$
$$z_m = b_{m1}x_1 + \cdots + b_{mn}x_n.$$

Let us add these forms,

$$y_1 + z_1 = (a_{11} + b_{11})x_1 + \cdots + (a_{1n} + b_{1n})x_n,$$
$$\cdots\cdots\cdots\cdots\cdots\cdots\cdots\cdots\cdots\cdots\cdots\cdots\cdots$$
$$y_m + z_m = (a_{m1} + b_{m1})x_1 + \cdots + (a_{mn} + b_{mn})x_n.$$

§1. THE SCOPE AND APPARATUS

It is natural to say that the matrix of the system of forms so obtained

$$\begin{bmatrix} a_{11} + b_{11} & \cdots & a_{1n} + b_{1n} \\ \cdots & \cdots & \cdots \\ a_{m1} + b_{m1} & \cdots & a_{mn} + b_{mn} \end{bmatrix}$$

is the sum of the matrices

$$\begin{bmatrix} a_{11} & \cdots & a_{1n} \\ \cdots & \cdots & \cdots \\ a_{m1} & \cdots & a_{mn} \end{bmatrix} \quad \text{and} \quad \begin{bmatrix} b_{11} & \cdots & b_{1n} \\ \cdots & \cdots & \cdots \\ b_{m1} & \cdots & b_{mn} \end{bmatrix}.$$

Similarly, the product of the matrix

$$\begin{bmatrix} a_{11} & \cdots & a_{1n} \\ \cdots & \cdots & \cdots \\ a_{m1} & \cdots & a_{mn} \end{bmatrix}$$

by the number c is defined as the matrix of the coefficients in the system of forms $cy_1, cy_2, \cdots, cy_m$, where $y_1, y_2, \cdots, y_m$ are the forms whose coefficients constitute the matrix

$$\begin{bmatrix} a_{11} & \cdots & a_{1n} \\ \cdots & \cdots & \cdots \\ a_{m1} & \cdots & a_{mn} \end{bmatrix}.$$

From this definition it is clear that

$$c \begin{bmatrix} a_{11} & \cdots & a_{1n} \\ \cdots & \cdots & \cdots \\ a_{m1} & \cdots & a_{mn} \end{bmatrix} = \begin{bmatrix} ca_{11} & \cdots & ca_{1n} \\ \cdots & \cdots & \cdots \\ ca_{m1} & \cdots & ca_{mn} \end{bmatrix}.$$

Finally, the operation of multiplication of a matrix by a matrix is defined as follows. Suppose that

$$\begin{aligned} z_1 &= a_{11} y_1 + \cdots + a_{1m} y_m, \\ &\cdots \cdots \cdots \cdots \cdots \cdots \cdots \cdots \cdots \\ z_k &= a_{k1} y_1 + \cdots + a_{km} y_m \end{aligned} \tag{1}$$

and

$$\begin{aligned} y_1 &= b_{11} x_1 + \cdots + b_{1n} x_n, \\ &\cdots \cdots \cdots \cdots \cdots \cdots \cdots \cdots \\ y_m &= b_{m1} x_1 + \cdots + b_{mn} x_n. \end{aligned}$$

When we substitute in (1) the expressions of $y_1, y_2, \cdots, y_m$ in terms of

$x_1, x_2, \cdots, x_n$, we find that $z_1, z_2, \cdots, z_k$ can also be expressed in terms of $x_1, x_2, \cdots, x_n$ by linear forms

$$z_1 = c_{11}x_1 + \cdots + c_{1n}x_n,$$
$$\cdots\cdots\cdots\cdots\cdots\cdots\cdots\cdots$$
$$z_k = c_{k1}x_1 + \cdots + c_{kn}x_n.$$

The matrix of coefficients

$$\begin{bmatrix} c_{11} & \cdots & c_{1n} \\ \cdots & \cdots & \cdots \\ c_{k1} & \cdots & c_{kn} \end{bmatrix}$$

is called the product of the matrices

$$\begin{bmatrix} a_{11} & \cdots & a_{1m} \\ \cdots & \cdots & \cdots \\ a_{k1} & \cdots & a_{km} \end{bmatrix} \text{ and } \begin{bmatrix} b_{11} & \cdots & b_{1n} \\ \cdots & \cdots & \cdots \\ b_{m1} & \cdots & b_{mn} \end{bmatrix}$$

and is denoted by

$$\begin{bmatrix} a_{11} & \cdots & a_{1m} \\ \cdots & \cdots & \cdots \\ a_{k1} & \cdots & a_{km} \end{bmatrix} \begin{bmatrix} b_{11} & \cdots & b_{1n} \\ \cdots & \cdots & \cdots \\ b_{m1} & \cdots & b_{mn} \end{bmatrix}.$$

It is easy to calculate how the elements of the product of two matrices are expressed in terms of the elements of its factors. The element c_{ij} is the coefficient of x_j in the expression for z_i in terms of $x_1, x_2, \cdots, x_n$.

But $z_i = a_{i1}y_1 + \cdots + a_{im}y_m$ and

$$y_1 = \cdots + b_{1j}x_j + \cdots,$$
$$\cdots\cdots\cdots\cdots\cdots\cdots\cdots\cdots$$
$$y_m = \cdots + b_{mj}x_j + \cdots.$$

Therefore,

$$z_i = \cdots + (a_{i1}b_{1j} + \cdots + a_{im}b_{mj})x_j + \cdots,$$

hence

$$c_{ij} = a_{i1}b_{1j} + \cdots + a_{im}b_{mj}.$$

Thus, the element in the ith row and the jth column of the product of two matrices is equal to the sum of the products of the elements of the ith row of the first factor by the corresponding elements of the jth column of the second factor. For example:

$$\begin{pmatrix} 2 & 1 & 3 \\ 3 & 1 & 1 \end{pmatrix} \begin{pmatrix} 3 & 1 \\ 1 & 2 \\ 2 & 4 \end{pmatrix} = \begin{pmatrix} 2\cdot 3 + 1\cdot 1 + 3\cdot 2 & 2\cdot 1 + 1\cdot 2 + 3\cdot 4 \\ 3\cdot 3 + 1\cdot 1 + 1\cdot 2 & 3\cdot 1 + 1\cdot 2 + 1\cdot 4 \end{pmatrix} = \begin{pmatrix} 13 & 16 \\ 12 & 9 \end{pmatrix}.$$

§1. THE SCOPE AND APPARATUS

Although a matrix is, so to speak, a "composite" object and many elements enter into its formation, it is useful and convenient to denote it by a single letter and to preserve the usual notation for operations of addition and multiplication. We shall use the capital letters of the Latin alphabet to denote matrices. The application of such a concise notation brings simplicity and lucidity into the theory of matrices by embracing in short formulas, that remind us of the formulas of ordinary algebra, complicated relations connecting a set of numbers, namely the elements of the matrix that occur in these formulas. Thus, for example, the set of linear forms

$$a_{11}x_1 + \cdots + a_{1n}x_n,$$
$$\dots\dots\dots\dots\dots\dots\dots\dots\dots$$
$$a_{m1}x_1 + \cdots + a_{mn}x_n$$

appears in matrix notation as AX, where A is the coefficient matrix and X the "column" formed by the variables $x_1, x_2, \cdots, x_n$. The system of linear equations

$$a_{11}x_1 + \cdots + a_{1n}x_n = b_1,$$
$$\dots\dots\dots\dots\dots\dots\dots\dots\dots$$
$$a_{m1}x_1 + \cdots + a_{mn}x_n = b_m$$

is written as

$$AX = B,$$

where A is the coefficient matrix, X the column of the unknowns, and B the column of the absolute terms.

The fundamental operations on matrices, namely addition and multiplication, are, of course, not always defined. The operation of addition makes sense for matrices of *equal structure*, i.e., having the same number of rows and of columns. As the result of addition, we obtain a matrix of the same structure. The operation of multiplication makes sense if the number of columns of the first matrix is equal to the number of rows of the second. As the result we obtain a matrix in which the number of rows is equal to the number of rows of the first factor and the number of columns is equal to the number of columns of the second factor.

The operations on square matrices are subject to most of the laws for operations on number, but some of the laws turn out to be violated.

Let us enumerate the fundamental properties for operations on matrices:

1. $A + B = B + A$ (commutative law for addition).
2. $(A + B) + C = A + (B + C)$ (associative law for addition).
3. $c(A + B) = cA + cB$ ⎫ (distributive laws for multiplication by a number. Here c, c_1, c_2 are
3'. $(c_1 + c_2)A = c_1A + c_2A$ ⎭ numbers and not matrices).

4. $(c_1 c_2) A = c_1(c_2 A)$ (associative law for multiplication by a number).

5. There exists a "null" matrix

$$O = \begin{pmatrix} 0 & \cdots & 0 \\ \cdots & \cdots & \cdots \\ 0 & \cdots & 0 \end{pmatrix}$$

such that $A + O = A$ for every matrix A.

6. $c \cdot O = 0 \cdot A = O$; conversely, if $cA = O$, then $c = 0$ or $A = O$ (here c is a number).

7. For every matrix A there exists an opposite matrix $-A$, i.e., such that $A + (-A) = O$.

8. $(A + B) \cdot C = AC + BC$ } (distributive laws for addition and
8'. $C(A + B) = CA + CB$ } multiplication of matrices).

9. $(AB) C = A(BC)$ (associative law for multiplication).

10. $(cA) B = A(cB) = c(AB)$.

These properties hold not only for square matrices but also for arbitrary rectangular matrices with the sole proviso that the operations that occur in each of the numbered formulas must be defined. For square matrices of equal order this proviso is automatically fulfilled.

All these properties of the operations are similar to the properties of operations on numbers.

We shall now point out two peculiarities of the operations on matrices. First, the commutative law for the multiplication of matrices, even square ones, need not hold; i.e., AB is not always equal to BA. For example:

$$\begin{pmatrix} 3 & -2 \\ -1 & 4 \end{pmatrix} \begin{pmatrix} 1 & 2 \\ 3 & 2 \end{pmatrix} = \begin{pmatrix} -3 & 2 \\ 11 & 6 \end{pmatrix};$$

$$\begin{pmatrix} 1 & 2 \\ 3 & 2 \end{pmatrix} \begin{pmatrix} 3 & -2 \\ -1 & 4 \end{pmatrix} = \begin{pmatrix} 1 & 6 \\ 7 & 2 \end{pmatrix}.$$

Second, the product of two numbers is, of course, equal to zero if and only if one of the factors is equal to zero. This theorem is well known to be fundamental in the theory of algebraic equations. But under multiplication of matrices it turns out to be false. For the product of two matrices may be equal to the null matrix, although neither factor is equal to the null matrix. For example:

$$\begin{pmatrix} 1 & 1 \\ 1 & 1 \end{pmatrix} \begin{pmatrix} 1 & 1 \\ -1 & -1 \end{pmatrix} = \begin{pmatrix} 0 & 0 \\ 0 & 0 \end{pmatrix}.$$

§1. THE SCOPE AND APPARATUS

Let us mention yet another property of the multilplication of matrices. The matrix $\bar{A}$ is called the *transpose* of A if in every row of $\bar{A}$ there stand the elements of the corresponding column of A in the same order. For example, for the matrix

$$A = \begin{bmatrix} 1 & 2 \\ 3 & 4 \\ 5 & 6 \end{bmatrix}$$

the transpose is the matrix

$$\bar{A} = \begin{bmatrix} 1 & 3 & 5 \\ 2 & 4 & 6 \end{bmatrix}.$$

The operation of multiplication is connected with that of transposition by the formula

$$\overline{AB} = \bar{B}\bar{A},$$

which is easily verified on the basis of the multiplication rule for matrices.

The theory of matrices forms an indispensable part of linear algebra in that it plays the role of an apparatus for stating and solving its problems.

Geometric analogies in linear algebra. Apart from the earlier described source for the emergence of the ideas and problems of linear algebra, there are also the needs of mathematical analysis and geometry, in particular analytic geometry, that lead to the development of linear algebra and, in turn, enrich it by important ideas and analogies. It is well known that the analytic geometry of the plane, and, in an even greater measure, of the space, as far as the theory of straight lines and planes is concerned, makes use of the apparatus of linear algebra in its simplest form. For a straight line in the plane is given by a linear equation in two variables that links the two coordinates of an arbitrary point of the line. A plane in space is given by a linear equation in three variables (the coordinates of an arbitrary point of this plane), a line in space by two linear equations.

A special simplicity and clarity is, of course, brought into analytical geometry and consequently into the theory of the simplest systems of linear equations by the use of a concept of a vector. Now a similar simplicity and clarity is brought into linear algebra, in particular into the general theory of systems of linear equations, by the use of the concept of a vector in a generalized sense. The way to this generalization is the following. A vector (in space) is given by three numbers, namely its three projections on the coordinate axes. Every triplet of real numbers in turn can be represented geometrically in the form of a vector (in space).

For vectors the operations of addition ("by the parallelogram rule") and multiplication by a number are defined. These operations are defined in accordance with similar operations on forces, velocities, accelerations, and other physical quantities that can be represented by means of vectors.

If vectors are given by their coordinates (i.e., their projections on the coordinate axes), then the operations of addition and multiplication by a number performed on vectors correspond to the analogous operations on the rows (or columns) of their coordinates.

Thus, it is convenient to interpret a row or column of three elements geometrically as a vector in three-dimensional space, and then the basic operations on "rows" (or "columns") are interpreted by the corresponding operations on vectors in space, so that the algebra of rows (or columns) of three elements formally does not differ at all from the algebra of the vectors of three-dimensional space. This circumstance makes it natural to introduce a geometric terminology into linear algebra.

A column (or row) of n numbers

$$\begin{pmatrix} x_1 \\ x_2 \\ \vdots \\ x_n \end{pmatrix}$$

is regarded as a "vector", i.e., as an element of some "n-dimensional vector space." The sum of the vectors

$$\begin{pmatrix} x_1 \\ x_2 \\ \vdots \\ x_n \end{pmatrix} \text{ and } \begin{pmatrix} y_1 \\ y_2 \\ \vdots \\ y_n \end{pmatrix}$$

is taken to be the vector

$$\begin{pmatrix} x_1 + y_1 \\ x_2 + y_2 \\ \cdots \\ x_n + y_n \end{pmatrix};$$

the product of the vector

$$\begin{pmatrix} x_1 \\ x_2 \\ \vdots \\ x_n \end{pmatrix}$$

§1. THE SCOPE AND APPARATUS 47

by the number c is taken to be the vector

$$\begin{pmatrix} cx_1 \\ cx_2 \\ \vdots \\ cx_n \end{pmatrix}.$$

The set of all vectors (columns) forms, by definition, the n-dimensional arithmetical vector space.

Together with the n-dimensional arithmetical vector space we can introduce the concept of an n-dimensional point space, by associating with each column of n real numbers a geometrical image, namely a point. Then the n-dimensional vector space is defined in the following way.

With every pair of points A and B we associate the vector $\overrightarrow{AB}$ leading from A to B by taking as its coordinates (its projections on the coordinate axes), by definition, the difference of the corresponding coordinates of the points B and A. Two vectors are taken to be equal if their corresponding coordinates are equal, just as in three-dimensional space we regard vectors as equal if one of them is obtained from the other by a parallel shift.

Between the vectors of an n-dimensional vector space and the points of an n-dimensional point space, there exists a natural one-to-one correspondence.

The point

$$\begin{pmatrix} 0 \\ 0 \\ \vdots \\ 0 \end{pmatrix}$$

is taken as the "origin of coordinates," and to every point there corresponds the vector that joins the origin to that point. Then we associate with every vector the point that is the end point of this vector, assuming that its beginning coincides with the origin. The introduction of the point space creates new analogies that enable us to "see" better in n-dimensional space.

However, in the further generalizations (§2) a rigorous definition of a point space becomes rather more complicated, and we shall, therefore, not make use of this concept. The reader who wishes to use the analogies arising from the investigation of a point space should visualize the elements of a vector space as vectors emanating from the origin of coordinates.

The introduction of a geometric terminology enables us to use in linear algebra analogies based on the geometric intuition which originates in the study of the geometry of three-dimensional space.

Of course, these analogies must be used with a certain care, bearing in mind that every intuitive-geometric argument can be checked in a strictly logical way applying only precise definitions of "geometric" concepts and rigorous proofs of theorems.

A characteristic feature of the elements of an n-dimensional vector space is the existence of the operations of addition and multiplication by a number, with properties reminiscent of the operations on numbers. Namely, as we have already mentioned in the account of the properties of operations on matrices, for the operation of addition the commutative and associative laws are satisfied, the distributive laws (for multiplication by a number) hold, the operation of addition has a unique inverse, and the product of a vector by a number gives the null vector if and only if either the vector is the null vector or the number is zero.

However, not only these columns (and rows) possess the features referred to. Such features also belong to the set of matrices of equal structure and to physical vector quantities: forces, velocities, accelerations, etc. They also belong to some mathematical objects of an altogether different nature, for example: the set of all polynomials in one variable, the set of all continuous functions on a given interval $[a, b]$, the set of all solutions of a linear homogeneous differential equation, etc.

This circumstance motivates a further generalization of a vector space, namely the introduction of general linear spaces. The elements of such generalized spaces may be arbitrary mathematical or physical objects for which the operations of addition and multiplication by a number are defined in a natural fashion. Such a very general and abstract approach to the concept of a linear space does not bring any complications into the theory, as we have seen earlier: Every linear space (of course, n-dimensional; the meaning of this will be clarified in the next section) does not differ in its structure and its properties from the arithmetical linear space, but the field of applicability is considerably extended by this generalization and it becomes possible to apply the methods of linear algebra to a very wide range of problems of theoretical science.

§2. Linear Spaces

Definition of a linear space. We now proceed to a rigorous definition of a linear space.

A *linear space* is a collection of objects of an arbitrary nature for which

§2. LINEAR SPACES

the concepts of a sum and of a product by a number make sense and which satisfy the following postulates:

1. $(X + Y) + Z = X + (Y + Z)$.
2. There exists a "null" element 0 such that $X + 0 = X$ for every X.
3. For every element X there exists an opposite $-X$ such that $X + (-X) = 0$.
4. $X + Y = Y + X$.
5. $1 \cdot X = X$.
6. $c_1(c_2 X) = c_1 c_2 X$.
7. $(c_1 + c_2) X = c_1 X + c_2 X$.
8. $c(X + Y) = cX + cY$.

Here X, Y, Z are elements of the linear space; 1, c_1, c_2, c are numbers.

These postulates (whibh are also called the *axioms of a linear space*) are very natural and constitute a formal account of those properties of the operations of addition and multiplication by a number that are necessarily linked with the concept of these operations in whatever generalized sense they are to be understood. Operations having one physical meaning or another are, in fact, treated as addition and multiplication by a number in all cases when these operations satisfy the postulates 1–8.

We mention some consequences of these axioms:

a. The null element 0 of the space is unique, i.e., there exists only one element satisfying axiom 2.

b. The opposite element of a given element X is unique.

c. "Subtraction" has a meaning; i.e., when a sum and one of the summands is given, the other summand is always defined, in fact, uniquely: If $X + Z = Y$, then $Z = Y + (-X)$.

d. $0 \cdot X = c \cdot 0 = 0$.

e. If $cX = 0$, then either $c = 0$, or $X = 0$.

f. $-X = (-1) X$.

The proof of these consequences are very simple and will be omitted. In what follows the elements of a linear space will be called vectors.

Linear dependence and independence of vectors. We now proceed to the important concept of linear dependence and independence of vectors.

The vector $c_1 X_1 + c_2 X_2 + \cdots + c_m X_m$ with arbitrary numerical values of the coefficients $c_1, c_2, \cdots, c_m$ is called a linear combination of the vectors $X_1, X_2, \cdots, X_m$. If among the vectors $X_1, X_2, \cdots, X_m$ there is at least one that is a linear combination of the remaining ones, then the vectors $X_1, X_2, \cdots, X_m$ are called linearly dependent. But if none of the vectors $X_1, X_2, \cdots, X_m$ is a linear combination of the remaining ones, then the vectors $X_1, X_2, \cdots, X_m$ are called linearly independent.

It is easy to see that for linear independence of the vectors $X_1, X_2, \cdots, X_m$ it is necessary and sufficient that the relation $c_1X_1 + c_2X_2 + \cdots + c_mX_m = 0$ should hold for $c_1 = c_2 = \cdots = c_m = 0$ only.

For vectors of the ordinary three-dimensional space the concepts of linear dependence and independence have a simple geometrical meaning.

Suppose two vectors X_1 and X_2 are given. Their linear dependence means that one of the vectors is a "linear combination" of the other, i.e., that they differ simply by a numerical factor. This means that both vectors belong to a common straight line, i.e., that they have equal or opposite direction.

Conversely, if two vectors are contained in one straight line, then they are linearly dependent. Consequently linear independence of two vectors X_1 and X_2 means that these vectors cannot be placed on one straight line; their directions are essentially distinct.

Let us now investigate what linear dependence and independence of three vectors means. Suppose that the vectors X_1, X_2 and X_3 are linearly dependent and, for the sake of definiteness, that the vector X_3 is a linear combination of the vectors X_1 and X_2. Then X_3 obviously lies in a plane containing the vectors X_1 and X_2; i.e., all three vectors X_1, X_2, X_3 belong to one plane. It is easy to see that if the vectors X_1, X_2, X_3 lie in one plane, then they are linearly dependent. For if the vectors X_1 and X_2 do not lie on one line, then X_3 can be decomposed with respect to X_1 and X_2, i.e., represented as a linear combination of X_1 and X_2. But if X_1 and X_2 lie on one line, then already X_1 and X_2 are linearly dependent.

Thus, linear dependence of three vectors X_1, X_2, X_3 is equivalent to the fact that they lie in one plane. Therefore X_1, X_2, X_3 are linearly independent if and only if they do not belong to one plane.

Four vectors in three-dimensional space are always linearly dependent. For if the vectors X_1, X_2, X_3 are linearly dependent, then the vectors X_1, X_2, X_3, X_4 are also linearly dependent for any X_4. But if X_1, X_2, X_3 are linearly independent, then they do not lie in one plane and every vector X_4 can be decomposed with respect to X_1, X_2, X_3, i.e., represented as a linear combination of them.

The preceding arguments can be generalized in the following way.

In three-dimensional space the vectors $X_1, X_2, \cdots, X_k$ $(k \geqslant 3)$ are linearly dependent if and only if they belong to a space (straight line or plane) of a dimension less than k.

In what follows we shall see after a rigorous definition of subspace and dimension that also in the general case linear dependence of the vectors $X_1, X_2, \cdots, X_k$ is equivalent to the fact that they belong to a space whose dimension is less than k; i.e., the "geometrical" meaning of linear depen-

§2. LINEAR SPACES

dence remains the same as for vectors in three-dimensional space.

The following theorem plays a fundamental role in the theory of linear spaces. If the vectors $X_1, X_2, \cdots, X_m$ are linear combinations of the vectors $Y_1, Y_2, \cdots, Y_k$ and $m > k$, then $X_1, X_2, \cdots, X_m$ are linearly dependent (theorem on the linear dependence of linear combinations).

For $k = 1$ the theorem is obvious. For $k > 1$ it is easily proved by the method of mathematical induction with respect to k.

Basis and dimension of a space. In three-dimensional space any three vectors X_1, X_2, X_3 that do not lie in one plane (i.e., that are linearly independent) form a *basis* of the space, which means that every vector of the space can be decomposed with respect to X_1, X_2, X_3, i.e., represented as a linear combination of them.

General linear vector spaces can be divided into two types.

It can happen that a space contains an arbitrarily large number of linearly independent vectors. Such spaces are called infinite-dimensional and their study leads to a branch of linear algebra that is the topic of a special mathematical discipline, functional analysis (see Chapter XIX).

A linear space is called finite-dimensional if there exists a finite bound for the number of linearly independent vectors, i.e., a number n such that there exist in the space n linearly independent vectors, but that any vectors more than n in number are linearly dependent. The number n is called the *dimension* of the space.

Thus, the space of vectors of the ordinary geometrical three-dimensional space is three-dimensional also in the sense of the general definition we have given. For in the three-dimensional geometric space, there exist many triplets of linearly independent vectors, but any four vectors are linearly dependent.

The space of n-term columns is n-dimensional in the sense of our definition. For there are n linearly independent vectors in the space, for example

$$e_1 = \begin{pmatrix} 1 \\ 0 \\ \vdots \\ 0 \end{pmatrix}, \quad e_2 = \begin{pmatrix} 0 \\ 1 \\ \vdots \\ 0 \end{pmatrix}, \quad \cdots, \quad e_n = \begin{pmatrix} 0 \\ 0 \\ \vdots \\ 1 \end{pmatrix} \tag{2}$$

but every vector

$$\begin{pmatrix} x_1 \\ x_2 \\ \vdots \\ x_n \end{pmatrix}$$

of the space is a linear combination of them, namely: $x_1 e_1 + x_2 e_2 + \cdots + x_n e_n$. Therefore by the theorem of linear dependence of linear combinations any vectors more than n in number are linearly dependent.

Polynomials in one variable form a linear space. For there is a natural definition of the operations of addition and of multiplication by a number for polynomials, and they satisfy the axioms 1–8. However, this space is infinite-dimensional, since the vectors $1, x, \cdots, x^N$ are linearly independent for any N. But the set of polynomials whose degree does not exceed a given number N form a finite-dimensional space whose dimension is $N + 1$. For the vectors $1, x, \cdots, x^N$ are linearly independent and their number is $N + 1$. Now every polynomial whose degree does not exceed N is a linear combination of $1, x, \cdots, x^N$ so that by the theorem on linear independence any polynomials of degree $\leqslant N$, if they are more than $N + 1$ in number, are linearly dependent.

We now introduce the important concept of a *basis* for an n-dimensional space. A basis is defined as a set of linearly independent vectors of the space such that every vector of the space is a linear combination of vectors of this set. Thus, in the space of columns a basis is, for example, the set of vectors (2). In the space of polynomials of degree $\leqslant N$ the "vectors" $1, x, \cdots, x^N$ can be taken as a basis. In the three-dimensional geometrical space any triplet of linearly independent vectors plays the role of a basis.

In an n-dimensional linear space, every set of n linearly independent vectors (and the existence of at least one such set is part of the definition of an n-dimensional space) form a basis of the space. For let $e_1, e_2, \cdots, e_n$ be linearly independent vectors of an n-dimensional linear space and X an arbitrary vector of the space. Then the vectors $X, e_1, \cdots, e_n$ are linearly dependent (since their number is more than n), i.e., there are numbers $c, c_1, c_2, \cdots, c_n$, not all equal to zero, such that $cX + c_1 e_1 + \cdots + c_n e_n = 0$. Here $c \neq 0$, because if we had $c = 0$, then the vectors $e_1, e_2, \cdots, e_n$ would be linearly dependent. Therefore $X = -(c_1/c) e_1 - \cdots - (c_n/c) e_n$; i.e., every vector of the space is a linear combination of the vectors $e_1, e_2, \cdots, e_n$.

Every basis of an n-dimensional linear space consists of exactly n vectors. For the vectors of a basis are linearly independent and therefore their number cannot be larger than n. On the other hand, let $e_1, e_2, \cdots, e_k$ be an arbitrary basis of an n-dimensional space. We have already established that $k \leqslant n$. But every vector of the space, by definition of a basis, is a linear combination of the vectors $e_1, e_2, \cdots, e_k$ and by the theorem on linear dependence of linear combinations any vectors, more than k in number, are linearly dependent, from which it follows that the dimension n of the space is not larger than the number k of vectors of a basis. Thus, $k = n$, and this is what we had to prove.

§2. LINEAR SPACES

We now introduce *coordinates* of a vector with respect to a given basis $e_1, e_2, \cdots, e_n$. As we have shown earlier, every vector X is a linear combination of the vectors of the basis. This representation is unique. For suppose that the vector X is expressed in terms of the basis $e_1, e_2, \cdots, e_n$ in two ways:

$$X = x_1 e_1 + x_2 e_2 + \cdots + x_n e_n,$$

$$X = x'_1 e_1 + x'_2 e_2 + \cdots + x'_n e_n.$$

Then $(x_1 - x'_1) e_1 + (x_2 - x'_2) e_2 + \cdots + (x_n - x'_n) e_n = 0$, and from this it follows by the linear independence of $e_1, e_2, \cdots, e_n$ that $x = x'_1, \cdots, x_n = x'_n$.

The coefficients $x_1, x_2, \cdots, x_n$ in the decomposition of an arbitrary vector X in terms of the vectors of a basis are called the *coordinates* of X in this basis. In this way every vector, *once a basis of the space is chosen*, can in a natural manner be associated with the row (or column) of its coordinates and vice versa: every row (or column) of n numbers can be regarded as the set of coordinates of a certain vector.

The operations of addition of vectors and multiplication of a vector by a number correspond to the similar operations on the rows (or columns) of their coordinates.

Therefore every n-dimensional linear space, irrespective of the nature of its elements (they may be functions, matrices, any physical quantities whatsoever, etc.), does not differ at all from the space of rows (or columns) with respect to these operations. Thus, as we have already mentioned, the generalized axiomatic approach to the concept of a linear space does not lead to any complications in comparison with the treatment of the space as a space of rows, but it extends the domain of applicability of this concept considerably.

An identity of properties of two sets of objects in relation to a given system of operations (or arbitrary other relations between their elements) is called in mathematics an isomorphism. An exact definition of isomorphism of algebraic systems will be given in Chapter XX. Using this term we can say that all n-dimensional linear spaces, irrespective of the nature of their elements, are isomorphic to one another and isomorphic to a single model, namely the space of rows.

Subspaces. A set of vectors of an n-dimensional linear space R_n, satisfying the condition that every linear combination of arbitrary vectors of the set under consideration also belongs to it, is called a *subspace* of the space. Obviously a subspace of the space R_n is itself a linear space and has, therefore, bases and a dimension. It is also obvious that the dimension of the subspace does not exceed the dimension of the whole space and can be equal to it if and only if the subspace coincides with the whole space.

Examples of subspaces of the three-dimensional vector space are the planes and lines we have studied, to within a translation, more accurately, the sets of all vectors that lie in a plane or on a line.

Very frequently we have to investigate the subspaces "spanned" by a system of vectors. These subspaces are defined as follows. Suppose that a system of linearly independent or dependent vectors $X_1, X_2, \cdots, X_m$ of the space R_n is given. Then the set of all linear combinations of these vectors $\{c_1 X_1 + c_2 X_2 + \cdots + c_m X_m\}$ forms a subspace of R_n which is called the subspace spanned by the vectors $X_1, X_2, \cdots, X_m$.

The dimension of this subspace is called the *rank* of the system of vectors $X_1, X_2, \cdots, X_m$. It is easy to see that the rank of a system of vectors is equal to the maximal number of linearly independent vectors contained in the system.

The "set" consisting only of the null vector formally satisfies the conditions imposed on a subspace. The dimension of this subspace is taken to be zero.

If two subspaces of a space R_n are given, then we can form from them in a natural manner two other subspaces, their vector sum (or union) and their intersection.

The *vector sum* of two subspaces P and Q is defined as the set of all sums of vectors belonging to the subspaces P and Q. The vector sum can also be regarded as the subspace spanned by the union of the bases of the subspaces P and Q.

The *intersection* of two subspaces is defined as the set of all vectors that belong to both subspaces. For example, the vector sum of two planes (i.e., two-dimensional vector subspaces) in the ordinary three-dimensional space is the whole space (provided only that the planes do not coincide) and their intersection is a straight line (under the same proviso).

The dimensions p and q of the two given subspaces, the dimension t of their vector sum, and the dimension s of their intersection satisfy the following interesting relation:

$$p + q = t + s.$$

We omit the proof of this statement.

From this relation we can make certain deductions concerning the intersection of subspaces in special cases. For example, two noncoincident planes (i.e., two-dimensional subspaces) in a space of four dimensions intersect in general only in a point (the dimension of their intersection is zero) and two planes intersect in a straight line only if their vector sum is three-dimensional, i.e., if both planes belong to some three-dimensional subspace. For in this case $t + s = 2 + 2 = 4$, from which it follows that $s = 1$ only when $t = 3$.

2§. LINEAR SPACES

Complex linear spaces. In the description of the space of rows and of the general linear space, we have not specified what sort of numbers we are dealing with in the definition of the operation of multiplication of a vector by a number. Since we have started out from a generalization of the ordinary vectors, i.e., the directed segments in the geometrical three-dimensional space, we have had in mind arbitrary real numbers. The so-constructed linear spaces, which are called real linear spaces, generalize in the most natural way the three-dimensional space of ordinary vectors. However, in many problems of contemporary mathematics it turns out to be useful to consider a *complex linear space*. By this we mean a collection of objects for which the operations of addition and of multiplication by an arbitrary complex number are defined so that these operations satisfy all the axioms 1–8. As an example of a complex space, we can take the space of rows whose elements are arbitrary complex numbers.

Formally the theory of complex spaces does not differ essentially from the theory of real spaces.

However, even a two-dimensional complex space does not have an intuitive geometric interpretation. The fact is that an n-dimensional complex space can also be regarded as a real one, in view of the fact that since the operation of multiplication by an arbitrary complex number is defined for it, the operation of multiplication by a real number is defined just as well. But the dimension of a complex n-dimensional space regarded as a real one is equal to $2n$, i.e., twice as much. For if $e_1, e_2, \cdots, e_n$ is a basis of the complex space, then we can take as a basis of the same space, regarded as a real one, for example the vectors

$$e_1, ie_1, e_2, ie_2, \cdots, e_n, ie_n, \quad \text{where} \quad i = \sqrt{-1}.$$

Therefore a two-dimensional complex space can be interpreted as a real one, but four-dimensional.

Furthermore, the theory of linear spaces does not undergo any changes formally if as the collection of numbers by which the "vectors" of the space may be multiplied we take an arbitrary set of numbers, other than that of all real or all complex numbers, provided only that the results of the basic arithmetical operations (addition, subtraction, multiplication, and division) performed on numbers of the set again belong to the set. A set of numbers satisfying these postulates is called a number field. (This concept will be studied in more detail in Chapter XX.) As an example of a number field, we can take the field of rational numbers.

In some parts of algebra that are close to the theory of numbers, the theory of linear spaces over an arbitrary field is successfully applied.

The n-dimensional Euclidean space. Some important concepts of the ordinary vector space have not yet been generalized in the preceding account, in particular, the concept of the length of a vector and the angle between vectors. It is well known that in analytic geometry problems relating to the intersection of lines and planes, parallelism, and many others make no use of these concepts. The properties of a space whose description does not require the concepts of the length of a vector and of angle can be characterized as the properties that remain unchanged under arbitrary affine transformations [see Chapter III, §11]. For this reason linear spaces in which the concept of the length of a vector is not defined are called *affine spaces*.

However, many problems of mathematics require generalizations of the concepts of the length of a vector and of an angle to n-dimensional spaces. These generalizations proceed by means of an analogy with the theory of vectors in a plane or in space.

Let us consider, first of all, the real space of rows. The length of the vector $X = (x_1, x_2, \cdots, x_n)$ is defined to be the number

$$|X| = \sqrt{x_1^2 + x_2^2 + \cdots + x_n^2}.$$

This is quite natural, since for $n = 2$ and $n = 3$ the length of a vector is computed precisely by this formula in terms of its coordinates with respect to Cartesian coordinate axes.

The concept of the angle between vectors is introduced in a natural way by the following considerations. In a plane and in space the angle between the vectors X and Y is the angle at the vertex A in the triangle with the sides $AB = |X|$, $AC = |Y|$ and $BC = |X - Y|$.

In an n-dimensional space it is natural to take this as the definition of the angle between vectors, i.e., to proceed as if we could "draw" a pair of vectors in an n-dimensional space and "place" them in a plane preserving their lengths and the angle between them. However, such a definition would lack rigor; the existence of a triangle ABC with the lengths of the vectors $|X|$, $|Y|$, and $|X - Y|$ is needed in the proof.

Disregarding this inaccuracy, we introduce a formula for the computation of an angle. By a well-known formula of trigonometry we have

$$BC^2 = AB^2 + AC^2 - 2AB \cdot AC \cos \phi;$$

hence,

$$\cos \phi = \frac{|X|^2 + |Y|^2 - |X - Y|^2}{2|X| \cdot |Y|}$$

$$= \frac{x_1^2 + \cdots + x_n^2 + y_1^2 + \cdots + y_n^2 - (x_1 - y_1)^2 - \cdots - (x_n - y_n)^2}{2|X| \cdot |Y|}$$

$$= \frac{x_1 y_1 + \cdots + x_n y_n}{|X| \cdot |Y|}.$$

§2. LINEAR SPACES

If we retain the term "scalar product," as in three-dimensional space, for the product of the lengths of vectors by the cosine of the angle between them, we find that the scalar product of the vectors is computed by the formula

$$X \cdot Y = x_1 y_1 + \cdots + x_n y_n,$$

which for $n = 2$ and $n = 3$ coincides with the well-known formulas for the scalar product of ordinary vectors.

Strictly speaking, the expression $x_1 y_1 + \cdots + x_n y_n$ should be taken as the definition of the *scalar product* (because there is a lack of rigor in the definition of the scalar product by means of the angle) and then the angle between vectors can be defined by the formula

$$\cos \phi = \frac{X \cdot Y}{|X| \cdot |Y|}. \tag{3}$$

This is what we shall do.

To justify this definition of an angle we have to show that the absolute value of the right-hand side of formula (3) does not exceed 1, i.e., that $(X \cdot Y)^2 \leqslant |X|^2 \cdot |Y|^2$.

In expanded form this inequality becomes

$$(x_1 y_1 + \cdots + x_n y_n)^2 < (x_1^2 + \cdots + x_n^2)(y_1^2 + \cdots + y_n^2).$$

It is known as the Cauchy-Bunjakovskiĭ inequality and can be proved directly, by a fairly tedious computation. We shall prove it by the following indirect argument.

First of all we mention that the scalar multiplication of vectors has the following properties:

1'. $X \cdot X = |X|^2 > 0$ for $X \neq 0$.
2'. $X \cdot Y = Y \cdot X$.
3'. $(cX) \cdot Y = c(X \cdot Y)$.
4'. $(X_1 + X_2) \cdot Y = X_1 \cdot Y + X_2 \cdot Y$.

That these properties hold follows immediately from the expression of the scalar product in terms of the coordinates.

We now introduce the vector $Y + tX$, where t is an arbitrary real number. We have $|Y + tX|^2 \geqslant 0$, because the square of the length of a vector cannot be negative. But by the properties of the scalar product $|Y + tX|^2 = |Y|^2 + 2tX \cdot Y + t^2 |X|^2$. Moreover, it is known that a quadratic trinomial is nonnegative for all values of the real variable t if and only if its roots are imaginary or equal, i.e., if its discriminant is negative or zero. But the discriminant of the trinomial $|Y|^2 + 2tX \cdot Y + t^2|X|^2$ is equal to $4(X \cdot Y)^2 - 4|X|^2|Y|^2$ so that $(X \cdot Y)^2 - |X|^2|Y|^2 \leqslant 0$, and this is equivalent to the Cauchy-Bunjakovskiĭ inequality.

From this inequality it follows that

$$\frac{|X \cdot Y|}{|X||Y|} \leqslant 1$$

and therefore the definition of an angle by the formula (3) is justified.

Furthermore, it is easy to deduce the inequalities

$$|X| - |Y| \leqslant |X \pm Y| \leqslant |X| + |Y|,$$

which imply, in particular, the existence of a triangle with the sides $|X|$, $|Y|$ and $|X - Y|$, so that the nonrigorous definition of an angle given previously, which was based on geometric intuition, now also becomes valid.

Axiomatic definition of an n-dimensional Euclidean space. In the preceding section we have introduced the concepts of the length of a vector, of angle, and of the scalar product in the space of rows. In the general axiomatic definition of an n-dimensional real linear space, these concepts are also defined axiomatically, and the concept of a scalar product comes first.

Scalar multiplication of vectors of a linear real space is the name for an operation which associates with every pair of vectors X and Y a real number, their so-called scalar product $X \cdot Y$, where this operation must satisfy the following postulates (axioms):

1'. $X \cdot X > 0$ for $X \neq 0$, $0 \cdot 0 = 0$.
2'. $X \cdot Y = Y \cdot X$.
3'. $(cX) \cdot Y = c(X \cdot Y)$.
4'. $(X_1 + X_2) \cdot Y = X_1 \cdot Y + X_2 \cdot Y$.

Furthermore, by the length of a vector we mean the number $\sqrt{X \cdot X}$, by the cosine of the angle between the vectors X and Y the number $X \cdot Y/|X| \cdot |Y|$. To justify this latter inequality, it is necessary to establish the Cauchy-Bunjakovskiĭ inequality $(X \cdot Y)^2 \leqslant |X|^2|Y|^2$. But this can be done exactly as we have done it in the preceding section. In our proof we have made use only of the properties 1', 2', 3' and 4' of the scalar product, the specific nature of the space of rows playing no role in the proof. A real linear space in which a scalar multiplication satisfying the axioms 1'-4' has been introduced is called a *Euclidean space*.

§2. LINEAR SPACES

In various concrete linear spaces that are studied in mathematics, scalar products are introduced by various methods whose choice is dictated by the nature of the problem. For example, in the spaces whose elements are the functions of one variable $X(t)$ defined on a given interval $a \leqslant t \leqslant b$, the scalar product of two elements $X(t)$ and $Y(t)$ is often taken to be the number $\int_a^b X(t) Y(t) dt$ or $\int_a^b X(t) Y(t) p(t) dt$, where $p(t)$ is some positive function. It is easy to see that all the axioms 1'- 4' are satisfied for either of these definitions.

Orthogonality; orthonormal bases. Two vectors of a Euclidean space are called *orthogonal* (or perpendicular) if their scalar product is zero. It is easy to see that pairwise orthogonal nonzero vectors are always linearly independent. For suppose that $X_1, X_2, \cdots, X_m$ are pairwise orthogonal nonzero vectors and that $c_1 X_1 + c_2 X_2 + \cdots + c_m X_m = \mathbf{0}$.

By the property of the scalar product $X_1(c_1 X_1 + c_2 X_2 + \cdots + c_m X_m) = c_1 |X_1|^2 = 0$; hence $c_1 = 0$. In the same way we can show that $c_2 = \cdots = c_m = 0$. Therefore $X_1, \cdots, X_m$ are linearly independent.

From what we have proved, it follows that in an n-dimensional space there cannot be more than n pairwise orthogonal nonzero vectors and that every set of n pairwise orthogonal vectors forms a basis of the space. If, moreover, the lengths of all the n pairwise orthogonal vectors are 1, then the basis they form is called *orthonormal*.

It is not difficult to show, but we shall omit the proof, that a Euclidean space has orthonormal bases, in fact infinitely many. Moreover, if in a space R some subspace P is chosen, then an orthonormal basis of the subspace can be extended to an orthonormal basis of the whole space by adjoining certain vectors.

It is often convenient to define vectors in a Euclidean space by their coordinates in an arbitrary orthonormal basis, because in this case we obtain a particularly simple expression for the scalar product. For if a vector X has the coordinates $(x_1, x_2, \cdots, x_n)$ in the orthonormal basis $e_1, e_2, \cdots, e_n$ and the vector Y the coordinates $(y_1, y_2, \cdots, y_n)$, i.e.,

$$X = x_1 e_1 + x_2 e_2 + \cdots + x_n e_n \quad \text{and} \quad Y = y_1 e_1 + y_2 e_2 + \cdots + y_n e_n,$$

then by the property of the scalar product

$$\begin{aligned} X \cdot Y &= x_1 y_1 e_1 e_1 + x_1 y_2 e_1 e_2 + \cdots + x_1 y_n e_1 e_n \\ &\quad + x_2 y_1 e_2 e_1 + x_2 y_2 e_2 e_2 + \cdots + x_2 y_n e_2 e_n \\ &\quad \cdots\cdots\cdots\cdots\cdots\cdots\cdots\cdots\cdots\cdots\cdots\cdots\cdots\cdots\cdots \\ &\quad + x_n y_1 e_n e_1 + x_n y_2 e_n e_2 + \cdots + x_n y_n e_n e_n \\ &= x_1 y_1 + x_2 y_2 + \cdots + x_n y_n, \end{aligned}$$

since $e_i e_k = 0$ for $i \neq k$ and $e_i e_i = |e_i|^2 = 1$ for every $i = 1, 2, \cdots, n$. In particular, $X \cdot X = x^2_1 + x^2_2 + \cdots + x^2_n$.

Thus, the length of a vector and the scalar product are expressed in terms of the coordinates of an orthonormal basis by the same formulas as in the space of rows.

The transition from one of the models of a Euclidean space, namely the space of rows, to the general axiomatically defined Euclidean space does not introduce any complications, but extends the domain of applicability of the theory.

Now let us deal with the problem of orthogonal projection of vectors on a subspace. Let R_n be an n-dimensional Euclidean space and P_m an m-dimensional subspace of it. Further, let $e_1, e_2, \cdots, e_m, f_1, \cdots, f_{n-m}$ be an orthonormal basis of R_n including an orthonormal basis of the subspace P_m. The subspace Q_{n-m} spanned by the vectors $f_1, f_2, \cdots, f_{n-m}$ is called the *orthogonal complement* of the subspace P_m. Its dimension is $n - m$. The orthogonal complement Q_{n-m} can be characterized as the subspace consisting of all vectors that are orthogonal to every vector of the subspace P_m.

Every vector Z belonging to R_n can be expressed uniquely as a sum of vectors X and Y of which one belongs to P_m, the other to Q_{n-m}. This is clear, because the vector Z can be expressed uniquely in the form

$$Z = x_1 e_1 + \cdots + x_m e_m + y_1 f_1 + \cdots + y_{n-m} f_{n-m},$$

so that $X = x_1 e_1 + \cdots + x_m e_m$, $Y = y_1 f_1 + \cdots + y_{n-m} f_{n-m}$.

The vector X is called the *orthogonal projection* of Z onto P_m.

Unitary spaces. The concepts of the length of a vector and the scalar product of vectors can also be defined in a complex space. As before, the concept of the scalar multiplication is put first, and this is defined as follows. With every pair X and Y of vectors of a complex space we associate a complex (not necessarily real) number, their so-called scalar product $X \cdot Y$. The operation of scalar multiplication must satisfy the following axioms:

1″. $X \cdot X$ is real and positive for $X \neq 0$, $0 \cdot 0 = 0$.

2″. $Y \cdot X = (X \cdot Y)'$. Here the prime denotes transition to the conjugate complex number.

3″. $(cX) \cdot Y = c(X \cdot Y)$ for an arbitrary complex c.

4″. $(X_1 + X_2) \cdot Y = X_1 \cdot Y + X_2 \cdot Y$ (distributive law).

In the space of rows with complex elements the scalar product of the

§3. SYSTEMS OF LINEAR EQUATIONS

vectors $(x_1, \cdots, x_n)$ and $(y_1, \cdots, y_n)$ can be taken to be the number $x_1 y_1' + \cdots + x_n y_n'$. It is easy to verify that all the axioms 1″-4″ are satisfied for this definition.

The length of a vector is defined as the number $\sqrt{X \cdot X}$. The concept of angle between vectors is not defined.

A complex linear space with a scalar product satisfying the axioms 1″-4″ is called a *unitary space*.

§3. Systems of Linear Equations

Systems of two equations with two unknowns and of three equations with three unknowns. A system of two linear equations with two unknowns appears in the following general form

$$a_1 x + b_1 y = c_1,$$
$$a_2 x + b_2 y = c_2.$$

To solve this system we multiply the first equation by b_2, the second by $-b_1$ and add. We obtain

$$(a_1 b_2 - a_2 b_1) x = c_1 b_2 - c_2 b_1.$$

Similarly, by multiplying the first equation by $-a_2$, the second by a_1, and adding, we obtain

$$(a_1 b_2 - a_2 b_1) y = a_1 c_2 - a_2 c_1.$$

From these equations it is easy to determine x and y, if only the expression $a_1 b_2 - a_2 b_1$ formed from the coefficients of the unknown x and y is different from zero. This expression is called the *determinant* of the matrix

$$\begin{bmatrix} a_1 & b_1 \\ a_2 & b_2 \end{bmatrix}$$

formed from the coefficients of the system. The determinant is denoted:

$$\begin{vmatrix} a_1 & b_1 \\ a_2 & b_2 \end{vmatrix}.$$

From the definition it follows that the determinant is computed by the scheme

$$\overset{+}{\diagdown}\overset{-}{\diagup},$$

which requires no further explanation.

Let us return to the solution of the system. The expressions $c_1b_2 - c_2b_1$ and $a_1c_2 - a_2c_1$ also appear as determinants in accordance with our definition, namely,
$$\begin{vmatrix} c_1 & b_1 \\ c_2 & b_2 \end{vmatrix} \quad \text{and} \quad \begin{vmatrix} a_1 & c_1 \\ a_2 & c_2 \end{vmatrix}.$$

Thus, if the determinant
$$\begin{vmatrix} a_1 & b_1 \\ a_2 & b_2 \end{vmatrix}$$
is different from zero, then we obtain the following formulas for the solution of the system:

$$x = \frac{\begin{vmatrix} c_1 & b_1 \\ c_2 & b_2 \end{vmatrix}}{\begin{vmatrix} a_1 & b_1 \\ a_2 & b_2 \end{vmatrix}}; \quad y = \frac{\begin{vmatrix} a_1 & c_1 \\ a_2 & c_2 \end{vmatrix}}{\begin{vmatrix} a_1 & b_1 \\ a_2 & b_2 \end{vmatrix}}. \tag{4}$$

Strictly speaking, these arguments are not complete. The operations on the equations that we have carried out to deduce the formulas for the solution of the system make sense only under the assumption that x and y are in fact numbers that form a solution of the system. The logical substance of our argument is the following: If the determinant of the coefficients of the system is not zero and the solution of the system exists, then it can be computed by the formulas (4). Therefore it is still necessary to verify that the values of the unknown that we have found do in fact satisfy both equations of the system. This can be done without any difficulty.

Thus, if the determinant of the matrix of the coefficients of the system is different from zero, then the system has a unique solution given by the formulas (4).

For a system of three equations with three unknowns
$$a_1x + b_1y + c_1z = d_1,$$
$$a_2x + b_2y + c_2z = d_2,$$
$$a_3x + b_3y + c_3z = d_3,$$

it is easy to carry out similar arguments and computations; for this purpose it is sufficient to add up the equations after multiplying them by factors such that after addition two of the unknowns disappear. To make the unknown y and z disappear, we have to take for these factors $b_2c_3 - b_3c_2$, $b_3c_1 - b_1c_3$ and $b_1c_2 - b_2c_1$, as is easy to verify by computation.

§3. SYSTEMS OF LINEAR EQUATIONS

We obtain the result that if the expression

$$\Delta = a_1 b_2 c_3 - a_1 b_3 c_2 + a_2 b_3 c_1 - a_2 b_1 c_3 + a_3 b_1 c_2 - a_3 b_2 c_1$$

is different from zero, then the system has a unique solution obtained by the formulas

$$x = \frac{\Delta_1}{\Delta}, \quad y = \frac{\Delta_2}{\Delta}, \quad z = \frac{\Delta_3}{\Delta},$$

where $\Delta_1, \Delta_2, \Delta_3$ are the expressions obtained from Δ by replacing the coefficients of the corresponding unknown by the absolute terms.

The expression Δ is called the determinant of the matrix

$$\begin{bmatrix} a_1 & b_1 & c_1 \\ a_2 & b_2 & c_2 \\ a_3 & b_3 & c_3 \end{bmatrix}$$

and is denoted by

$$\begin{vmatrix} a_1 & b_1 & c_1 \\ a_2 & b_2 & c_2 \\ a_3 & b_3 & c_3 \end{vmatrix}.$$

For the computation of a determinant the following scheme is useful:

In the first of these schemes, the lines (a diagonal and two triangles) connect the positions of the elements whose product occurs in the composition of the determinant with a plus sign; and in the second scheme, they connect the terms occuring in the determinant with a minus sign.

For systems of two equations with two unknowns and of three equations with three unknowns, we have obtained entirely similar results. In both cases the system has a unique solution, provided the determinant of the matrix of the coefficients is different from zero. The formulas for the solution are also similar: In the denominator of each of the unknowns stands the determinant of the matrix of coefficients, and in the numerators the determinants of the matrices that arise from the matrix of coefficients by replacing the coefficients of the unknown to be computed by the absolute terms.

An immediate generalization of these results to systems of n equations in n unknowns for arbitrary n is somewhat difficult. It becomes comparatively easy by an indirect method: First we generalize the concept of a determinant to square matrices of arbitrary order, and having studied the properties of determinants we apply their theory to the investigation of systems of equations.

Determinants of the nth order. When we consider the explicit expression for the determinants

$$\begin{vmatrix} a_1 \ b_1 \\ a_2 \ b_2 \end{vmatrix} = a_1 b_2 - a_2 b_1$$

and

$$\begin{vmatrix} a_1 \ b_1 \ c_1 \\ a_2 \ b_2 \ c_2 \\ a_3 \ b_3 \ c_3 \end{vmatrix} = a_1 b_2 c_3 - a_1 b_3 c_2 + a_2 b_3 c_1 - a_2 b_1 c_3 + a_3 b_1 c_2 - a_3 b_2 c_1,$$

we notice that in every term there occurs as a factor exactly one element from each row and one from each column of the determinant, and that all possible products of this form occur in the determinant with a plus or a minus sign. This property is at the bottom of the generalization of the concept of the determinant to square matrices of arbitrary order. In fact, the determinant of a square matrix of order n or, briefly, a *determinant of the nth order* is defined as the algebraic sum of all possible products of the elements of the matrix, precisely one from each row and one from each column; these products are given plus or minus signs by a certain well-defined rule. This rule is somewhat complicated to explain, and we shall not dwell on its formulation. It is sufficient to mention that it is arranged in such a way that the following important basic properties of a determinant are secured:

1. When two rows are interchanged, the determinant changes its sign.

For determinants of order 2 and 3, this property is easy to verify by an immediate computation. In the general case it is proved on the basis of the rule for the signs that we have not formulated here.

Determinants have quite a number of other remarkable properties that enable us to apply determinants successfully in diverse theoretical and numerical calculations, notwithstanding the fact that determinants are extraordinarily cumbersome: A determinant of order n contains, as is easy to see, $n!$ terms, each term consists of n factors, and the factors are provided with their signs according to a complicated rule.

We now proceed to enumerate the basic properties of determinants but omit their detailed proofs. The first of these properties has been formulated.

2. A determinant does not change when its matrix is transposed, i.e.,

§3. SYSTEMS OF LINEAR EQUATIONS

when the rows are replaced by the columns, preserving their order. The proof is based on a detailed study of the rule of the distribution of signs in the terms of the determinant. This property enables us to transfer every statement concerning the rows of the determinant to a statement on columns.

3. A determinant is a linear function of the elements of each row (or column). In detail,

$$\begin{vmatrix} a_{11} & \cdots & a_{1n} \\ \cdots & \cdots & \cdots \\ a_{i1} & \cdots & a_{in} \\ \cdots & \cdots & \cdots \\ a_{n1} & \cdots & a_{nn} \end{vmatrix} = a_{i1}A_{i1} + a_{i2}A_{i2} + \cdots + a_{in}A_{in}, \tag{5}$$

where $A_{i1}, A_{i2}, \cdots, A_{in}$ are expressions that do not depend on the elements of the ith row.

This property follows evidently from the fact that every term contains one and only one factor from each row, in particular the ith row.

The equation (5) is called the expansion of the determinant with respect to the elements of the ith row, and the coefficients $A_{i1}, A_{i2}, \cdots, A_{in}$ are called the algebraic complements of the elements $a_{i1}, a_{i2}, \cdots, a_{in}$ of the determinant.

4. The algebraic complement A_{ij} of the element a_{ij} is equal, apart from the sign, to the so-called minor Δ_{ij} of the determinant, i.e., the determinant of order $(n-1)$ that arises from the given one by crossing out the ith row and jth column. To obtain the algebraic complement the minor must be taken with the sign $(-1)^{i+j}$. The properties 3 and 4 reduce the computation of a determinant of order n to the computation of n determinants of order $n-1$.

The fundamental properties that we have enumerated have a number of interesting consequences. We now mention some of these.

5. A determinant with two equal rows is zero.

For if a determinant has two equal rows, then the determinant does not change when they are interchanged, because the rows are identical; but on the other hand, by our first property it should change its sign. Therefore it is equal to zero.

6. The sum of the products of the elements of any row by the algebraic complements of another row is zero.

For such a sum is the result of expanding a determinant with two equal rows with respect to one of them.

7. A common factor of the elements of any row can be taken before the determinant sign.

This follows from property 3.

8. A determinant with two proportional rows is zero.

It is sufficient to take out the factor of proportionality and then we have a determinant with two equal rows.

9. A determinant does not change if we add to the elements of any one row numbers that are proportional to the elements of another row.

For by property 3 the modified determinant is equal to the sum of the original determinant and a determinant with two proportional rows, which is zero.

The last property gives us a good method for computing determinants. Without changing the value of a determinant, we can transform its matrix by applying this rule so that in one row (or column) all the elements except one become zero. Then by expanding the determinant with respect to the elements of this row (column) we reduce the computation of a determinant of order n to that of a *single* determinant of order $n - 1$, namely, the algebraic complement of the only nonzero element of the row in question.

For example, suppose we have to compute the determinant

$$\Delta = \begin{vmatrix} 1 & 1 & -1 & 2 \\ 2 & -1 & 1 & 1 \\ -1 & 2 & 0 & 1 \\ 1 & 1 & -2 & 1 \end{vmatrix}.$$

We add to the second column the first, multiplied by -1, to the third, the first, and to the fourth, the first, multiplied by -2 and obtain

$$\Delta = \begin{vmatrix} 1 & 0 & 0 & 0 \\ 2 & -3 & 3 & -3 \\ -1 & 3 & -1 & 3 \\ 1 & 0 & -1 & -1 \end{vmatrix}.$$

Expanding Δ by the elements of the first row, we obtain

$$\Delta = 1 \cdot (-1)^{1+1} \begin{vmatrix} -3 & 3 & -3 \\ 3 & -1 & 3 \\ 0 & -1 & -1 \end{vmatrix}.$$

Finally, we add to the first row the second and expand with respect to the elements of the first column; we obtain

$$\Delta = \begin{vmatrix} 0 & 2 & 0 \\ 3 & -1 & 3 \\ 0 & -1 & -1 \end{vmatrix} = 3 \cdot (-1)^{1+2} \begin{vmatrix} 2 & 0 \\ -1 & -1 \end{vmatrix} = -3 \cdot (-2) = 6.$$

§3. SYSTEMS OF LINEAR EQUATIONS

The determinant of a matrix A is denoted by $|A|$.

In conclusion, we mention a further very important property of determinants.

The determinant of the product of two square matrices is equal to the product of the determinants of the factors, i.e., in short notation, $|AB| = |A||B|$.

This property enables us in particular to multiply determinants of equal order by the rule for multiplication of matrices.

Systems of n linear equations in n unknowns. With the apparatus of determinants, it is now easy to generalize the results obtained earlier for systems of two equations with two unknowns and of three equations with three unknowns to systems of n equations with n unknowns, under the assumption that the determinant of the coefficient matrix is different from zero.

Let

$$a_{11}x_1 + \cdots + a_{1j}x_j + \cdots + a_{1n}x_n = b_1,$$
$$\dots\dots\dots\dots\dots\dots\dots\dots\dots\dots\dots\dots\dots\dots\dots\dots$$
$$a_{n1}x_1 + \cdots + a_{nj}x_j + \cdots + a_{nn}x_n = b_n$$

be such a system. We denote by Δ the determinant of the coefficient matrix of the system. By assumption it is different from zero. Furthermore, we denote by A_{ij} the algebraic complement of the element a_{ij}. We multiply the first equation by A_{1j}, the second by A_{2j}, $\cdots$, the nth by A_{nj} and add. We obtain

$$\Delta x_j = b_1 A_{1j} + \cdots + b_n A_{nj}.$$

For the coefficients of all the unknowns except x_j vanish, because they appear as sums of the products of the algebraic complements of the elements of the jth column with the elements of other columns (property 6, applied to columns); but the coefficients of the unknown x_j is equal to the sum of the products of the elements of the jth column with their algebraic complements, i.e., is equal to Δ.

Thus,

$$x_j = \frac{b_1 A_{1j} + \cdots + b_n A_{nj}}{\Delta} \quad \text{for all} \quad j = 1, 2, \cdots, n. \tag{6}$$

As we have said, these arguments are valid only if we understand by $x_1, x_2, \cdots, x_n$ a solution of the system, the existence of which must be assumed in the first instance.

Hence the result of the argument is the following.

If a solution of the system exists, then it is given by the formulas (6) and is therefore unique.

To complete the exposition, it is necessary to prove the existence of a solution, and this can be done by substituting the values we have found for the unknowns into all the equations of the original system. It is easy to verify by using the same property of the determinant (but this time applied to the rows) that these values in fact satisfy all the equations.

Thus the following theorem is true: If the determinant of the coefficient matrix of a system of n equations with n unknowns is different from zero, then the system has a unique solution given by the formulas (6).

These formulas can be transformed by remarking that the sum $b_1 A_{1j} + \cdots + b_n A_{nj}$ can be written in the form of a determinant, namely:

$$\Delta_j = b_1 A_{1j} + \cdots + b_n A_{nj} = \begin{vmatrix} a_{11} \cdots b_1 \cdots a_{1n} \\ \cdots\cdots\cdots\cdots\cdots\cdots \\ a_{n1} \cdots b_n \cdots a_{nn} \end{vmatrix}$$

(the absolute terms occur in the jth column).

Hence the results we have stated for systems of equations with two and three unknowns have been completely generalized to a system of n equations, and even the formulas for the solution are formally exactly the same.

We mention one corollary of this theorem: If it is known of a system of equations that it has no solution at all or that the solution is not unique, then the determinant of the coefficient matrix is equal to zero.

This corollary is particularly often applied to homogeneous systems, i.e., those in which the absolute terms $b_1, b_2, \cdots, b_n$ are all zero. Homogeneous systems always have the obvious "trivial" solution $x_1 = x_2 = \cdots = x_n = 0$.

If a homogeneous system has, apart from the trivial one, also a nontrivial solution, then its determinant is zero.

This statement opens up the possibility of using the theory of determinants in other branches of mathematics and its applications.

Let us consider, for example, a problem in analytical geometry: to find the equation of the plane passing through three given points (x_1, y_1, z_1), (x_2, y_2, z_2) and (x_3, y_3, z_3) that do not lie on one line.

From elementary geometry it is known that such a plane exists. Suppose that its equation is of the form $Ax + By + Cz + D = 0$. Then

$$Ax_1 + By_1 + Cz_1 + D = 0,$$
$$Ax_2 + By_2 + Cz_2 + D = 0,$$
$$Ax_3 + By_3 + Cz_3 + D = 0.$$

Let x, y, z be the coordinates of an arbitrary point in that plane. Then we also have

$$Ax + By + Cz + D = 0.$$

§3. SYSTEMS OF LINEAR EQUATIONS

We regard these four equations as a system of linear homogeneous equations for the coefficients A, B, C, D of the required plane. This system has a nontrivial solution, because the required plane exists. Therefore the determinant of the system is zero; i.e.,

$$\begin{vmatrix} x_1 & y_1 & z_1 & 1 \\ x_2 & y_2 & z_2 & 1 \\ x_3 & y_3 & z_3 & 1 \\ x & y & z & 1 \end{vmatrix} = 0. \tag{7}$$

Now this is the equation of the required plane. For it is an equation of the first degree in x, y, z, a fact which follows from the linearity of the determinant with respect to the elements of the last row.

By making use of the fact that the given points do not lie on one line, it is easy to verify that not all the coefficients of this equation are zero. Consequently the equation (7) is indeed the equation of a plane. This plane passes through the given points, because their coordinates obviously satisfy the equation.

Matrix notation for a system of n equations in n unknowns. A system of n linear equations in n unknowns

$$a_{11}x_1 + \cdots + a_{1n}x_n = b_1,$$
$$\dotfill$$
$$a_{n1}x_1 + \cdots + a_{nn}x_n = b_n$$

can be written in matrix notation in the form of a single equation

$$AX = B.$$

Here A denotes the coefficient matrix, X the column formed by the unknowns, and B the column of the absolute terms.

The solution of the system (if the determinant of the matrix A is different from zero) can be written explicitly as follows [see formula (6)]:

$$x_1 = \frac{A_{11}}{\Delta}b_1 + \frac{A_{21}}{\Delta}b_2 + \cdots + \frac{A_{n1}}{\Delta}b_n,$$

$$x_2 = \frac{A_{12}}{\Delta}b_1 + \frac{A_{22}}{\Delta}b_2 + \cdots + \frac{A_{n2}}{\Delta}b_n,$$

$$\dotfill$$

$$x_n = \frac{A_{1n}}{\Delta}b_1 + \frac{A_{2n}}{\Delta}b_2 + \cdots + \frac{A_{nn}}{\Delta}b_n$$

or in matrix form

$$X = \begin{bmatrix} \dfrac{A_{11}}{\Delta} & \dfrac{A_{21}}{\Delta} & \cdots & \dfrac{A_{n1}}{\Delta} \\ \dfrac{A_{12}}{\Delta} & \dfrac{A_{22}}{\Delta} & \cdots & \dfrac{A_{n2}}{\Delta} \\ \multicolumn{4}{c}{\dotfill} \\ \dfrac{A_{1n}}{\Delta} & \dfrac{A_{2n}}{\Delta} & \cdots & \dfrac{A_{nn}}{\Delta} \end{bmatrix} B.$$

The matrix standing as first factor on the right-hand side of the equation is called the *inverse* to the matrix A and is denoted by A^{-1}. Using this notation we obtain the solution of the system $AX = B$ in the following simple and natural form, which recalls the formula for the solution of a single equation in one unknown:

$$X = A^{-1}B.$$

We can easily give another proof of the result obtained, in terms of the algebra of matrices.

For this purpose we must first of all mention the special role of the matrix

$$E = \begin{pmatrix} 1 & 0 & 0 & \cdots & 0 \\ 0 & 1 & 0 & \cdots & 0 \\ \multicolumn{5}{c}{\dotfill} \\ 0 & 0 & 0 & \cdots & 1 \end{pmatrix},$$

the so-called *unit matrix*.

The unit matrix plays among square matrices the same role as the number 1 among numbers. In fact, for every matrix A the following equations hold: $AE = A$ and $EA = A$. This is easy to verify by the rule for the multiplication of matrices.

The matrix A^{-1} defined previously, the inverse to A, plays in relation to it a similar role to that played by the inverse of a given number:

$$AA^{-1} = A^{-1}A = E.$$

The validity of these equations can be verified by the rule for the multiplication of matrices and by the properties 3 and 6 of a determinant.

Knowing these properties of the unit and the inverse matrix we can obtain the solution of the system $AX = B$ in the following way.

Suppose that $AX = B$. Then $A^{-1}(AX) = A^{-1}B$. But $A^{-1}(AX) = (A^{-1}A)X = EX = X$ and therefore $X = A^{-1}B$.

Suppose now that $X = A^{-1}B$, then $AX = AA^{-1}B = EB = B$.

§3. SYSTEMS OF LINEAR EQUATIONS

Thus the "equation" $AX = B$ has the unique solution $X = A^{-1}B$, provided only that A^{-1} exists.

We have established the existence of the inverse matrix A^{-1} for A under the assumption that the determinant of A is different from zero. This condition is not only sufficient but also necessary for the existence of the inverse matrix. For suppose that the matrix A has an inverse A^{-1} such that $AA^{-1} = E$. Then by the property of the determinant of the product of two matrices

$$|A||A^{-1}| = |E| = 1,$$

and from this it follows that the determinant of A is not zero.

A matrix whose determinant is different from zero is called *nondegenerate* or *nonsingular*. We have therefore established that an inverse matrix always exists for nondegenerate matrices and only for them.

The introduction of the concept of an inverse matrix turns out to be useful not only in the theory of systems of linear equations but also in many other problems of linear algebra.

In conclusion, we mention that the formulas we have derived for the solution of linear systems are an irreplaceable tool in theoretical considerations but are not convenient for the numerical solution of systems. As we have already mentioned, various methods and computational schemes have been worked out for the numerical solution of systems, and in view of the great importance of this problem for practical investigations, the work of simplifying the numerical solution of systems (especially with large numbers of unknowns) is intensively pursued even at present.

The general case of systems of linear equations. We now turn to the investigation of systems of linear equations in the most general case when it is not assumed that the number of equations is equal to the number of unknowns. In such a general setting it cannot be expected, naturally, that a solution of the system always exists or that, in case it exists, it turns out to be unique. It is natural to assume that if the number of equations is less than the number of unknowns the system has infinitely many solutions. For example, two equations of the first degree in three unknowns are satisfied by the coordinates of every point on the straight line that is the intersection of the planes defined by the equations. However it can happen in this case that the system has no solution at all, namely when the planes are parallel. And if the number of equations is greater than the number of unknowns, then as a rule the system has no solution. However, in this case it is possible that the system has solutions, even infinitely many.

In order to investigate the existence and the character of the set of solutions of a system in this general setting, we turn to a "geometrical" interpretation of the system.

We interpret the system of equations

$$a_{11}x_1 + a_{12}x_2 + \cdots + a_{1n}x_n = b_1,$$
$$a_{21}x_1 + a_{22}x_2 + \cdots + a_{2n}x_n = b_2, \qquad (8)$$
$$\cdots\cdots\cdots\cdots\cdots\cdots\cdots\cdots\cdots\cdots\cdots\cdots$$
$$a_{m1}x_1 + a_{m2}x_2 + \cdots + a_{mn}x_n = b_m$$

in the m-dimensional space of columns in the form

$$x_1 A_1 + x_2 A_2 + \cdots + x_n A_n = B.$$

Here $A_1, A_2, \cdots, A_n$ denote the columns of the coefficients of the corresponding unknowns and B the column of the absolute terms.

In this interpretation the problem of the existence of a solution of the system turns into the problem of whether the given vector B is a linear combination of the vectors $A_1, A_2, \cdots, A_n$.

The answer to this problem is almost obvious. For the vector B to be a linear combination of the vectors $A_1, A_2, \cdots, A_n$ it is necessary and sufficient that B should be contained in the subspace spanned by $A_1, A_2, \cdots, A_n$ or, in other words, that the subspaces spanned by the systems of vectors $A_1, A_2, \cdots, A_n$ and $A_1, A_2, \cdots, A_n, B$ should coincide.

Since the first of these subspaces is contained in the second, they coincide if and only if their dimensions are equal. We recall that the dimension of the subspace spanned by a given system of vectors is called the rank of this system. Thus, a necessary and sufficient condition for the existence of a solution of the system $x_1 A_1 + x_2 A_2 + \cdots + x_n A_n = B$ is the equality of the ranks of the vector systems $A_1, A_2, \cdots, A_n$ and $A_1, A_2, \cdots, A_n, B$.

It can be proved, but we shall not do it here, that the rank of a system of vectors is equal to the rank of the matrix formed from the coordinates of these vectors. Here we understand by the *rank of a matrix* the largest order of a nonzero determinant that can be formed from the given matrix by omitting part of its rows and columns.

Since the coordinates of the vectors $A_1, A_2, \cdots, A_n$ (in the natural basis for the space of columns) are the coefficients of the system and the coordinates of the vector B its absolute terms, we obtain the following final formulation of the condition for the existence of a solution of a system.

For the existence of at least one solution of the system of linear equations

$$a_{11}x_1 + \cdots + a_{1n}x_n = b_1,$$
$$\cdots\cdots\cdots\cdots\cdots\cdots\cdots\cdots\cdots\cdots$$
$$a_{m1}x_1 + \cdots + a_{mn}x_n = b_m$$

§3. SYSTEMS OF LINEAR EQUATIONS

it is necessary and sufficient that the rank of the matrix formed from the coefficients of the system should be equal to the rank of the matrix formed from the coefficients and the absolute terms.

Now let us investigate the character of the set of solutions if they exist. Let $x_1^0, x_2^0, \cdots, x_n^0$ be any solution of the system (8). We set $x_1 = x_1^0 + y_1$, $x_2 = x_2^0 + y_2, \cdots, x_n = x_n^0 + y_n$. Then in view of the fact that x_1^0, $x_2^0, \cdots, x_n^0$ form a solution of the system (8), the new unknowns $y_1, y_2, \cdots, y_n$ must satisfy the homogeneous system

$$a_{11}y_1 + \cdots + a_{1n}y_n = 0,$$
$$\cdots\cdots\cdots\cdots\cdots\cdots\cdots\cdots\cdots\cdots\cdots \qquad (9)$$
$$a_{m1}y_1 + \cdots + a_{mn}y_n = 0$$

with the same coefficient matrix. Conversely, if we add to the original solution $x_1^0, x_2^0, \cdots, x_n^0$ of the system (8) an arbitrary solution of the homogeneous system (9), then we obtain another solution of the system (8).

Thus, in order to obtain the general solution of the system (8), it is only necessary to take an arbitrary particular solution of it and to add it to the general solution of the homogeneous system (9).

In this way the problem of the character of the set of solutions of the system (8) is reduced to the same problem for the homogeneous system (9). We shall consider this problem in the next section.

Homogeneous systems. We shall interpret the homogeneous system of linear equations

$$a_{11}y_1 + a_{12}y_2 + \cdots + a_{1n}y_n = 0,$$
$$a_{21}y_1 + a_{22}y_2 + \cdots + a_{2n}y_n = 0,$$
$$\cdots\cdots\cdots\cdots\cdots\cdots\cdots\cdots\cdots\cdots\cdots$$
$$a_{m1}y_1 + a_{m2}y_2 + \cdots + a_{mn}y_n = 0$$

in the n-dimensional Euclidean space. (Here we assume that the coefficients of the system are real. For systems with complex coefficients, we can give a similar interpretation in unitary space and obtain similar results.)

Let $A_1', A_2', \cdots, A_m', Y$ be the vectors of a Euclidean space whose coordinates in an orthonormal basis are, respectively,

$$(a_{11}, a_{12}, \cdots, a_{1n}), (a_{21}, a_{22}, \cdots, a_{2n}), \cdots, (a_{m1}, a_{m2}, \cdots, a_{mn}),$$
$$(y_1, y_2, \cdots, y_n).$$

Then the system assumes the form

$$A_1'Y = 0, A_2'Y = 0, \cdots, A_m'Y = 0;$$

i.e., every solution of the system determines a vector orthogonal to all the vectors formed by the coefficients of the various equations.

Therefore, the set of solutions forms the subspace that is the orthogonal complement to the subspace spanned by the vectors $A_1', A_2', \cdots, A_m'$. The dimension of the latter subspace is equal to the rank r of the matrix formed from the coefficients of the system. The dimension of the orthogonal complement, i.e., of the "solution space," is then equal to $n - r$.

Now every subspace has a basis, i.e., a system of linearly independent vectors equal in number to the dimension of the subspace and such that their linear combinations fill the whole subspace. Therefore, among the solutions of a homogeneous system there exist $n - r$ linearly independent solutions such that all the solutions of the system are linear combinations of them. Here n denotes the number of unknowns and r the rank of the coefficient matrix.

Thus, the structure of the solutions of a homogeneous system and, consequently, also of an inhomogeneous system is completely clarified. In particular, a homogeneous system has the unique trivial solution $x_1 = x_2 = \cdots = x_n = 0$ if and only if the rank of the coefficient matrix is equal to the number of unknowns. By what we have said at the end of the preceding paragraph, the same condition is also the condition for uniqueness of the solution for systems of inhomogeneous equations (providing the consistency condition is satisfied).

Our investigation of these systems shows clearly how the introduction of generalized geometrical concepts leads to simplicity and lucidity in a complicated algebraic problem.

§4. Linear Transformations

Definition and examples. In many mathematical investigations it becomes necessary to change the variables, i.e., to go over from one system of variables $x_1, x_2, \cdots, x_n$ to another $y_1, y_2, \cdots, y_n$, connected with the first by means of a functional dependence:

$$y_1 = \phi_1(x_1, x_2, \cdots, x_n),$$
$$y_2 = \phi_2(x_1, x_2, \cdots, x_n),$$
$$\cdots\cdots\cdots\cdots\cdots\cdots\cdots\cdots\cdots$$
$$y_n = \phi_n(x_1, x_2, \cdots, x_n).$$

For example, if the variables are the coordinates of a point in a plane or in space, then the transition from one system of coordinates to another system gives rise to a transformation of coordinates that is defined by the

§4. LINEAR TRANSFORMATIONS

expressions for the original coordinates in terms of the new ones or vice versa.

Moreover, a transformation of variables arises in studying the changes due to a transition from one position or configuration to another for objects whose position or configuration is described by the values of the variables. As a typical example of this kind of tranformation, we can take the change of coordinates of the points of some body under deformations.

An abstractly given transformation of a system of n variables is usually interpreted as a transformation (deformation) of an n-dimensional space, i.e., as an association between each vector of the space (or part of it) with coordinates $x_1, x_2, \cdots, x_n$ and a corresponding vector with coordinates $y_1, y_2, \cdots, y_n$.

As we have said previously, every "smooth" function (having continuous partial derivatives) of several variables is close to a linear function for small changes of these variables. Therefore every "smooth" transformation (i.e., one for which the functions $\phi_1, \phi_2, \cdots, \phi_n$ in its analytical expression have continuous partial derivatives) is close to a linear transformation in a small part of the space:

$$
\begin{aligned}
y_1 &= a_{11}x_1 + a_{12}x_2 + \cdots + a_{1n}x_n + b_1, \\
y_2 &= a_{21}x_1 + a_{22}x_2 + \cdots + a_{2n}x_n + b_2, \\
&\cdots\cdots\cdots\cdots\cdots\cdots\cdots\cdots\cdots\cdots\cdots\cdots \\
y_n &= a_{n1}x_1 + a_{n2}x_2 + \cdots + a_{nn}x_n + b_n.
\end{aligned}
\qquad (10)
$$

This circumstance alone makes the study of the properties of linear transformations one of the most important problems of mathematics. For example, from the theory of n linear equations in n unknowns, we know that a necessary and sufficient condition for the system of equations (10) with respect to $x_1, x_2, \cdots, x_n$ to have a unique solution, i.e., for the corresponding linear transformation to be invertible, is the nonvanishing of the determinant of the coefficients. This circumstance is the foundation of an important theorem of analysis: For a transformation

$$
\begin{aligned}
y_1 &= \phi_1(x_1, x_2, \cdots, x_n), \\
y_2 &= \phi_2(x_1, x_2, \cdots, x_n), \\
&\cdots\cdots\cdots\cdots\cdots\cdots\cdots\cdots \\
y_n &= \phi_n(x_1, x_2, \cdots, x_n),
\end{aligned}
$$

which is smooth in the neighborhood of a given point, to have a smooth inverse transformation it is necessary and sufficient that at the given point the determinant

$$\begin{vmatrix} \dfrac{\partial \phi_1}{\partial x_1} & \cdots & \dfrac{\partial \phi_1}{\partial x_n} \\ \dfrac{\partial \phi_2}{\partial x_1} & \cdots & \dfrac{\partial \phi_2}{\partial x_n} \\ \cdots & \cdots & \cdots \\ \dfrac{\partial \phi_n}{\partial x_1} & \cdots & \dfrac{\partial \phi_n}{\partial x_n} \end{vmatrix}$$

should be different from zero.

The study of the general linear transformation (10) essentially reduces to the study of the homogeneous transformation with the same coefficients

$$\begin{aligned} y_1 &= a_{11}x_1 + \cdots + a_{1n}x_n, \\ y_2 &= a_{21}x_1 + \cdots + a_{2n}x_n, \\ &\cdots \\ y_n &= a_{n1}x_1 + \cdots + a_{nn}x_n, \end{aligned} \qquad (11)$$

and in what follows, when speaking of linear tranformations, we shall always have homogeneous transformations in mind.

Linear transformations of an n-dimensional space can also be defined by their intrinsic properties, apart from the formulas (11) that connect the coordinates of corresponding points. Such a coordinate-free definition of the concept of a linear transformation is useful in that it does not depend on the choice of a basis. This definition is as follows.

A linear transformation of an n-dimensional linear space is a function $Y = A(X)$ where X and Y are vectors. This function satisfies the postulate of linearity

$$A(c_1X_1 + c_2X_2) = c_1A(X_1) + c_2A(X_2). \qquad (12)$$

In what follows, when speaking of a linear transformation of a space, we shall understand it in the sense of this definition.

This definition is equivalent to the preceding one in terms of coordinates. For the function $Y = A(X)$, which to the vector X with the coordinates $x_1, x_2, \cdots, x_n$ associates the vector Y with the coordinates $y_1, y_2, \cdots, y_n$ in such a way that the coordinates $y_1, y_2, \cdots, y_n$ are expressed in terms of the coordinates $x_1, x_2, \cdots, x_n$ in the form of linear homogeneous functions, obviously satisfies the postulate (12). Conversely, if the function $Y = A(X)$ satisfies the postulate (12) and if $e_1, e_2, \cdots, e_n$ is an arbitrary basis of the space, then

$$A(x_1e_1 + x_2e_2 + \cdots + x_ne_n) = x_1A(e_1) + x_2A(e_2) + \cdots + x_nA(e_n).$$

§4. LINEAR TRANSFORMATIONS

We denote the coordinates (in the same basis) of the vector $A(e_j)$ by $a_{1j}, \cdots, a_{nj}; j = 1, \cdots, n$. Then the coordinates of the vector $Y = A(X)$ are

$$y_1 = a_{11}x_1 + a_{12}x_2 + \cdots + a_{1n}x_n,$$
$$y_2 = a_{21}x_1 + a_{22}x_2 + \cdots + a_{2n}x_n,$$
$$\cdots\cdots\cdots\cdots\cdots\cdots\cdots\cdots\cdots\cdots\cdots\cdots$$
$$y_n = a_{n1}x_1 + a_{n2}x_2 + \cdots + a_{nn}x_n.$$

Thus, to every linear transformation of a linear space there corresponds a certain square matrix with respect to a given basis. This transformation can be written in matrix language in the form $Y = AX$. Here X is the column of the coordinates of the original vector, Y the column of the coordinates of the transformed vector, and A the coefficient matrix of the transformation. The columns of the matrix A are formed by the coordinates of those vectors into which the vectors of the basis are transformed. In accordance with the matrix notation, we shall subsequently often write a linear transformation itself in the form $Y = AX$, omitting the parentheses.

From the formula

$$A(x_1e_1 + x_2e_2 + \cdots + x_ne_n) = x_1A(e_1) + x_2A(e_2) + \cdots + x_nA(e_n)$$

it follows that the whole space is mapped under a linear transformation into the subspace spanned by the vectors $A(e_1), \cdots, A(e_n)$. The dimension of this subspace is equal to the rank of the system of vectors $A(e_1)$, $A(e_2), \cdots, A(e_n)$, or, what is the same, to the rank of the matrix formed by their coordinates, i.e., the rank of the matrix A associated with the transformation. This subspace coincides with the whole space if and only if the rank of the matrix A is equal to n, i.e., if the determinant of A is different from zero. In this case the linear transformation is called *nonsingular* or *nondegenerate*.

From the theory of systems of linear equations, we know that nondegenerate transformations are uniquely invertible and that the coordinates of the original vector are expressed in terms of the coordinates of the transformed vector by the formula $\mathbf{X} = \mathbf{A}^{-1}\mathbf{Y}$.

A transformation whose matrix has the determinant zero is called *singular* or *degenerate*. A degenerate transformation is not invertible. This follows from the theory of linear transformations or more intuitively from the fact that it transforms the whole space into part of it.

As a first example of a nondegenerate transformation, we take the identity transformation that maps every vector into itself. The matrix of the identity transformation in any basis is the unit matrix E. A nonsingular transformation is also given by a similarity that consists in multiplying all the vectors of the space by one and the same number. The matrix of a similarity transformation does not depend on the choice of a basis and has the form aE, where a is the similarity factor.

An important special case of nondegenerate transformation are the orthogonal transformations. The concept of an *orthogonal transformation* has a meaning when applied to a Euclidean space and is defined as a linear transformation preserving the lengths of vectors. An orthogonal transformation is a generalization to n-dimensional space of a rotation of the space around the fixed origin of coordinates or a rotation combined with a reflection in an arbitrary plane passing through the origin.

It is easy to see that under an orthogonal transformation not only the lengths of vectors are preserved but also scalar products and that, consequently, orthogonal transformations carry an orthogonal basis of the space into a system of pairwise orthogonal unit vectors which in turn is then necessarily also a basis.

The matrix connected with an orthogonal transformation with respect to an orthonormal basis has the following specific properties.

First, the sum of the squares of the elements of each column is 1, since these sums are the squares of the lengths of the vectors into which the vectors of the given basis are transformed. Second, the sums of the products of corresponding elements taken from distinct columns are zero, since these sums are the scalar products of the vectors into which the vectors of the basis are transformed.

In matrix notation both these properties can be written by the single formula

$$\tilde{P}P = E.$$

Here P is the matrix of the orthogonal transformation (with respect to an orthonormal basis), and $\tilde{P}$ is its transposed matrix, i.e., the matrix whose rows are the columns of P in the same order.

For the diagonal elements of the matrix $\tilde{P}P$ are by the rule for the multiplication of matrices equal to the sum of the squares of the elements of the corresponding column of P, and the nondiagonal elements are equal to the sum of the products of the corresponding elements taken from distinct columns of P.

As an example for a degenerate transformation, we can take the orthogonal projection of all vectors of a Euclidean space onto some subspace (see §2). For in this transformation the whole space is mapped onto part of it.

Transformation of coordinates. We now consider the problem of the transformation of coordinates in n-dimensional space, i.e., the problem

§4. LINEAR TRANSFORMATIONS

how the coordinates of vectors are changed on transition from one basis to another.

Let the original basis be $e_1, e_2, \cdots, e_n$ and let $f_1, f_2, \cdots, f_n$ be an arbitrary other basis of the space. Suppose further that

$$C = \begin{bmatrix} c_{11}, & \cdots, & c_{1n} \\ & \cdots & \\ c_{n1}, & \cdots, & c_{nn} \end{bmatrix}$$

is the matrix whose columns are the coordinates of the vectors of the new basis $f_1, f_2, \cdots, f_n$ with respect to the original one. The matrix C is obviously nondegenerate, because the vectors $f_1, f_2, \cdots, f_n$ are linearly independent. It is called the *matrix of the coordinate transformation*.

We denote by $x_1, x_2, \cdots, x_n$ the coordinates of a certain vector X with respect to the basis $e_1, e_2, \cdots, e_n$ and by $x'_1, x'_2, \ldots, x'_n$ the coordinates of the same vector with respect to the basis $f_1, f_2, \cdots, f_n$. Then $X = x'_1 f_1 + x'_2 f_2 + \cdots + x'_n f_n$ and therefore the coordinates of the vector X with respect to the original basis form the column

$$\begin{bmatrix} x_1 \\ x_2 \\ \vdots \\ x_n \end{bmatrix} = \begin{bmatrix} c_{11}x'_1 + c_{12}x'_2 + \cdots + c_{1n}x'_n \\ c_{21}x'_1 + c_{22}x'_2 + \cdots + c_{2n}x'_n \\ \cdots \\ c_{n1}x'_1 + c_{n2}x'_2 + \cdots + c_{nn}x'_n \end{bmatrix} = \begin{bmatrix} c_{11} & \cdots & c_{1n} \\ c_{21} & \cdots & c_{2n} \\ & \cdots & \\ c_{n1} & \cdots & c_{nn} \end{bmatrix} \begin{bmatrix} x'_1 \\ x'_2 \\ \vdots \\ x'_n \end{bmatrix}.$$

Thus, the original coordinates are expressed linearly and homogeneously in terms of the transformation with the matrix C.

The formulas that express the connection between the coordinates with respect to the original and the transformed basis coincide formally with the formulas that link the coordinates of corresponding vectors in a nondegenerate linear transformation of the space. This circumstance enables us to interpret an abstractly given linear homogeneous transformation of variables with a nondegenerate matrix either as a transformation of coordinates or as a transformation of the space. In each concrete case the choice of one of these two interpretations is determined by the context of the problem under consideration.

Let us now deal with the question how the matrix of a linear transformation of the space is changed under a coordinate transformation.

Suppose that the given linear transformation has the matrix A in the basis $e_1, e_2, \cdots, e_n$ so that the column Y of the coordinates of the transformed vector is linked with the column X of the original one by the formula
$$Y = AX.$$

Suppose now that a transformation of coordinates with the matrix C is made; X', Y' denote, respectively, the columns of the coordinates of the original and the transformed vectors with respect to the new basis. Then $X = CX'$, $Y = CY'$ and hence

$$Y' = C^{-1}Y = C^{-1}AX = C^{-1}ACX'.$$

Thus, the matrix of our transformation with respect to the new basis is $C^{-1}AC$.

Two matrices A and B connected by the relation $B = C^{-1}AC$, where C is a nonsingular matrix, are called *similar*. One and the same linear transformation corresponds with respect to various bases to a class of pairwise similar matrices.

Eigenvectors and eigenvalues of a linear transformation. A very important class of linear transformations consists of the transformations that come about in the following way.

Let e_1, e_2, $\cdots$, e_n be arbitrary linearly independent vectors of the space. Suppose that under the transformation they are multiplied by certain numbers λ_1, λ_2, $\cdots$, λ_n. If the vectors e_1, e_2, $\cdots$, e_n are taken as a basis of the space, then the transformation can be described by

$$\begin{bmatrix} \lambda_1 & 0 & \cdots & 0 \\ 0 & \lambda_2 & \cdots & 0 \\ \multicolumn{4}{c}{\cdots\cdots\cdots\cdots\cdots} \\ 0 & 0 & \cdots & \lambda_n \end{bmatrix}.$$

The transformations of this class have a simple and intuitive geometrical meaning (of course, only for real spaces and for $n = 2$ or $n = 3$). Namely, if all the numbers λ_i are positive, then the transformation that we describe consists in a stretching (or compressing) of the space in the directions of the vectors e_1, e_2, $\cdots$, e_n with coefficients λ_1, λ_2, $\cdots$, λ_n. If some of the λ_i are negative, then the deformation of the space is accompanied by a change of direction of some of the vectors e_1, e_2, $\cdots$, e_n into the opposite. Finally, if for example $\lambda_1 = 0$, then a projection of the space parallel to e_1 takes place onto the subspace spanned by e_2, $\cdots$, e_n with a subsequent deformation in these directions.

The class of transformations we have considered is important, because in spite of its simplicity it is very general. In fact, it can be established that every linear transformation satisfying certain not very severe restrictions belongs to our class; i.e., we can find for it a basis in which it is described by a diagonal matrix.

The restrictions to be imposed on the transformation become particularly

§4. LINEAR TRANSFORMATIONS

clear if we consider linear transformation of a complex space. In what follows this will be assumed.

We introduce the following definition.

A nonzero vector X which under a linear transformation A of the space goes into a collinear vector λX is called an *eigenvector of the transformation*. In other words, a nonzero vector X is an eigenvector of the transformation A if and only if $AX = \lambda X$. The number λ is called an *eigenvalue* of the transformation A.

It is obvious that if in some basis a transformation has a diagonal matrix, then this basis consists of eigenvectors and the diagonal elements are eigenvalues. Conversely, if there exists in the space a basis consisting of eigenvectors of the transformation A, then in this basis the matrix of the transformation A is diagonal and consists of the eigenvalues corresponding to the vectors of the basis.

We now proceed to study the properties of eigenvectors and eigenvalues. With this aim we write the definition of an eigenvector in coordinate notation. Let A be the matrix corresponding to the transformation A with respect to a certain basis and X the column of coordinates of the vector X in the same basis. The equation $AX = \lambda X$ in coordinate notation is written as $AX = \lambda X$ or

$$(A - \lambda E) X = 0.$$

In expanded form this equation turns into the system

$$\begin{aligned}
(a_{11} - \lambda) x_1 + a_{12} x_2 + \cdots + a_{1n} x_n &= 0, \\
a_{21} x_1 + (a_{22} - \lambda) x_2 + \cdots + a_{2n} x_n &= 0, \\
&\cdots \\
a_{n1} x_1 + a_{n2} x_2 + \cdots + (a_{nn} - \lambda) x_n &= 0.
\end{aligned}$$

We can regard this system of equations as a system of linear and homogeneous equations for $x_1, x_2, \cdots, x_n$. We are interested in the case when this system has a nontrivial solution, because the coordinates of an eigenvector must not all be equal to zero. Now we know that a necessary and sufficient condition for the existence of a nontrivial solution of a system of linear homogeneous equations is that the rank of the coefficient matrix should be less than the number of unknowns, and this is equivalent to the vanishing of the determinant of the system

$$\begin{vmatrix} a_{11} - \lambda & & \cdots & a_{1n} \\ a_{21} & a_{22} - \lambda & \cdots & a_{2n} \\ \cdots & & & \\ a_{n1} & a_{n2} & \cdots & a_{nn} - \lambda \end{vmatrix} = 0.$$

Thus, all eigenvalues of the transformation A are roots of the polynomial $|A - \lambda E|$ and, conversely, every root of this polynomial is an eigenvalue of the transformation since to every root there corresponds at least one eigenvector. The polynomial $|A - \lambda E|$ is called the *characteristic polynomial* of the matrix A. The equation $|A - \lambda E| = 0$ is called the *characteristic* or *secular equation* and its roots *characteristic numbers* of the matrix.*

By the fundamental theorem of higher algebra (Chapter IV), every polynomial has at least one root; therefore every linear transformation has at least one eigenvalue and hence at least one eigenvector. But, of course, it is quite possible that even in the case when the transformation can be expressed by a real matrix it turns out that all or some of its eigenvalues are complex. Consequently, the theorem on the existence of (real) eigenvalues and eigenvectors for an arbitrary linear transformation is not true in a real space. For example, the transformation of the plane that consists in a rotation around the origin of coordinates by any angle other than 180° changes the directions of all the vectors of the plane so that there are no eigenvectors for this transformation.

The roots of the characteristic polynomial of a matrix A are the eigenvalues of the transformation A; therefore matrices that correspond to one and the same transformation in distinct bases have identical sets of roots of the characteristic polynomial. This leads to the plausible asssertion that the characteristic polynomial of a linear transformation also depends on the transformation only and not on the choice of a basis. This can be verified by the following elegant calculation, which is based on the properties of operations on matrices and determinants.

We know that if a matrix A corresponds to a transformation A in some basis, then in any other basis the transformation A has a similar matrix $C^{-1}AC$, where C is some nonsingular matrix. But

$$|C^{-1}AC - \lambda E| = |C^{-1}AC - C^{-1}\lambda EC| = |C^{-1}(A - \lambda E)C|$$
$$= |C^{-1}||C||A - \lambda E| = |C^{-1}C||A - \lambda E| = |A - \lambda E|.$$

Thus, matrices corresponding to one and the same transformation A in distinct bases have in fact one and the same characteristic polynomial, which can therefore be called the polynomial of the transformation.

We shall now make the assumption that all the eigenvalues of the transformation A are distinct. Let us prove that the eigenvectors, one for each eigenvalue, are linearly independent. For if we suppose that some

---

* The name "secular equation" has arisen in celestial mechanics in connection with the problem of the so-called secular disturbances in the motions of the planets.

§4. LINEAR TRANSFORMATIONS

of them, say $e_1, \cdots, e_k$, are linearly independent and the remaining ones, among them e_{k+1}, are linear combinations of these, then

$$e_{k+1} = c_1 e_1 + c_2 e_2 + \cdots + c_k e_k. \tag{13}$$

When we apply the linear transformation to both sides of this equation, we obtain

$$A e_{k+1} = c_1 A e_1 + c_2 A e_2 + \cdots + c_k A e_k,$$

from which it follows by the definition of an eigenvector that

$$\lambda_{k+1} e_{k+1} = c_1 \lambda_1 e_1 + c_2 \lambda_2 e_2 + \cdots + c_k \lambda_k e_k.$$

When we multiply equation (13) by λ_{k+1} and subtract from it the equation just obtained, we have

$$c_1(\lambda_{k+1} - \lambda_1) e_1 + c_2(\lambda_{k+1} - \lambda_2) e_2 + \cdots + c_k(\lambda_{k+1} - \lambda_k) e_k = 0,$$

hence it follows by the linear independence of $e_1, e_2, \cdots, e_k$ that

$$c_1(\lambda_{k+1} - \lambda_1) = c_2(\lambda_{k+1} - \lambda_2) = \cdots = c_k(\lambda_{k+1} - \lambda_k) = 0.$$

But we had assumed that all the eigenvalues are distinct and the vectors are chosen one for each eigenvalue. Therefore $\lambda_{k+1} - \lambda_1 \neq 0$, $\lambda_{k+1} - \lambda_2 \neq 0, \cdots, \lambda_{k+1} - \lambda_k \neq 0$ and the equation (13) is impossible, since the coefficients $c_1, c_2, \cdots, c_k$ cannot all be zero.

Now it is clear that if all the eigenvalues of a linear transformation are distinct, then there exists a basis in which the matrix of the transformation has diagonal form. For we can choose as such a basis a system of eigenvectors, one for each eigenvalue. As we have shown, they are linearly independent and their number is equal to the dimension of the space; i.e., they do in fact form a basis.

The theorem we have proved can be stated in terms of the theory of matrices as follows. If all the eigenvalues of a matrix are distinct, then the matrix is similar to the diagonal matrix whose diagonal elements are these eigenvalues.

The problem of transforming the matrix of a linear transformation to its simplest form is considerably more complicated if there are equal ones among the roots of the characteristic polynomial. We shall confine ourselves to a short account of the final result.

A "canonical box" of order m is defined as a matrix of the form

$$I_{m,\lambda_i} = \begin{bmatrix} \lambda_i & 1 & & & \\ & \lambda_i & 1 & & \\ & & \ddots & \ddots & \\ & & & \ddots & 1 \\ & & & & \lambda_i \end{bmatrix}.$$

All the unnamed elements are equal to zero.

A canonical Jordan matrix is defined as a matrix in which there are "canonical boxes" along the main diagonal and all the remaining elements are zero:

$$\begin{bmatrix} I_{m_1,\lambda_1} & & & \\ & I_{m_2,\lambda_2} & & \\ & & \ddots & \\ & & & I_{m_k,\lambda_k} \end{bmatrix}$$

The numbers λ_i in the distinct "boxes" are not necessarily pairwise distinct. Every matrix can be reduced to a canonical Jordan matrix similar to it. The proof of this theorem is rather complicated. We ought to mention that this theorem plays an important role in many applications of algebra to other problems of mathematics, in particular in the theory of systems of linear differential equations.

A matrix can be reduced to diagonal form if and only if the orders m_i of all boxes are equal to 1.

§5. Quadratic Forms

Definition and simplest properties. A *quadratic form* is a homogeneous polynomial of degree 2 in several variables.

A quadratic form in n variables $x_1, x_2, \cdots, x_n$ consists of terms of two types: squares of the variables and products of two variables, both with certain coefficients. A quadratic form can be written in the following quadratic scheme:

$$\begin{aligned} f(x_1, x_2, \cdots, x_n) = \quad & a_{11}x_1^2 + a_{12}x_1x_2 + \cdots + a_{1n}x_1x_n \\ & + a_{21}x_2x_1 + a_{22}x_2^2 + \cdots + a_{2n}x_2x_n \\ & \cdots\cdots\cdots\cdots\cdots\cdots\cdots\cdots\cdots\cdots\cdots\cdots \\ & + a_{n1}x_nx_1 + a_{n2}x_nx_2 + \cdots + a_{nn}x_n^2. \end{aligned}$$

Pairs of similar terms $a_{12}x_1x_2$ and $a_{21}x_2x_1$, etc., are written with equal coefficients so that each of them gets half the coefficient of the corresponding product of the variables. Thus, every quadratic form is uniquely connected with its coefficient matrix, which is symmetric.

A quadratic form can conveniently be represented in the following

§5. QUADRATIC FORMS

matrix notation. We denote by X the column of the variables $x_1, x_2, \cdots, x_n$ by $\tilde{X}$ the row $(x_1, x_2, \cdots, x_n)$, i.e., the transposed matrix of X. Then

$$
\begin{aligned}
f(x_1, x_2, \cdots, x_n) &= x_1(a_{11}x_1 + a_{12}x_2 + \cdots + a_{1n}x_n) \\
&\quad + x_2(a_{21}x_1 + a_{22}x_2 + \cdots + a_{2n}x_n) + \cdots \\
&\quad + x_n(a_{n1}x_1 + a_{n2}x_2 + \cdots + a_{nn}x_n) \\
&= (x_1, x_2, \cdots, x_n) \begin{bmatrix} a_{11}x_1 + a_{12}x_2 + \cdots + a_{1n}x_n \\ a_{21}x_1 + a_{22}x_2 + \cdots + a_{2n}x_n \\ \cdots \\ a_{n1}x_1 + a_{n2}x_2 + \cdots + a_{nn}x_n \end{bmatrix} \\
&= (x_1, x_2, \cdots, x_n) \begin{bmatrix} a_{11} a_{12} \cdots a_{1n} \\ a_{21} a_{22} \cdots a_{2n} \\ \cdots \\ a_{n1} a_{n2} \cdots a_{nn} \end{bmatrix} \begin{bmatrix} x_1 \\ x_2 \\ \vdots \\ x_n \end{bmatrix} = \tilde{X}AX.
\end{aligned}
$$

Quadratic forms occur in many branches of mathematics and its applications.

In the theory of numbers and in crystallography, we consider quadratic forms under the proviso that the variables $x_1, x_2, \cdots, x_n$ assume only integral values. In analytical geometry a quadratic form arises in setting up the equation of a curve (or surface) of the second order. In mechanics and physics a quadratic form appears in the expression for the kinetic energy of a system in terms of the components of the generalized velocities, etc. Furthermore, a study of quadratic forms is necessary even in analysis for the investigation of functions of several variables in connection with problems where it is important to clarify how a given function in the neighborhood of a given point deviates from its approximating linear function. An example of problems of this type is the investigation of maxima and minima of a function.

Let us consider, for example, the problem of finding the maxima and minima of a function of two variables $w = f(x, y)$ having continuous partial derivatives of the third order. A necessary condition for the point (x_0, y_0) to give a maximum or minimum of the function w is that the partial derivatives of the first order should vanish at the point (x_0, y_0). Let us assume that this condition is satisfied. We give to the variables x and y small increments h and k and consider the corresponding increment of the function $\Delta w = f(x_0 + h, y_0 + k) - f(x_0, y_0)$. By Taylor's formula this increment is equal, to within terms of higher order of smallness, to the quadratic form $\frac{1}{2}(rh^2 + 2shk + tk^2)$, where r, s and t are the values of the second derivatives $\partial^2 w/\partial x^2$, $\partial^2 w/\partial x \partial y$, $\partial^2 w/\partial y^2$, computed at the point (x_0, y_0). If this quadratic form is positive for all values of h and k (except

$h = k = 0$), then the function w has a minimum at the point (x_0, y_0); if it is negative, then a maximum. Finally, if the form assumes positive as well as negative values, then there is neither a maximum nor a minimum. Functions with a larger number of variables can be investigated in a similar way.

The study of quadratic forms essentially consists in the investigation of the problem of the equivalence of forms under one set or another of linear transformations of the variables. Two quadratic forms are called *equivalent* if one of them can be carried into the other by one of the transformations of the given set. Closely connected with the problem of equivalence is the problem of *reduction* of a form, i.e., its transformation into a certain form, as simple as possible.

In various problems connected with quadratic forms, we consider various sets of admissible transformations of the variables.

In problems of analysis arbitrary nonsingular transformations of the variables are admitted; for the purposes of analytical geometry, the greatest interest lies in orthogonal transformations, i.e., those that correspond to the transition from one system of variable Cartesian coordinates to another. Finally, in the theory of numbers and in crystallography, we consider linear transformations with integer coefficients and with a determinant equal to 1.

Here we shall consider two of these problems: the problem of the reduction of a quadratic form to the simplest possible form by means of arbitrary nonsingular transformations and the same problem for orthogonal transformations. Above all, we shall find out how the matrix of the quadratic form is changed under a linear transformation of the variables.

Suppose that $f(x_1, x_2, \cdots, x_n) = \bar{X}AX$, where A is the symmetric matrix of the coefficients of the form and X is the column of the variables.

We take a linear transformation of the variables, writing it briefly $X = CX'$. Here C denotes the coefficient matrix of this transformation and X' the column of the new variables. Then $\bar{X} = \bar{X}'\bar{C}$ and consequently $\bar{X}AX = \bar{X}'(\bar{C}AC)X'$ so that the matrix of the transformed quadratic form is $\bar{C}AC$.

The matrix $\bar{C}AC$ is automatically symmetric, as is easy to verify. Thus, the problem of reducing a quadratic form to the simplest possible form is equivalent to the task of reducing a symmetric matrix to the simplest form by multiplying it on the left and on the right by transposed matrices.

Transformation of a quadratic form to the canonical form by successive completion of squares. We shall show that every (real) quadratic form can be reduced to a sum of squares of new variables with certain coefficients by a real nonsingular linear transformation.

§5. QUADRATIC FORMS

To prove this we shall show first of all that if the form is not identically zero, then we can make the coefficient of the square of the first variable different from zero by the application of a nonsingular transformation of the variables.

For suppose that

$$
\begin{aligned}
f(x_1, x_2, \cdots, x_n) = \quad & a_{11}x_1^2 + a_{12}x_1x_2 + \cdots + a_{1n}x_1x_n \\
+ \; & a_{21}x_2x_1 + a_{22}x_2^2 + \cdots + a_{2n}x_2x_n \\
& \cdots\cdots\cdots\cdots\cdots\cdots\cdots\cdots\cdots\cdots \\
+ \; & a_{n1}x_nx_1 + a_{n2}x_nx_2 + \cdots + a_{nn}x_n^2.
\end{aligned}
$$

If $a_{11} \neq 0$, then no transformation is required. If $a_{11} = 0$, but some one of the diagonal coefficients $a_{kk} \neq 0$, then we set $x_1 = x_k'$, $x_k = x_1'$ equating the remaining original variables to the corresponding new ones. This nonsingular transformation achieves our object. Finally, if all the diagonal coefficients are equal to zero, then at least one of the nondiagonal coefficients is different from zero, for example a_{12}.

By taking the nonsingular transformation

$$
x_1 = x_1',
$$
$$
x_2 = x_1' + x_2'
$$

and equating the remaining original variables to the new ones, we achieve our aim.

Thus, without loss of generality we can assume that $a_{11} \neq 0$.

Let us now separate out the square of a linear function such that all the terms containing x_1 occur in this square.

This can easily be done. For

$$
\begin{aligned}
f(x_1, x_2, \cdots, x_n) = \quad & a_{11}x_1^2 + a_{12}x_1x_2 + \cdots + a_{1n}x_1x_n \\
+ \; & a_{21}x_2x_1 + a_{22}x_2^2 + \cdots + a_{2n}x_2x_n \\
& \cdots\cdots\cdots\cdots\cdots\cdots\cdots\cdots\cdots\cdots \\
+ \; & a_{n1}x_nx_1 + a_{n2}x_nx_2 + \cdots + a_{nn}x_n^2 \\
= a_{11}\left[x_1 + \frac{a_{12}}{a_{11}}x_2 + \cdots + \frac{a_{1n}}{a_{11}}x_n\right]^2 & - a_{11}\left[\frac{a_{12}}{a_{11}}x_2 + \cdots + \frac{a_{1n}}{a_{11}}x_n\right]^2 \\
+ \; a_{22}x_2^2 + \cdots + a_{2n}x_2x_n & \\
\cdots\cdots\cdots\cdots\cdots\cdots\cdots\cdots\cdots & \\
+ a_{n2}x_nx_2 + \cdots + a_{nn}x_n^2. &
\end{aligned}
$$

By removing the parentheses in the second summand and collecting similar terms, we obtain

$$f(x_1, x_2, \cdots, x_n) = a_{11} \left[x_1 + \frac{a_{12}}{a_{11}} x_2 + \cdots + \frac{a_{1n}}{a_{11}} x_n \right]^2 + f_1(x_2, \cdots, x_n),$$

where f_1 is a form in $n - 1$ variables.

The transformation

$$x_1 + \frac{a_{12}}{a_{11}} x_2 + \cdots + \frac{a_{1n}}{a_{11}} x_n = x_1',$$
$$x_2 = x_2',$$
$$\cdots\cdots\cdots\cdots\cdots$$
$$x_n = x_n',$$

is obviously nonsingular. By making this transformation we reduce our form to

$$a_{11} x_1'^2 + f_1(x_2', \cdots, x_n').$$

Continuing the process in a similar manner we reduce the form to the required "canonical" form

$$\alpha_1 z_1^2 + \alpha_2 z_2^2 + \cdots + \alpha_n z_n^2.$$

Here $z_1, z_2, \cdots, z_n$ are the new variables introduced at the last step.

The law of inertia of quadratic forms. In the reduction of a quadratic form to canonical form there always is a considerable arbitrariness in the choice of the transformation of variables that brings about this reduction. This arbitrariness comes in, among other things, from the fact that we can precede the previous method of successive separation of squares by an arbitrary nonsingular transformation of the variables.

However, notwithstanding this arbitrariness, as a result of the reduction of the given form we obtain an almost unique canonical quadratic form independent of the choice of the reducing transformation. Namely the number of squares of the new variables that occur with positive, negative, and zero coefficients is always one and the same, irrespective of the method of reduction. This theorem is known as the *law of inertia* of quadratic forms. We shall not prove it here.

The law of interia of a quadratic form solves the problem of equivalence of a real quadratic form under all nonsingular transformations. For two forms are equivalent if and only if their reduction to canonical form leads to forms with the same number of squares with positive, negative, and zero coefficients.

§5. QUADRATIC FORMS

Of special interest for the applications are the quadratic forms that under reduction to canonical form turn into a sum of squares of the new variables with all the coefficients *positive*. Such forms are called *positive definite*.

Positive-definite quadratic forms are characterized by the property that their values for real values of the variables not all equal to zero are always positive.

Orthogonal transformation of quadratic forms to canonical form. Of special interest among all possible methods of reducing a quadratic form to canonical form are the orthogonal transformations, i.e., those that can be obtained by a linear transformation of the variables with an orthogonal matrix. Such transformations are of interest, for example, in analytical geometry, in the problem of reducing the general equation of a curve or surface of the second order to canonical form.

In order to convince ourselves of the possibility of such a transformation, it is convenient to regard the quadratic form as a function of vectors in a Euclidean space by considering the variables $x_1, x_2, \cdots, x_n$ as the coordinates of a variable vector in an orthonormal basis. Then an orthogonal transformation of the variables can be interpreted as a transition from one orthonormal basis to another.

With the quadratic form

$$f(x_1, x_2, \cdots, x_n) = a_{11}x_1^2 + \cdots + a_{1n}x_1x_n$$
$$\cdots\cdots\cdots\cdots\cdots\cdots\cdots\cdots\cdots\cdots\cdots$$
$$+ a_{n1}x_nx_1 + \cdots + a_{nn}x_n^2$$

we connect the linear transformation A that has with respect to the chosen basis the matrix

$$A = \begin{bmatrix} a_{11} & \cdots & a_{1n} \\ \cdots & \cdots & \cdots \\ a_{n1} & \cdots & a_{nn} \end{bmatrix}$$

Then our quadratic form can be regarded as the scalar product $AX \cdot X$ (where X is the vector with the coordinates $x_1, x_2, \cdots, x_n$), and its coefficients a_{ij} are the scalar product $Ae_i \cdot e_j$, where $e_1, e_2, \cdots, e_n$ is the chosen orthonormal basis.

It is easy to see that in consequence of the symmetry of the matrix A the following equation holds for arbitrary vectors X and Y

$$AX \cdot Y = X \cdot AY.$$

We shall show first of all that the transformation A has at least one real eigenvalue and eigenvector corresponding to it.

For this purpose we consider the values of the form $AX \cdot X$ under the assumption that the vector X ranges over the unit sphere, i.e., the set of all unit vectors. Under these conditions the form $AX \cdot X$ will have a maximum. Let us show that this maximum $AX \cdot X$ is an eigenvalue of the tranformation A and the vector X_0 for which this maximum is assumed is a corresponding eigenvector; i.e., $AX_0 = \lambda_1 X_0$.

The proof of this statement is by an indirect method, namely, by establishing that the vector AX_0 is orthogonal to all vectors orthogonal to X_0.

We note that for an arbitrary vector Z we have the inequality $AZ \cdot Z \leqslant \lambda_1 \mid Z \mid^2$. This is obvious from the fact that $X = Z/\mid Z \mid$ is a unit vector and λ_1 the maximum of the values of the form $AX \cdot X$ on the unit sphere. We consider $Z = X_0 + \epsilon Y$, where ϵ is a real number and Y an arbitrary vector orthogonal to the vector X_0. Then

$$AZ \cdot Z = (AX_0 + \epsilon AY) \cdot (X_0 + \epsilon Y) = AX_0 \cdot X_0 + 2\epsilon AX_0 \cdot Y + \epsilon^2 AY \cdot Y$$
$$= \lambda_1 + 2\epsilon AX_0 \cdot Y + \epsilon^2 AY \cdot Y.$$

Moreover,

$$\mid Z \mid^2 = (X_0 + \epsilon Y) \cdot (X_0 + \epsilon Y) = \mid X_0 \mid^2 + \epsilon^2 \mid Y \mid^2 = 1 + \epsilon^2 \mid Y \mid^2,$$

because
$$X_0 \cdot Y = 0, \quad \mid X_0 \mid^2 = 1.$$

Therefore,
$$\lambda_1 + 2\epsilon AX_0 \cdot Y + \epsilon^2 AY \cdot Y \leqslant \lambda_1 + \epsilon^2 \lambda_1 \mid Y \mid^2,$$

and from this we obtain on dividing by ϵ^2

$$\frac{2}{\epsilon} AX_0 \cdot Y < \lambda_1 \mid Y \mid^2 - AY \cdot Y. \tag{14}$$

This last inequality must be satisfied for arbitrary real ϵ of sufficiently small absolute value.

But it can only be satisfied under the condition $AX_0 \cdot Y = 0$, because if $AX_0 \cdot Y > 0$, then the inequality (14) is impossible for sufficiently small positive ϵ, and if $AX_0 \cdot Y < 0$, then it is impossible for negative ϵ of sufficiently small absolute value. Thus, $AX_0 \cdot Y = 0$, i.e., AX_0 is in fact orthogonal to every vector orthogonal to X_0. Therefore AX_0 and X_0 are collinear; i.e., $AX_0 = \lambda' X_0$, where λ' is a real number. The fact that $\lambda' = \lambda_1$ is easy to verify, for

$$\lambda_1 = AX_0 \cdot X_0 = \lambda' X_0 \cdot X_0 = \lambda'.$$

§6. FUNCTIONS OF MATRICES

Now it is easy to show that every quadratic form can in fact be reduced to canonical form by an orthogonal transformation.

Let $e_1, e_2, \cdots, e_n$ be the original orthonormal basis of the space and $f_1, f_2, \cdots, f_n$ a new orthonormal basis in which the first vector f_1 is an eigenvector X_0 of the transformation A. Let $x_1, x_2, \cdots, x_n$ be the coordinates of the vector X in the original basis and $x'_1, x'_2, \cdots, x'_n$ its coordinates in the new basis. Then

$$\begin{bmatrix} x_1 \\ x_2 \\ \vdots \\ x_n \end{bmatrix} = P \begin{bmatrix} x'_1 \\ x'_2 \\ \vdots \\ x'_n \end{bmatrix}$$

where P is an orthogonal matrix.

Let us carry out the transition to the new variables in the quadratic form $AX \cdot X$. In the new variables the quadratic form has the coefficients $a'_{ij} = Af_i \cdot f_j$. Therefore

$$a'_{11} = Af_1 \cdot f_1 = \lambda_1 f_1 \cdot f_1 = \lambda_1,$$

$$a'_{1j} = a'_{j1} = Af_1 \cdot f_j = \lambda_1 f_1 \cdot f_j = 0 \quad \text{for} \quad j \neq 1;$$

i.e., the form is now

$$\lambda_1 x'^2_1 + \phi(x'_2, \cdots, x'_n).$$

Thus, by means of an orthogonal transformation we have succeeded in separating out one square of a new variable.

By repeating the same arguments with the new form $\phi(x'_2, \cdots, x'_n)$, etc., we eventually arrive at the conclusion that the form turns out to be reduced to canonical form by a chain of orthogonal transformations. But it is obvious that a chain of orthogonal transformations is equivalent to a single orthogonal transformation. This concludes the proof of the theorem.

§6. Functions of Matrices and Some of Their Applications

Functions of matrices. The applications of linear algebra to other branches of mathematics are very numerous and diverse. It is not an exaggeration to say that the ideas and results of linear algebra are used in a large part of contemporary mathematics and theoretical physics in one form or another, particularly in the form of matrix calculus.

We shall deal briefly with one of the methods of applying the matrix calculus to the theory of ordinary differential equations. Here functions of matrices play an important role.

First of all we define the powers of a square matrix A. We set $A^0 = E$, $A^1 = A$, $A^2 = AA$, $A^3 = A^2 A$, $A^4 = A^3 A$, etc. By the associative law it is easy to show that $A^m A^n = A^{m+n}$ for arbitrary natural numbers m and n. The operations of addition and of multiplication by a number have been defined for matrices. This enables us to define in a natural way the meaning of a polynomial (in one variable) of a matrix. For if $\phi(x) = a_0 x^n + a_1 x^{n-1} + \cdots + a_n$, then we set (by definition) $\phi(A) = a_0 A^n + a_1 A^{n+1} + \cdots + a_n E$. Thus, the concept of the simplest function of a matrix argument, namely the polynomial, is defined.

By means of a limit process it is easy to generalize the concept of a function of a matrix argument to a considerably wider class of functions than polynomials in one variable. Without treating this problem in all its generality we shall restrict ourselves here to the examination of analytic functions.

First of all we introduce the concept of the limit of a sequence of matrices. A sequence of matrices

$$A_1 = \begin{bmatrix} a_{11}^{(1)} & \cdots & a_{1n}^{(1)} \\ \cdots & \cdots & \cdots \\ a_{n1}^{(1)} & \cdots & a_{nn}^{(1)} \end{bmatrix}, \quad A_2 = \begin{bmatrix} a_{11}^{(2)} & \cdots & a_{1n}^{(2)} \\ \cdots & \cdots & \cdots \\ a_{n1}^{(2)} & \cdots & a_{nn}^{(2)} \end{bmatrix}, \cdots$$

is said to *converge* to the matrix

$$A = \begin{bmatrix} a_{11} & \cdots & a_{1n} \\ \cdots & \cdots & \cdots \\ a_{n1} & \cdots & a_{nn} \end{bmatrix}$$

(or to have the matrix A as its limit) if $\lim_{k \to \infty} a_{ij}^{(k)} = a_{ij}$ for all i, j. Further, the sum of a series $A_1 + A_2 + \cdots + A_k + \cdots$ is defined as the limit of the sums of its segments $\lim_{k \to \infty} (A_1 + A_2 + \cdots + A_k)$ if this limit exists.

Let $f(z)$ be an analytic function regular in a neighborhood of $z = 0$. Then $f(z)$ can be expanded in a power series

$$f(z) = a_0 + a_1 z + a_2 z^2 + \cdots + a_k z^k + \cdots.$$

For every square matrix A it is natural to set

$$f(A) = a_0 E + a_1 A + a_2 A^2 + \cdots + a_k A^k + \cdots.$$

Now it turns out that this series converges for all matrices A whose eigenvalues lie within the circle of convergence of the power series $a_0 + a_1 z + \cdots + a_k z^k + \cdots$.

§6. FUNCTIONS OF MATRICES

Of interest for the applications are the elementary functions of matrices. For example, the geometric series $E + A + A^2 + \cdots + A^k + \cdots$ is convergent for matrices whose eigenvalues are of absolute value less than 1, and the sum of the series is the matrix $(E - A)^{-1}$, in complete accordance with the formula

$$1 + x + \cdots + x^k + \cdots = \frac{1}{1-x}.$$

The representation of $(E - A)^{-1}$ in the form of an infinite series gives an effective method for the approximate solution of systems of linear equations whose coefficient matrix is close to the unit matrix.

For when we write such a system in the form

$$(E - A) X = B,$$

we obtain

$$X = (E - A)^{-1} B = B + AB + A^2 B + \cdots \qquad (15)$$

and this gives a convenient formula for the solution of the system, if only the series (15) converges sufficiently fast.

It is useful to consider the binomial series

$$(E + A)^m = E + \frac{m}{1} A + \frac{m(m-1)}{2!} A^2 + \cdots$$

which can be applied (provided the eigenvalues of A are less than 1 in modulus) not only for natural exponents m but also for fractional and negative exponents.

Particularly important for the applications is the exponential matrix function

$$e^A = E + A + \frac{A^2}{2!} + \frac{A^3}{3!} + \cdots.$$

The series defining the exponential function converges for every matrix A. The exponential matrix function has properties reminding us of properties of the ordinary exponential function. For example, if A and B commute under multiplication, i.e., $AB = BA$, then $e^{A+B} = e^A \cdot e^B$. However, for noncommuting A and B the formula ceases to be true.

Application to the theory of systems of ordinary linear differential equations. In the theory of systems of ordinary linear differential equations it is appropriate to consider matrices whose elements are functions of an independent variable:

$$U(t) = \begin{bmatrix} a_{11}(t) & \cdots & a_{1n}(t) \\ \cdots & & \cdots \\ a_{m1}(t) & \cdots & a_{mn}(t) \end{bmatrix}.$$

For such matrices the concept of the derivative with respect to the argument t is defined in a natural way, namely:

$$\frac{dU(t)}{dt} = \begin{bmatrix} a'_{11}(t) \cdots a'_{1n}(t) \\ \cdots\cdots\cdots\cdots\cdots \\ a'_{m1}(t) \cdots a'_{mn}(t) \end{bmatrix}.$$

It is not difficult to verify that some elementary formulas of differentiation are valid for matrices. For example,

$$\frac{d(U+V)}{dt} = \frac{dU}{dt} + \frac{dV}{dt},$$

$$\frac{d(cU)}{dt} = c\frac{dU}{dt},$$

$$\frac{d(UV)}{dt} = \frac{dU}{dt}V + U\frac{dV}{dt}.$$

(The multiplication must be carried out strictly in the order as given in the formula!)

A system of ordinary linear homogeneous differential equations

$$\frac{dy_1}{dt} = a_{11}(t)\,y_1 + a_{12}(t)\,y_2 + \cdots + a_{1n}(t)\,y_n,$$

$$\frac{dy_2}{dt} = a_{21}(t)\,y_1 + a_{22}(t)\,y_2 + \cdots + a_{2n}(t)\,y_n$$

$$\cdots\cdots\cdots\cdots\cdots\cdots\cdots\cdots\cdots\cdots\cdots\cdots\cdots$$

$$\frac{dy_n}{dt} = a_{n1}(t)\,y_1 + a_{n2}(t)\,y_2 + \cdots + a_{nn}(t)\,y_n$$

can be written in this notation in the form

$$\frac{dY}{dt} = A(t)\,Y,$$

where

$$Y = \begin{bmatrix} y_1 \\ \vdots \\ y_n \end{bmatrix}, \quad A(t) = \begin{bmatrix} a_{11}(t) \cdots a_{1n}(t) \\ \cdots\cdots\cdots\cdots\cdots \\ a_{n1}(t) \cdots a_{nn}(t) \end{bmatrix},$$

i.e, in a form similar to a single linear homogeneous differential equation.

If the coefficients of the system are constants, i.e., if the matrix A is constant, then the solution of the system also looks outwardly like the solution of the equation $y' = ay$. For in this case $Y = e^{At}C$, where C is a column of arbitrary constants.

The solution in this form is very convenient for computations. The fact is that for an arbitrary analytic function $f(z)$ we have the equation

$$f(B^{-1}LB) = B^{-1}f(L)B.$$

Since every matrix can be reduced to the canonical Jordan form (see §4), the computation of a function of an arbitrary matrix reduces to the computation of a function of a canonical matrix, which is easy to carry out. Therefore, if $A = B^{-1}LB$, where L is a canonical matrix, then

$$Y = e^{At}C = B^{-1}e^{Lt}BC = B^{-1}e^{Lt}C',$$

where $C' = BC$ is a column of arbitrary constants.

From this formula it is easy to derive an explicit expression for all the components of the required column Y.

The Soviet mathematician I. A. Lappo-Danilevskiĭ has successfully developed the apparatus of the theory of matrix functions and was the first to apply it to the investigation of systems of differential equations, including those with variable coefficients. His results count among the most brilliant achievements of mathematics in the last fifty years.

Suggested Reading

G. Birkhoff and S. MacLane, *A survey of modern algebra*, Macmillan, New York, 1941.

V. N. Faddeeva, *Computational methods of linear algebra.* Translated by C. D. Benster, Dover, New York, 1959.

F. R. Gantmacher, *Applications of the theory of matrices.* Translated by J. L. Brenner, Interscience, New York, 1959.

——, *The theory of matrices*, Vols. 1, 2. Translated by K. A. Hirsch, Chelsea New York, 1959.

  [This book and the one above are translations from the Russian original, published in 1953, but with various additions, some of which were supplied by the author. The two volumes contain a great deal of material not available in other texts.]

I. M. Gel'fand, *Lectures on linear algebra*, Interscience, New York, 1961.

F. Klein, *Elementary mathematics from an advanced standpoint*. Vol. I. *Arithmetic, algebra, analysis.* Translated by E. R. Hedrick and C. A. Noble, Dover, New York, 1953.

O. Schreier and E. Sperner, *Introduction to modern algebra and matrix theory*, Chelsea, New York, 1951.

V. I. Smirnov, *Linear algebra and group theory.* Translated and revised by R. A. Silverman, McGraw-Hill, New York, 1961.

H. W. Turnbull and A. C. Aitken, *An introduction to the theory of canonical matrices*, Dover, New York, 1961.

B. L. van der Waerden, *Modern algebra*, Vol. I. Translated from the second revised German edition by Fred Blum. With revisions and additions by the author. Frederick Ungar, New York, 1949.

CHAPTER XVII

NON-EUCLIDEAN GEOMETRY

Ever since N. I. Lobačevskiĭ first demonstrated the possibility of a non-Euclidean geometry and put forward a new notion concerning the relationship of geometry to the material reality, the scope of geometry, its methods and applications, have been enlarged exceedingly. Nowadays mathematicians study several "spaces": apart from the Euclidean space they deal with Lobačevskiĭ space, projective space, various n-dimensional and even infinite-dimensional spaces, Riemannian, topological, and other spaces; the number of such spaces is boundless and each of them has its own properties, its own "geometry." In physics we use the concept of the so-called phase and configuration "spaces"; the theory of relativity utilizes the notion of a curved space and other results of abstract geometrical theories.

How and whence have these mathematical abstractions arisen? What real basis, what real value and application do they have? What is their relation to reality? How are they defined and how are they studied in mathematics? What is the significance of the general ideas of contemporary geometry in mathematics?

These questions will be answered in the present chapter. We shall not give an account of the theory of abstract mathematical spaces as such; that would require a far longer explanation and far more attention to the specific mathematical apparatus. Our task is to throw light on the essence of the new ideas in geometry, i.e., to answer the questions raised here, and this can be done without complicated proofs and formulas.

The history of our problem goes right back to Euclid's "Elements" and the axiom or, as one also says, the postulate on parallel lines.

XVII. NON-EUCLIDEAN GEOMETRY

§1. History of Euclid's Postulate

In his "Elements" Euclid formulates the fundamental premises of geometry in the form of so-called postulates and axioms. Among these there is Postulate V (in other copies of the "Elements" Axiom XI) which is now usually stated as follows: "Through a point not lying on a given line not more than one line parallel to the given line can be drawn." We recall that a line is called parallel to a given line if both lie in one plane and do not intersect; we think here of the infinite lines, so to speak, not of their finite segments.

It is easy to prove that we can always draw at least one parallel to a given line a through a point A not lying on it.

For let us drop a perpendicular b from A to a and draw through A a line c perpendicular to b (figure 1). The figure so obtained is completely

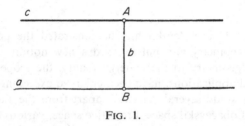

Fig. 1.

symmetric with respect to b, since the angles formed by b with a and c at both its ends are equal. Therefore, when we turn the plane about b, we bring the half lines of a and c into coincidence. Hence it is clear that if a and c were to intersect on one side of AB, then they would have to intersect also on the other side. It would then come out that the lines a and c have two points in common; but this is impossible, because by a fundamental property of the line only one line can be drawn through two points (so that lines having two common points must necessarily coincide).

Thus, from the basic properties of the line and of the motion of a figure (inasmuch as a turn about the line AB is a rotation of a semiplane with this line as axis) it follows that at least one parallel to a given line can always be drawn through a given point. Now Euclid's postulate supplements this result by the statement that this parallel must be the only one and no other can exist.

Among the other postulates (axioms) of geometry this one occupies a somewhat special place. Euclid's own formulation of it is rather complicated, but even in its usual form cited here it contains a certain difficulty. This difficulty is already inherent in the very concept of parallel

§1. HISTORY OF EUCLID'S POSTULATE

lines, here we deal with the whole line. But how are we to convince ourselves that two given lines are parallel? For this purpose we ought to produce them on both sides "to infinity" and convince ourselves that they do not intersect anywhere *on their whole infinite extent*. This notion clearly has its difficulties. And all this was apparently the reason why the parallel postulate occupied a somewhat special position even for Euclid: In his "Elements" this postulate is used beginning with the 29th proposition only, whereas in the first 28 propositions he dispenses with it. In view of the complicated nature of the postulate the wish to do without it is quite natural, and already in antiquity attempts were made to change the definition of parallel lines, to change the formulation of the postulate itself or, better still, to deduce it as a theorem from other axioms and basic concepts of geometry.

Thus, the theory of parallel lines, founded on the Fifth Postulate, became an object of comment and criticism in the works of many geometers, from the days of antiquity. In the course of these investigations it was the principal aim to get rid of the Fifth Postulate altogether, by deducing it as a theorem from other basic propositions of geometry.

This task attracted many geometers: the Greek Proclus (5th century A. D.) who wrote a commentary to Euclid, the Persian Nasir ed Din et Tusi* (13th century), the Englishman Wallis (1616-1703), the Italian Saccheri (1667-1733), the German philosopher and mathematician Lambert (1728-1777), the Frenchman Legendre (1752-1833), and many others; in the span of more than two thousand years since Euclid's "Elements" appeared they all outdid themselves in subtlety and geometrical ingenuity trying to prove the Fifth Postulate.

However, the result of these attempts invariably remained negative. Every time it became clear that the author of one proof or another had in fact relied on some proposition, no matter how obvious, that did not at all follow with logical necessity from the other premises of geometry. In other words, each time what they had done amounted to replacing the Fifth Postulate by some other statement from which in fact this postulate followed, but which itself required a proof.†

---

* Translator's note: Nasiraddin

† Of such statements, equivalent to the Fifth Postulate, quite a large number were set up. Here are some examples: (1) a line parallel to a given line has a constant distance from it (Proclus); (2) there exist similar (but not equal) triangles, i.e., triangles whose angles are equal but whose sides are unequal (Wallis); (3) there exists at least one rectangle, i.e., a quadrangle whose angles are all right angles (Saccheri); (4) a line perpendicular to one arm of an acute angle also intersects the other arm (Legendre); (5) the sum of the angles of a triangle is equal to two right angles (Legendre); (6) there exist triangles of arbitrarily large area (Gauss). This list could now be continued indefinitely.

Saccheri and Lambert penetrated deeper into the problem than the others. Saccheri was the first to attempt a proof of the Fifth Postulate by a *reductio ad absurdum*; i.e., he took as a starting point the opposite assertion and by developing its consequences hoped to come to a contradiction. When he arrived, in these inferences, at results that appeared entirely unimaginable, he thought that he had solved the problem. But he was mistaken, because a contradiction to intuitive ideas does not yet indicate a logical contradiction. After all, the problem was to prove the Euclidean postulate on the basis of other propositions of geometry and not to convince oneself once more of its intuitive truth. Intuitively, the postulate itself is convincing enough. But, let us repeat, intuitive conclusiveness and logical necessity are two different things.

Lambert proved to be a deeper thinker than Saccheri and his predecessors. Starting out on the same path he did not find a logical contradiction and did not make the mistakes of the others; he did not claim to have proved the Fifth Postulate. But after him, at the beginning of the 19th century, Legendre once more "proves" the Fifth Postulate by falling into the old mistake: Again he replaces the postulate by other assertions which require proof themselves.

Thus, at the beginning of the 19th century the problem of proving the Fifth Postulate remained as unsolved as it was in Euclid's time. The efforts remained in vain and the problem, it seemed, would not yield. Here, indeed, was a deep enigma of geometry; a problem whose solubility was not doubted by the best geometers did not yield a solution in two thousand years.

The theory of parallel lines became one of the central problems of geometry in the 19th century. It attracted many geometers: Gauss, Lagrange, d'Alembert, Legendre, Wachter, Schweikart, Taurinus, Farkas Bólyai, and others.

However, a proof of the postulate did not come forth. What, then, was the matter: Was it due to lack of ability or was the problem perhaps inaccurately stated? This question began to cross the minds of some geometers that surpassed the others in depth of thought. Gauss, the most famous German mathematician, occupied himself with the problem from 1792 onward and the correct statement of the question gradually dawned upon him. Finally he decided to abandon the Fifth Postulate and from 1813 on he developed a sequence of theorems that are consequences of the opposite assertion. A little later the German mathematician Schweikart, who was then a professor of Law at Kharkov, and Taurinus followed on the same path. But none of these arrived at a final answer to the problem. Gauss carefully concealed his investigations, Schweikart confined himself to private correspondence with Gauss, and only Taurinus

§2. THE SOLUTION OF LOBAČEVSKIĬ

went into print with the elements of a new geometry based on the negation of the Fifth Postulate. However, he himself ruled out the possibility of such a geometry. Thus, none of these solved the problem and the question whether the statement was altogether correctly put remained without an answer. The answer was first given by N. I. Lobačevskiĭ, a young professor at the University of Kazan: On February 23, 1826 he read a paper on the theory of parallels at a meeting of the physical-mathematical faculty and in 1829 he published its contents in the Journal of the University of Kazan.

§2. The Solution of Lobačevskiĭ

1. The ideas of Lobačevskiĭ. The essence of Lobačevskiĭ's solution of the problem of the Fifth Postulate was expressed by him in his work "New Elements of Geometry" (1835) in the following words:

"It is well known that in geometry the theory of parallels has so far remained incomplete. The futile efforts from Euclid's time on throughout two thousand years have compelled me to suspect that the concepts themselves do not contain the truth which we have wished to prove, but that it can only be verified like other physical laws by experiments, such as astronomical observations. Convinced, at last, of the truth of my conjecture and regarding the difficult problem as completely solved, I put down my arguments in 1826."

Let us examine what Lobačevskiĭ had in mind in this statement in which his new idea is so to speak focused, which not only gave the solution of the problem of the Fifth Postulate, but revolutionized our whole conception of geometry and of other branches of mathematics as well.

Already in 1815 I. N. Lobačevskiĭ had begun to work on the theory of parallels, trying at first, like the other geometers, to prove the Fifth Postulate. In 1823 he realized clearly that all the proofs, "no matter of what kind, can only be regarded as clarifications, but do not deserve to be called mathematical proofs in the full sense."* He was aware, then, that "the concepts themselves do not contain the truth which we have wished to prove"; i.e., in other words, the Fifth Postulate cannot be deduced from the fundamental propositions and concepts of geometry. How did he convince himself that this deduction is impossible?

He did this by going farther along the path on which Saccheri and Lambert had taken the first steps. As a hypothesis he took the statement contradicting the Euclidean postulate, namely: "through a point not

---

* So wrote Lobačevskiĭ in 1823 in his course of geometry, which was not published during his lifetime. This course "Geometry" was first edited in 1910.

lying on a given line not one, but at least two lines parallel to the given one can be drawn." Let us take this statement conditionally as an axiom and, adjoining it to the other propositions of geometry, let us develop further consequences of it. Then, if this assertion is incompatible with the other propositions of geometry, we shall arrive at a contradiction and so the Fifth Postulate will be proved indirectly: The opposite proposition leads to a contradiction. However, in view of the fact that no such contradiction is detected, we come to two conclusions which Lobačevskiĭ also reached.

The first conclusion is that the Fifth Postulate is not provable. The second conclusion is that on the basis of the opposite axiom just formulated we can develop a chain of consequences, i.e., theorems that do not contain a contradiction. These consequences form in their own right a certain logically possible noncontradictory theory that can be regarded as a new non-Euclidean geometry. Lobačevskiĭ cautiously called it "imaginary," because he could not yet find a real explanation for it. But its logical possibility was clear to him. By expressing and defending this strong conviction Lobačevskiĭ displayed the true grandeur of a genius who defends his convictions without wavering and does not hide them from public opinion for fear of misunderstanding and criticism.

Thus, the first two conclusions reached by Lobačevskiĭ consisted in the statements that the Fifth Postulate is not provable and that it is possible to develop on the basis of a contrary axiom a new geometry that is logically just as rich and perfect as the Euclidean, notwithstanding the fact that its results are at variance with the intuitive picture of space. Lobačevskiĭ in fact developed this new geometry, which now bears his name. A general result of enormous importance was involved here: *More than one geometry is logically conceivable.* The significance of this result in its full extent will be discussed later; in it is really contained not a small part of the solution of the problems concerning abstract mathematical spaces that were raised at the beginning of this chapter.

But let us return to Lobačevskiĭ's statement quoted previously. He says that geometrical truth like other physical laws can only be verified by experiments. This means, first, that we must interpret truth as a correspondence between abstract concepts and reality. This correspondence can only be established by experiment so that the verification of one result or another requires experimental investigations and that mere logical inference is insufficient for the purpose. Although Euclidean geometry reflects the real properties of space very accurately, it cannot be certain that further investigations might not reveal that Euclidean geometry is only approximately true as a theory of the properties of real space. Geometry as a science of real space (and not as a logical system) would

§2. THE SOLUTION OF LOBAČEVSKIĬ

then have to be changed and made more precise in accordance with the new experimental data.

This brilliant idea of Lobačevskiĭ has been completely vindicated in a new branch of physics, the theory of relativity.

Lobačevskiĭ himself undertook computations on the basis of astronomical observations with the object of verifying the accuracy of Euclidean geometry. These computations corroborated at the time its truth within the limits of the available accuracy. The situation has now changed, although it must be emphasized at once that Lobačevskiĭ geometry, too, did not prove to be more accurate in its application to space, whose properties turned out to be different and more complicated. But even before this the Lobačevskiĭ geometry had become well based and applicable in another connection, of which we shall speak in more detail later.

It must be emphasized that Lobačevskiĭ did not at all regard his geometry as a mere logical scheme constructed on arbitrarily chosen premises. The important task he saw not in the logical analysis of the foundations of geometry but in the investigation of its relationship to reality. Since an experiment cannot give an absolutely accurate solution of the problem of truth of Euclid's postulate, it makes sense to investigate those logical possibilities that are represented by the most fundamental premises of geometry. This mathematical investigation helps to mark a path on which the physical study of the properties of real space must proceed. Furthermore, Euclid's geometry is a limiting case of Lobačevskiĭ's geometry, so that the latter includes wider possibilities. From this point of view the restriction to Euclid's postulate would have been a hindrance to the development of the theory. The theory must go beyond the known frontiers so as to search for ways of disclosing new facts and laws. A deeper understanding of the links between mathematics and reality enables us to select from the diverse logical possibilities precisely those that have the best chance of being useful in the study of nature. If geometry after Lobačevskiĭ had not developed the mathematical doctrine of the possible properties of space, contemporary physics would not possess the mathematical tools that made it possible to formulate and develop the theory of relativity.

Thus, we can summarize Lobačevskiĭ's solution of the problem of the Fifth Postulate.

1. The postulate is not provable.

2. By adding the opposite axiom to the basic propositions of geometry a logically perfect and comprehensive geometry, different from the Euclidean, can be developed.

3. The truth of the results of one logically conceivable geometry or another in its application to real space can only be verified experimentally. A logically conceivable geometry must be elaborated not only as an arbitrary logical scheme, but also as a theory indicating possible ways and methods of developing physical theories.

The solution is altogether different from what the geometers wished to obtain when they tried to prove the Euclidean postulate. It went so much against the established ideas that it did not meet with much understanding among mathematicians. For them it was too new and radical. Lobačevskiĭ, as it were, cut the Gordian knot of the theory of parallels instead of trying to disentangle it, as others had expected to do.

2. Other geometers and philosophers. Almost simultaneously with Lobačevskiĭ the Hungarian geometer János Bólyai (1802-1860) also discovered the impossibility of proving the Fifth Postulate and the possibility of a non-Euclidean geometry; he published his results as an appendix to the geometric treatise of his father Farkas Bólyai of 1832. Previously the father had sent his son's paper for an opinion to Gauss and had received an encouraging reply, in which Gauss mentioned that he too had reached the same results long ago. However, Gauss refrained from publishing anything. In one of his letters he explains that he was afraid of being misunderstood.

In science it always happens that the time is ripe for some results and that they are obtained almost simultaneously and independently by several scholars. The integral and differential calculus was developed simultaneously by Newton and Leibnitz; Darwin's ideas were independently reached at the same time by Wallace; the elements of the theory of relativity were found simultaneously by Einstein and also by Poincaré, and of such examples there are many more. They show once more that science grows inevitably by the solution of problems for which it is ripe, and not by accidental discoveries and guesses. So it was also with the discovery of the possibility of non-Euclidean geometry, which was made simultaneously by several geometers: Lobačevskiĭ, Bólyai, Gauss, Schweikart, and Taurinus.

However, as it also happens constantly in science, not all the scholars who arrive at a new result play an equal role in its establishment and not all of them have an equal share in the service performed. Priority is of importance here, but also clarity and depth of results, and coherence and sound arguments in their derivation. Neither Schweikart nor Taurinus were convinced of the equal status of the new geometry, and this was a decisive feature of the case, all the more since partial results had already been obtained by Saccheri and Lambert. Gauss, although he apparently

§3. Lobačevskiĭ Geometry

had this conviction, was not resolute enough to risk coming out with it into the open.

Bólyai did not display any indecision, but he did not develop the new ideas as far and as deeply as Lobačevskiĭ. For Lobačevskiĭ was the first to express the new ideas openly, orally in 1826 and in print in 1829, and continued to develop and propagate them in a number of papers culminating in the "Pangeometry" of 1855, which he dictated as a blind old man in his declining years, retaining his strength of mind and his confidence in his work. And so the new geometry bears his name.

§3. Lobačevskiĭ Geometry

1. Some striking results. Thus, Lobačevskiĭ took as his starting point the statement contradictory to the Fifth Postulate: in a given plane at least two straight lines can be drawn through a point that do not intersect a given line. From this he derived a number of far-reaching consequences that formed the new geometry. This geometry was, therefore, constructed as a conceivable theory, as a collection of theorems that can be proved logically proceeding from the postulate, in conjuction with other* basic assumptions of Euclidean or, as Lobačevskiĭ used to say, "customary" geometry.

Among his deductions Lobačevskiĭ obtained all the results analogous to those of the "customary" elementary geometry, i.e., right up to non-Euclidean trigonometry and the solution of triangles, to the calculation of areas and volumes. We cannot here follow this chain of Lobačevskiĭ's deductions not because they are too complicated but merely for lack of space. After all, even a school course in "customary" geometry is rather long, and Lobačevskiĭ's deductions are neither simpler nor shorter than these "customary" deductions. Therefore we shall mention here only some striking results of Lobačevskiĭ, and refer the reader who is interested in a deeper study of non-Euclidean geometry to the special literature. Later on we shall explain a simple interpretation of non-Euclidean geometry in the actual world.

Let us begin with the theory of parallel lines. Suppose that a line a and a point A not on it are given. We drop the perpendicular AB from A to a. By the fundamental assumption there exist at least two lines passing through A and not intersecting our line a. Then every line in the angle between these two lines also does not intersect a. It is true that in figure 2 the lines b and b', if produced far enough, would actually intersect a,

---

* These so-called "remaining" propositions of geometry will also be formulated accurately in §5.

against Lobačevskiĭ's assumption. But there is nothing surprising in this. For of course Lobačevskiĭ did not argue from figures as we can draw them in the ordinary plane; he developed logical consequences from his

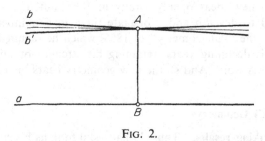

FIG. 2.

assumption, which contradicts what we are accustomed to see in diagrams. Figures play only an auxiliary role here; in them the facts of non-Euclidean geometry cannot be expressed accurately, because in a figure we draw ordinary lines in an ordinary plane, entirely Euclidean within the limits of accuracy of the figure.

This contradiction between logical possibility and visual representation was an important obstacle to the understanding of Lobačevskiĭ geometry. But if we are concerned with geometry as a logical theory, then we must look for logical rigor of the reasoning and not for agreement with customary figures.

2. The angle of parallelism. Let us turn again to our line a and point A. Through A we draw a half line x that does not intersect a (for example perpendicular to AB) and rotate it around A so that the angle ϕ between AB and x decreases, but without bringing it to an intersection with a. Then the half line x reaches a limiting position corresponding to the least value of ϕ. This limiting half line c also does not intersect a.

For if it did intersect a in some point X (figure 3), then we could take

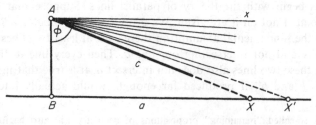

FIG. 3.

§3. LOBAČEVSKIĬ GEOMETRY

a point X' to the right and obtain a half line AX' intersecting a, but forming a larger angle with AB. But this is impossible, since by the construction of c every half line x that forms a larger angle with AB does not intersect a.

Therefore c does not intersect a and is, moreover, the extreme one of all the half lines passing through A and not intersecting a.

By symmetry, it is obvious that on the other side we can also draw a half line c' not intersecting a and also extreme among all such half lines. If c and c' were continuations of one another, then together they would form a single line $c + c'$. This line would then be the unique parallel to a through the given point A, so that under the slightest rotation either c or c' would intersect a. So once it has been assumed that the parallel is not unique, but that there are at least two, the half lines c and c' cannot be continuations of one another.

Thus, we have proved the first theorem of Lobačevskiĭ geometry:

Through a point A not lying on a given line a two half lines c and c' can be drawn such that they do not intersect a, but that every half line in the angle between them intersects a.

If the half lines c and c' are produced, then we obtain (figure 4) two lines not intersecting a with the additional property that every line passing through A in the angle α between these lines does not intersect a, but every line in the angle β intersects a. Lobačevskiĭ called these lines c, c' *parallels* to a: c parallel on the right, c' parallel on the left. Half of the angle β is called the *angle of parallelism* by Lobačevskiĭ; it is less than a right angle, because β is less than two right angles.

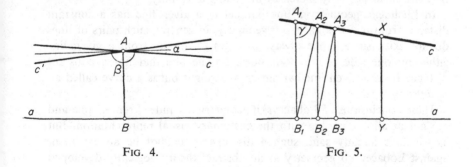

Fig. 4.   Fig. 5.

3. Convergence of parallel lines; the equidistant curve. Let us now investigate how the distance from a of a point X on c changes when X is shifted along c (figure 5). In Euclidean geometry the distance between parallel lines is constant. But here we can convince ourselves that when X

moves to the right, its distance from a (i.e., the length of the perpendicular XY) decreases.

We drop the perpendicular A_1B_1 from a point A_1 to a. From B_1 we drop the perpendicular B_1A_2 to c (A_2 lies to the right of A_1, since γ is an acute angle). Finally we drop the perpendicular A_2B_2 from A_2 to a. Let us show that A_2B_2 is less than A_1B_1.

The theorem that the perpendicular is shorter than a slant line is valid in Lobačevskiĭ geometry, because its proof (which can be found in every school book on geometry) does not depend on the concept of parallel lines nor on deductions connected with them. Now since the perpendicular is shorter than a slant line, B_1A_2 as a perpendicular to c is shorter than A_1B_1, and similarly A_2B_2 as a perpendicular to a is shorter than B_1A_2. Therefore A_2B_2 is shorter than A_1B_1.

When we now drop the perpendicular B_2A_3 to c from B_2 and repeat these arguments, we see that A_3B_3 is shorter than A_2B_2. Continuing this construction we obtain a sequence of shorter and shorter perpendiculars; i.e., the distances of A_1, A_2, $\cdots$ from a decrease. Furthermore, by supplementing our simple argument we could prove that, generally, if a point X'' on c lies to the right of X', then the perpendicular $X''Y''$ is shorter than $X'Y'$. We shall not dwell on this point. The preceding arguments, we trust, make the substance of the matter sufficiently clear and a rigorous proof is not one of our tasks.

But it is remarkable that, as can be proved, the distance XY not only decreases when X moves on c to the right, but actually tends to zero as X tends to infinity. That is, *the parallel lines a and c converge asymptotically*! Moreover, it can be proved that in the opposite direction the distance between them not only increases but tends to infinity.

In Euclidean geometry a line parallel to a given line has a constant distance from it. In Lobačevskiĭ geometry, in general, such pairs of lines do not exist, since a line always diverges to infinity from a given line either on one side or on both sides. So the line that has a constant distance from a given line can never be straight but is a curve called an *equidistant*.

These conclusions of Lobačevskiĭ geometry are indeed remarkable and are not at all compatible with the customary visual representation. But as we have already said, such a discrepancy cannot be an argument against Lobačevskiĭ geometry as an abstract theory, logically developed from the premises assumed.

4. The magnitude of the angle of parallelism. We shall now study the angle of parallelism, i.e., the angle γ that the line c parallel to a given line a forms with the perpendicular CA (figure 6). Let us show that this

§3. LOBAČEVSKIĬ GEOMETRY

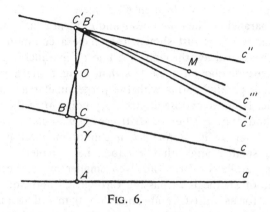

Fig. 6.

angle is smaller, the further C is from a. For this purpose we begin by proving the following. If two lines b and b' form equal angles α, α' with a secant BB', then they have a common perpendicular (figure 7).

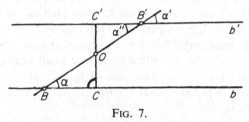

Fig. 7.

For the proof we draw through the midpoint O of BB' the line CC' perpendicular to B. We obtain two triangles OBC and $OB'C'$. Their sides OB and OB' are equal by construction. The angles at the common vertex O are equal as vertically opposite. The angle α'' is equal to α' since they are also vertically opposite. But α' is equal to α by assumption. Therefore α is equal to α''. Thus, in our triangles OBC and $OB'C'$ the sides OB and OB' and their adjacent angles are equal. But then, by a well-known theorem, the triangles are equal, in particular their angles at C and C'. But the angle at C is a right angle, since the line CC' is by construction perpendicular to b. Therefore the angle at C' is also a right angle; i.e., CC' is also perpendicular to b'. Thus, the segment CC' is a common perpendicular to both b and b'. This proves the existence of a common perpendicular.

Now let us prove that the angle of parallelism decreases with increasing distance from the line. That is, if the point C' lies further from a than C, then, as in figure 6, the parallel c' passing through C' forms with the perpendicular $C'A$ a smaller angle than the parallel c passing through C.

For the proof we draw through C' a line c'' under the same angle to $C'A$ as the parallel c. Then the lines c and c'' form equal angles with CC'. Therefore, as we have just shown, they have a common perpendicular BB'. Then we can draw through B' a line c''' parallel to c and forming with the perpendicular an angle less than a right angle, since we know already that a parallel forms with the perpendicular an angle less than a right angle. Now we choose an arbitrary point M in the angle between c'' and c''' and draw the line $C'M$. It lies in the angle between c'' and c''' and cannot intersect c'. A fortiori, it cannot intersect c. But it forms with AC' a smaller angle than c'' does, i.e., smaller than γ. Then, a fortiori, the parallel c' forms an even smaller angle, because it is the extreme one of all the lines passing through C' and not intersecting a. Therefore c' forms with $C'A$ an angle less than c does and this means that the angle of parallelism decreases on transition to a farther point C'; this is what we set out to prove.

We have shown, then, that the angle of parallelism decreases for increasing distance of C from a. Even more can be shown: *If the point C recedes to infinity, then this angle tends to zero.* That is, for a sufficiently large distance from the line a a parallel to it forms with the perpendicular *to it* an arbitrarily small angle.* In other words, if at a point very far from a the line perpendicular to a is tilted by a very small angle, the "tilted" line will no longer intersect a. This fact of Lobačevskiĭ geometry, too, makes an amazing impression. But further on we shall obtain other no less amazing results.

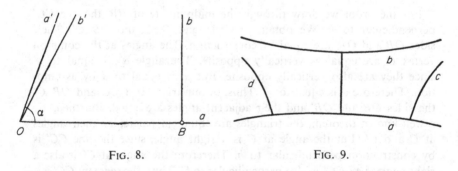

Fig. 8.   Fig. 9.

For example, let us take the acute angle α formed by two half lines a and a'. We erect the perpendicular b to a at a point sufficiently far from

---

* If h is the length of the perpendicular and γ is the angle of parallelism, then as Lobačevskiĭ proved $\tan \gamma/2 = e^{-h/k}$, where k is a constant depending on the unit of length and e is the base of natural logarithms. Obviously $e^{-h/k}$, and with it γ, tends to zero for $h \to \infty$.

§3. LOBAČEVSKIĬ GEOMETRY

the vertex O of α so that the angle of parallelism corresponding to the chosen distance OB (figure 8) is less than α. Once the angle α is larger than the angle of parallelism, the line b' through O parallel to b forms a smaller angle with a. But it does not intersect b. Therefore a', a fortiori, does not intersect b. Thus we have shown that the perpendicular to one arm of an acute angle erected sufficiently far from the vertex does not intersect the other arm.

5. More striking theorems. We have drawn all the preceding conclusions with a twofold aim. First of all, and this is the main point, we wanted to show by some simple examples how theorems of Lobačevskiĭ geometry can in fact be obtained from the premises assumed. This is a very simple instance of the way in which mathematicians reach conclusions in abstract geometry, of how conclusions can be reached at all that are not connected with the usual visual representation. Second, we wanted to show what peculiar results are obtained in Lobačevskiĭ geometry. Let us give a few more examples.

Two lines in a Lobačevskiĭ plane either intersect or they are parallel in the sense of Lobačevskiĭ, and then they converge asymptotically on the one side and on the other they diverge infinitely, or else they have a common perpendicular and diverge infinitely on both sides of it.

If the lines a, b have a common perpendicular (figure 9), then two perpendiculars c, d can be drawn to a that are parallel (in the sense of Lobačevskiĭ) to b and the whole line b lies in the strip between c and d.

The limit of a circle of infinitely increasing radius is not a line but a certain curve, a so-called *limiting circle*. It is not always possible to draw a circle through three points not on one line, but either a circle or a limiting circle or an equidistant (i.e., a line formed by the points that are equidistant from a certain line) can be drawn through the three points.

The sum of the angles of a triangle is always less than two right angles. If a triangle is increased so that all three heights grow without bound, then its three angles tend to zero.

There are no triangles of arbitrarily large area.

Two triangles are equal when their angles are equal.

The length l of the circumference of a circle is not proportional to the radius r but grows more rapidly (essentially by an exponential law). In fact, the following formula holds

$$l = \pi k (e^{r/k} - e^{-r/k}), \tag{1}$$

where k is a constant depending on the unit of length. Since

$$e^{r/k} = 1 + \frac{r}{k} + \frac{1}{2}\left(\frac{r}{k}\right)^2 + \cdots, \quad e^{-r/k} = 1 - \frac{r}{k} + \frac{1}{2}\left(\frac{r}{k}\right)^2 - \cdots,$$

we obtain from (1):

$$l = 2\pi r \left(1 + \frac{1}{6}\frac{r^2}{k^2} + \cdots\right). \qquad (2)$$

Only for small ratios r/k is it true with sufficient accuracy that $l = 2\pi r$.

All these conclusions are logical sequences of the premises assumed: "the axioms of Lobačevskiĭ" in conjunction with the basic propositions of the "customary" geometry.

6. Lobačevskiĭ's geometry compared with Euclid's.

An extremely important property of Lobačevskiĭ geometry consists in the fact that for sufficiently small domains it differs but little from Euclid's geometry; the smaller the domain, the less the difference. Thus, for sufficiently small triangles the connection between sides and angles is expressed with sufficient accuracy by the formulas of ordinary trigonometry, and the more accurately, the smaller the triangle.

The formula (2) shows that for small radii the length of a circle is proportional to the radius, with a good accuracy. Similarly the sum of the angles of a triangle differs by little from two right angles, etc.

In the formula for the length of the circumference of a circle, there occurs a constant k depending on the unit of length. If the radius is small in comparison with k, i.e., if r/k is small, then, as is clear from formula (2), the length l is nearly $2\pi r$. Generally, the smaller the ratio of the dimensions of a figure to this constant, the more accurately the properties of the figure approach the properties of the corresponding figure in Euclidean geometry.*

A measure for the deviation of the properties of a figure in Lobačevskiĭ geometry from the properties of a figure of Euclidean geometry is the

---

* For example, if a, b, c are the sides and the hypotenuse of a right-angled triangle then instead of Pythagoras' theorem the following relation holds

$$2(e^{c/k} + e^{-c/k}) = (e^{a/k} + e^{-a/k})(e^{b/k} + e^{-b/k}).$$

Expanding in series we obtain

$$c^2 + \frac{c^4}{12k^2} + \cdots = a^2 + b^2 + \frac{a^4 + 6a^2b^2 + b^4}{12k^2} + \cdots,$$

so that for large l we have the theorem of Pythagoras $c^2 = a^2 + b^2$. Furthermore, from Lobačevskiĭ's formula for the angle of parallelism γ (see the previous footnote), $\tan(\gamma/2) = e^{-h/k}$. If h/k is small, i.e., if the parallels are close together, then $\tan \gamma/2 = e^{-h/k} \approx 1$ and $\gamma \approx 90°$. Thus, for small distances parallels in a Lobačevskiĭ plane differ little from Euclidean parallels.

§3. LOBAČEVSKIĬ GEOMETRY

ratio r/k if r measures the dimensions of the figure (radius of a circle, sides of a triangle, etc.).

This has an important consequence.

Suppose we have to do with the actual space of the external world and measure distances in kilometers. Let us assume that the constant k is very large, say 10^{12}.

Then, for example, by the formula (2), for a circle with a radius of even 100 km the ratio of its length to the radius differs from 2π by less than 10^{-9}. Of the same order are the deviations from other ratios of Euclidean geometry. Within the limits of 1 kilometer they would even be of the order $1/k$, i.e., 10^{-12}, and within the limits of a meter of the order 10^{-15}; i.e., they would be altogether negligible. Such deviations from Euclidean geometry could not be observed, because the dimensions of an atom are a hundred times larger (they are of the order of 10^{-13} km). On the other hand, on the astronomical scale the ratio r/k could turn out to be not too small.

Therefore Lobačevskiĭ also assumed that, although on the ordinary scale Euclid's geometry is true with great accuracy, the deviation from it could be noted by astronomical observations. As we have already mentioned, this assumption has been justified, but the insignificant deviations from Euclidean geometry that have now been observed on the astronomical scale turn out to be even more complicated.

Finally, the arguments given have another important consequence. It is this: Since the deviation from Euclidean geometry becomes smaller for increasing values of the constant k, in the limit when k grows without bound, Lobačevskiĭ geometry goes over into Euclid's geometry. That is, *Euclid's geometry is just a limiting case of Lobačevskiĭ geometry.* Therefore, if this limiting case is added to Lobačevskiĭ geometry, then it comprises also Euclid's geometry and so it turns out, in this sense, to be a more general theory. In view of this situation Lobačevskiĭ called his theory "pangeometry," i.e., universal geometry. Such a relationship of theories constantly appears in the development of mathematics and the natural sciences: A new theory includes the old one as a limiting case, in accordance with the advance of our knowledge from more special to more general deductions.

However, all the reasonings and deductions we have made would remain, as it were, a hardly intelligible game of the mind if we could not establish a comparatively simple real meaning of Lobačevskiĭ geometry within the system of the usual concepts of Euclidean geometry. The solution of this problem was not finally reached by Lobačevskiĭ himself; it fell to the lot of his successors and was found almost forty years after his first paper had appeared. This solution is described in the next section.

§4. The Real Meaning of Lobačevskiĭ Geometry

1. Beltrami's interpretation on the pseudosphere. An intuitive interpretation of Lobačevskiĭ geometry was first given in 1868, when the Italian geometer Beltrami noted that the intrinsic geometry of a certain surface, namely the pseudosphere, coincided with the geometry on part of the Lobačevskiĭ plane. We recall that by the intrinsic geometry of a surface one understands the collection of properties of figures on it that can be determined by measuring lengths only on the surface itself. In figure 10 on the left we have drawn the so-called *tractrix*. This is the curve

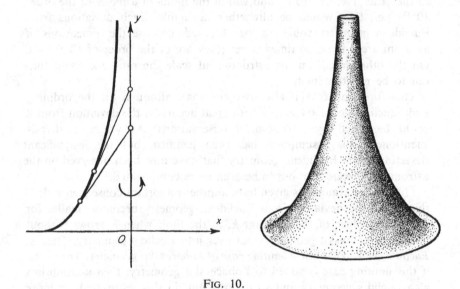

Fig. 10.

with the property that the length of the segment of its tangent from the point of contact to the intersection with the Oy-axis is constant for all points of the curve. The Oy-axis is its asymptote. By rotating the tractrix around its asymptote, we obtain a surface, which is called a *pseudosphere* and is illustrated in figure 10 on the right.

The interpretation of Lobačevskiĭ geometry by Beltrami comes to this, that all geometrical relations on part of the Lobačevskiĭ plane coincide with the geometrical relations on a suitable part of the pseudosphere provided the following convention is adopted. The role of straight line segments is taken over by shortest lines on the surface, the so-called geodesics. The distance between points is defined as the length of the shortest line joining them on the surface. Figures are called equal if their

§4. THE REAL MEANING OF LOBAČEVSKIĬ GEOMETRY

points can be put in correspondence with each other in such a way that intrinsic distances between corresponding points are equal. A motion of figures on the pseudosphere that preserves their dimensions from the point of view of the intrinsic geometry, although accompanied by bending, represents a motion in the Lobačevskiĭ plane. Lengths, angles, and areas are measured on the surface as usual and correspond to lengths, angles, and areas in Lobačevskiĭ geometry.

Beltrami's interpretation shows that, given these conditions, to every statement of Lobačevskiĭ geometry referring to part of the plane there corresponds an immediate fact of the intrinsic geometry of the pseudosphere. Lobačevskiĭ geometry consequently has a perfectly real meaning: It is nothing but an abstract account of the geometry on the pseudosphere.

We ought to mention that, thirty years before Beltrami's discovery, the intrinsic geometry of the pseudosphere had already been investigated by F. Minding, who had in fact established the properties that show that it coincides with Lobačevskiĭ geometry. However, neither he nor anyone else noted this, until Lobačevskiĭ's ideas had been sufficiently propagated. Beltrami had only to compare the results of Lobačevskiĭ and Minding to become aware of the connection between them.

Beltrami's discovery at once changed the attitude of mathematicians to Lobačevskiĭ geometry; from being "fictitious" it became real.*

2. Klein's interpretation in the circle and the sphere. However, as we have emphasized, not the whole Lobačevskiĭ plane is realized on the pseudosphere, but only part of it.†

---

* The history of the gradual establishment of the real meaning of Lobačevskiĭ geometry was in fact even more complicated. First of all, Lobačevskiĭ himself had the means of proving its noncontradictory character by a so-called analytical model, but he did not succeed in carrying the proof right through. This was done much later. Second, the German mathematician Riemann came forward in 1854 with a theory (see §10) in which Beltrami's results are already contained, although Riemann did not express them clearly; his paper was not understood and was not published until after his death in 1868 in the same year as the appearance of Beltrami's paper. The whole history of Lobačevskiĭ geometry from the attempts at proving Euclid's postulate to the complete clarification of the significance of non-Euclidean geometry is highly instructive in that it shows what struggles and roundabout ways are often needed to discover a truth which afterwards turns out to be simple and intelligible.

† The pseudosphere has everywhere the same negative Gaussian curvature. All surfaces of constant negative curvature have (at least in small parts) the same intrinsic geometry and can therefore serve as models of Lobačevskiĭ geometry. However, as Hilbert proved in 1901, none of these surfaces can be extended infinitely in all directions without singularities, so that they cannot serve as a model of the *whole* Lobačevskiĭ plane. On the other hand, the young Dutch mathematician Kuiper showed in 1955 that there exist smooth surfaces that represent in the sense of their intrinsic geometry the whole Lobačevskiĭ plane, but such surfaces although smooth cannot be bent continuously, they do not have a definite curvature.

Let us also note the following. When Lobačevskiĭ geometry is represented on a

Therefore the task of giving an actual interpretation of Lobačevskiĭ geometry on the whole plane, and all the more for his geometry in space, still remained unsolved. This was done later in 1870 by the German mathematician Klein. Let us explain what his solution was.

In an ordinary Euclidean plane we take a circle and consider only the interior of the circle; i.e., we exclude from our investigation its circumference and the domain outside. We agree to call this interior of the circle a "plane," since it will turn out to play the role of a Lobačevskiĭ plane. The chords of our circle will be called "lines" and in accordance with the agreement we have just made the end points of chords, as lying on the circumference, are excluded. Finally, a "motion" shall be any transformation of the circle that carries it into itself and carries lines into lines, i.e., does not distort its chords. The simplest example of such a transformation is a rotation of the circle around its center, but it turns out that there are far more of them. What these transformations are will be stated in the following paragraphs.

Now if we introduce these conventions of nomenclature, then the facts of the ordinary geometry within our circle are transformed into theorems of Lobačevskiĭ geometry. And conversely, every theorem of Lobačevskiĭ geometry is interpreted as a fact of the ordinary geometry within the circle.

For example, by Lobačevskiĭ's axiom at least two lines can be drawn through a point not lying on a given line that do not intersect the line. Let us translate this axiom into the language of ordinary geometry by our conventions, i.e., replace lines by chords. Then we obtain the statement: At least two chords can be drawn through a point inside the circle not lying on a given chord that do not intersect the chord. The truth of this statement is obvious from figure 11. Therefore Lobačevskiĭ's axiom is satisfied here.

We recall further that in Lobačevskiĭ geometry among the lines passing through the given point and not intersecting the given line there are two extreme ones, namely the ones that are called by Lobačevskiĭ parallels to the given line. This means that among the chords passing through the given point A and not intersecting the given chord BC there are two extreme chords. And indeed, these extreme chords are the ones that pass through B or C, respectively. They do not have common points with BC, because we exclude points on the circumference. Thus, this theorem of Lobačevskiĭ is satisfied here.

For a further translation of Lobačevskiĭ's theorems into the language

---

surface of negative constant curvature K, the constant k that figures in the formulas of the preceding section assumes a simple meaning: $k^2 = -1/K$.

§4. THE REAL MEANING OF LOBAČEVSKIĬ GEOMETRY

of ordinary geometry within the circle, it is necessary to explain how segments and angles are to be measured in the circle in such a way that these measurements correspond to Lobačevskiĭ geometry. Of course, the measuring cannot be the same as the usual one, because in the ordinary sense a chord has finite length and the line that the chord represents is infinite. This could perhaps be regarded even as a contradiction, but we shall see that no contradiction exists.

Let us recall first of all that in ordinary geometry lengths of segments are measured as follows. Some segment AB is chosen whose length is taken to be the unit and the length of any other segment XY is determined by comparing it with AB. For this purpose the segment AB is laid off along XY. If there remains a part of XY less than AB, then AB is divided into, say, 10 equal parts (equal in the sense that each is obtained from the other by a translation); these parts are laid off on the remaining portion of XY; if necessary, AB is then divided into 100 parts, etc. As a result the length of XY is expressed as a decimal fraction, which may be infinite. Consequently, lengths are measured by means of a motion of the whole or part of the segment chosen as unit; i.e., measurement is based on motion. And once motions are defined (in our case we have defined them as transformations of the circle that carry lines into lines), then it is known what segments are to be regarded as equal and how length has to be measured. The term that defines motion already contains, though in an implicit form, the rule for measuring lengths. Angles are measured in just the same way, by laying off the angle taken to be the unit. Thus, the rule for measuring angles is also contained in the definition of motion.

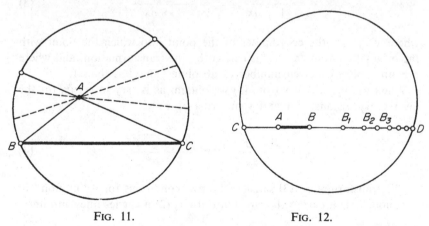

Fig. 11.   Fig. 12.

The corresponding rules for measuring lengths and angles in Lobačevskiĭ geometry are fairly simple, although essentially different from the ordinary rules. We shall not derive them here, because this is not a point of principal significance in our arguments.*

By the rule for measuring lengths it comes out that a chord has infinite length. And this is so because if by a transformation that we have taken to be a motion the segment AB is carried into BB_1, then into B_1B_2, etc., these segments B_kB_{k+1} become shorter and shorter in the usual sense (although equal in the sense of our model of Lobačevskiĭ geometry; figure 12). The points B_1, B_2, $\cdots$, B_k, $\cdots$ accumulate at the end point of the chord. But for us the chord has no end point; the end point is excluded by agreement, and in this sense it is "at infinity." In the sense of Lobačevskiĭ geometry the points B_1, B_2, $\cdots$ do not accumulate anywhere, they extend to infinity. By means of transformations that we have taken to be motions, laying off equal segments one after another, we cannot reach the circumference of the circle from within.

In order to understand clearly how segments are laid off in the model, let us consider the transformation that plays the role of a translation along a line.

Suppose that rectangular coordinates are introduced in the plane with the center of the circle as origin. To fix our ideas let us assume that our circle has unit radius so that its circumference is represented by the equation $x^2 + y^2 = 1$ and the points of the interior satisfy the inequality $x^2 + y^2 < 1$.

We consider the transformation given by the formulas

$$x' = \frac{x+a}{1+ax}, \quad y' = \frac{y\sqrt{1-a^2}}{1+ax}, \qquad (3)$$

where x', y' are the coordinates of the point into which the point with the original coordinates x, y is carried by the transformation, and where a is an arbitrarily given number of absolute value less than 1.

When we find x and y from (3) we obtain, as is easy to verify, for the inverse expressions of x and y in terms of x', y'

$$x = \frac{x'-a}{1-ax'}, \quad y = \frac{y'\sqrt{1-a^2}}{1-ax'}. \qquad (4)$$

The transformation (3) satisfies the two conditions for a "motion" in our model: (1) it carries the circle into itself; (2) it carries lines into lines.

---

* The rule for measuring length turns out to be the following. Let the segment AB lie on the chord CD (figure 12). Measuring segments in the usual way we form the so-called cross ratio $CB/CA : DB/DA$. Its logarithm is taken to be the length of AB.

§4. THE REAL MEANING OF LOBAČEVSKIĬ GEOMETRY

To prove the first property we have to convince ourselves, strictly speaking, that the inequality or equality $x^2 + y^2 \leqslant 1$ implies the corresponding relation $x'^2 + y'^2 \leqslant 1$ and vice versa. Let us show, for example, that when $x^2 + y^2 = 1$, then necessarily $x'^2 + y'^2 = 1$, i.e., that the points on the circumference of the given circle remain on it.

We compute $x'^2 + y'^2$, using (3) and taking into account that $x^2 + y^2 = 1$, i.e., $y^2 = 1 - x^2$,

$$x'^2 + y'^2 = \frac{(x+a)^2 + y^2(1-a^2)}{(1+ax)^2} = \frac{(x+a)^2 + (1-x^2)(1-a^2)}{(1+ax)^2}$$

$$= \frac{x^2 + 2ax + a^2 + 1 - x^2 - a^2 + a^2x^2}{(1+ax)^2} = \frac{1 + 2ax + a^2x^2}{1 + 2ax + a^2x^2}$$

$$= 1.$$

Therefore, when $x^2 + y^2 = 1$, then also $x'^2 + y'^2 = 1$. The remaining cases are verified similarly.

The second property of (3) is established very simply. For we know that every line is represented by a linear equation and, conversely, every linear equation represents a line. Suppose that the given line is

$$Ax + By + C = 0. \tag{5}$$

After the transformation (4) we obtain

$$A \frac{x' - a}{1 - ax'} + B \frac{y' \sqrt{1 - a^2}}{1 - ax'} + C = 0,$$

or, by reducing to the common denominator,

$$(A - aC) x' + B \sqrt{1 - a^2}\, y' + (C - aA) = 0.$$

This equation is linear and consequently represents a line. This is the line into which (5) is carried by the transformation. Let us also note that the transformation (3) carries the Ox-axis into itself, causing only a displacement of the points along it. This is clear, because on this axis $y = 0$ and by (3) we then also have $y' = 0$. On the Ox-axis the transformation is given by the single formula

$$x' = \frac{x + a}{1 + ax} \; (|a| < 1). \tag{3}$$

On this line the segment $x_1 x_2$ goes over into $x_1' x_2'$ by the formula (3′) and by our convention these arguments are to be regarded as equal. This is the manner in which the "laying off a segment" proceeds.

For the center O of the circle $x = 0$ so that $x' = a$; i.e., under (3′) the center goes over into the point A with the coordinate $x = a$.

Since a can be arbitrary subject only to $|a| < 1$, the center can be carried into any point on the diameter along the Ox-axis.

Under the same transformation, the point that was at A before is carried into the point A_1 with coordinate

$$x_1 = \frac{a + a}{1 + a^2} = \frac{2a}{1 + a^2}.$$

Thus, the segment OA goes over under (3) into AA_1 and so it is "laid off" on the "line" representing the diameter of the circle.

By repeating the same transformation, we can again lay off the same segment arbitrarily often. The point A_n with the coordinate x_n goes over into the point A_{n+1} with the coordinate

$$x_{n+1} = \frac{x_n + a}{1 + ax_n}.$$

So we obtain points $A, A_1, A_2, \cdots$ with the coordinates

$$x_0 = a, \quad x_1 = \frac{2a}{1 + a^2}, \quad x_2 = \frac{x_1 + a}{1 + ax_1} = \frac{3a + a^3}{1 + 3a^2}, \cdots.$$

Since all the segments $A_n A_{n+1}$ are obtained from OA by transformations expressing a motion, they are all "equal" to one another, equal in the sense of Lobačevskiĭ geometry as it is represented in the model. It is easy to show that the points A_n converge to the end point of the diameter. In the sense of the model they recede to infinity.

Since the Ox-axis can be given any direction, the same shift transformations are possible along any diameter. By combining them with rotations around the center of the circle and reflections in a diameter, we obtain all "motions" as they are to be understood in the model; they consist of shifts, rotations, and reflections. These transformations will be studied in more detail in the next section, where it will be proved rigorously that Lobačevskiĭ geometry is really satisfied in our model and that, in particular, the transformations we have defined as motions satisfy all the conditions (axioms) to which motions are subject in geometry.

Let us repeat once more what sort of a model of Lobačevskiĭ geometry Klein proposed. In a plane we have taken the interior of a circle; a point

§4. THE REAL MEANING OF LOBAČEVSKIĬ GEOMETRY

is regarded as a point, a line as a chord (with the end points excluded), a motion is taken to be a transformation carrying the circle into itself and chords into chords; the situation of points (a point lies on a line; a point lies between two others) is to be understood in the usual sense. The rule for measuring lengths and angles (and also areas) already follows from the way in which motion is defined, equality of segments and angles (and of arbitrary figures) is also defined and the same definition is applicable to the operation of laying off one segment along another.

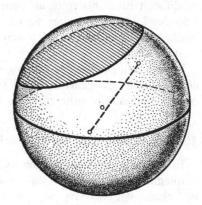

Fig. 13.

Under all these conditions to every theorem of Lobačevskiĭ geometry in the plane there corresponds a fact of Euclidean geometry within the circle, and vice versa: every such fact can be reinterpreted in the form of a theorem of Lobačevskiĭ geometry.

A model of Lobačevskiĭ geometry in space can be constructed similarly. For the space we take the interior of some sphere (figure 13), a line is interpreted as a chord, and a plane as a circle with its circumference on the sphere, but the surface of the sphere itself, and hence the end points of chords and the circumferences of these circles, are excluded; finally, a motion is defined as a transformation of the sphere into itself that carries chords into chords.

When this model of Lobačevskiĭ geometry was given, it was established incidentally that his geometry has a simple real meaning. Lobačevskiĭ geometry is valid, because it can be understood as a specific account of geometry in a circle or a sphere. At the same time its noncontradictory character was proved: Its results cannot lead to a contradiction, since every one can be translated into the language of ordinary Euclidean geometry within a circle (or a sphere, if we are concerned with Lobačevskiĭ geometry in space).*

3. Other interpretations. After Klein another model of Lobačevskiĭ geometry was given by the French mathematician Poincaré, who used it to derive important results in the theory of functions of a complex

---

* Mathematicians usually say that Lobačevskiĭ geometry can be represented within Euclid's geometry, and so it is just as noncontradictory as Euclid's geometry.

variable.* Thus, in his hands Lobačevskiĭ geometry led to the solution of difficult problems from an entirely different branch of mathematics. Lobačevskiĭ geometry has found a number of other applications in mathematics and theoretical physics; for example, in 1913 the physicist Varičak applied it in the theory of relativity.

Lobačevskiĭ geometry is growing successfully; the theory of geometrical constructions, the general theory of curves and surfaces, the theory of convex bodies, and other subjects are being developed in it.

§5. The Axioms of Geometry; Their Verification in the Present Case

1. Precise formulation of the axioms. In order to give a strict mathematical proof that Klein's model really provides an interpretation of Lobačevskiĭ geometry, we have first of all to state accurately what it is that has to be proved. To verify Lobačevskiĭ's theorems one after the other would be absurd; there are too many of them, in fact infinitely many, because one can prove more and more new theorems. However, it will be sufficient to show that in Klein's model the fundamental propositions of Lobačevskiĭ geometry are satisfied, since from these the remaining ones can be deduced. But in that case these fundamental propositions must be formulated precisely.

Thus, the problem of proving that Lobačevskiĭ geometry is noncontradictory reduces to the problem of stating its fundamental propositions, i.e., its axioms, accurately and completely. And since the assumptions of Lobačevskiĭ geometry differ from those of Euclid's geometry by the axiom of parallelism only, our task comes to a precise and complete formulation of the axioms of Euclidean geometry. In Euclid such a formulation does not yet exist; in particular, a definition of the properties of motion or superposition of figures is altogether absent, although of course, he makes use of them. The task of making Euclid's axioms accurate and complete came to the fore precisely in connection with the development of Lobačevskiĭ geometry; and also with the earlier mentioned general trend at the end of the last century toward making the foundations of mathematics more rigorous.

As a result of the investigations of a number of geometers, the problem of formulating the axioms of geometry was finally solved.

Generally speaking, the axioms can be chosen in various ways, taking

---

* Poincaré's model amounts to this: The Lobačevskiĭ plane is again taken to be the interior of a circle, but lines are interpreted as arcs of a circle perpendicular to the circumference of the given circle; a motion is defined as an arbitrary conformal transformation of the circle into itself. (The connection with conformal transformation also yields the connection with the theory of functions of a complex variable.)

§5. AXIOMS OF GEOMETRY; THEIR VERIFICATION

various concepts as starting points. Here we shall give an account of the axioms of geometry in a plane which is based on the concepts of point, straight line, motion, and such concepts as: The point X lies *on* the line a; the point B lies *between* the points A and C; a *motion* carries the point X into the point Y. (In our case other concepts can be defined in terms of these; for example, a segment is defined as the set of all points that lie between two given ones.)

The axioms fall into five groups.

I. Axioms of incidence.

1. One and only one straight line passes through any two points.
2. On every straight line there are at least two points.
3. There exist at least three points not lying on one straight line.

II. Axioms of order.

1. Of any three points on a straight line, just one lies between the other two.
2. If A, B are two points of a straight line, then there is at least one point C on the line such that B lies between A and C.
3. A straight line divides the plane into two half planes (i.e., it splits all the points of the plane not lying on the line into two classes such that points of one class can be joined by segments without intersecting the line, and points of distinct classes cannot).

III. Axioms of motion.

(A motion is to be understood as a transformation not of an individual figure, but of the whole plane.)

1. A motion carries straight lines into straight lines.
2. Two motions carried out one after the other are equivalent to a certain single motion.
3. Let A, A' and a, a' be two points and half lines going out from them, and α, α' half planes bounded by the lines a and a' produced; then there exists a unique motion that carries A into A', a into a' and α into α'. (Speaking intuitively, A is carried in A' by a translation, then the half line a is carried by a rotation into a', and finally the half plane α either coincides with α' or else it has to be subjected to a "revolution" around a as axis.)

IV. Axiom of continuity.

1. Let $X_1, X_2, X_3, \cdots$ be points situated on a straight line such that each succeeding one lies to the right of the preceding one, but that there *is* a point A lying to the right of them all.* Then there exists a point B that also lies to the right of all the points $X_1, X_2, \cdots$, but such that a point X_n is arbitrarily near to it (i.e., no matter what point C is taken to the left of B, there is a point X_n on the segment CB).

V. Axiom of parallelism (Euclid).

1. Only one straight line can pass through a given point that does not intersect a given straight line.

These axioms, then, are sufficient to construct Euclidean geometry in the plane. All the axioms of a school course of plane geometry can in fact be derived from them, though their derivation is very tedious.

The axioms of Lobačevskiĭ geometry differ only in the axiom of parallelism.

V'. Axiom of parallelism (Lobačevskiĭ).

1. At least two straight lines pass through a point not lying on a given straight line that do not intersect the line.

It may appear somewhat strange that in the list of axioms there is, for example, this one: "On every straight line there are at least two points." Surely in our idea of a line there are even infinitely many points on it. No wonder that neither to Euclid nor to any one of the mathematicians up to the end of the last century did it occur that such an axiom had to be stated: it was assumed tacitly. But now the situation has changed. When we give a new interpretation of geometry, we may understand by a straight line not the usual line, but something else: a geodesic on a surface, a chord of a circle, or what have you. Therefore the need clearly arises for stating accurately and exhaustively *everything* we have to postulate of those objects that will be described as straight lines. The same applies to all the other concepts and axioms.

So the appearance of various interpretations of geometry, as we have already said, was one of the important stimuli towards an accurate account of its fundamental statements. This is also the historical order: The precise formulation of the axioms came after the models of Beltrami, Klein, and Poincaré.

---

* "Right" can be replaced by "left."

§5. AXIOMS OF GEOMETRY; THEIR VERIFICATION

2. Verification of the axioms for Klein's model. We shall now show that in Klein's model all the enumerated axioms are satisfied except the Euclidean axiom of parallelism. As we have mentioned in the preceding section (figure 11), it is clear that here not this axiom but Lobačevskiĭ's axiom holds. It remains to verify the axioms I-IV.

The plane in the model is the interior of a circle (whose radius will be taken to be 1). Points play the role of points, chords the role of straight lines; the concepts "a point lies on a straight line" and "a point lies between two others" are understood in the usual sense. Hence it is obvious that the axioms of incidence, order, and continuity are satisfied. For example, the third axiom of order simply means that a chord divides the circle into two parts.

It remains to verify the axioms of motion. A motion is defined as a transformation that carries the circle into itself and straight lines into straight lines. From this definition it is obvious that these transformations fulfill the first two axioms of motion: The first, because straight lines are just chords and consequently preservation of chords means preservation of straight lines; the second, because if two transformations carrying the circle into itself and the chords into chords are carried out, then the resulting transformation also carries the circle into itself and chords into chords; i.e., it is one of those regarded as "motions."

Thus, only the third axiom of motion remains and the verification of it is the only point of difficulty here.

First of all we note that this axiom contains two statements.

Let A, A' be two points, a, a' two half lines going out from them, α, α' two half planes bounded by the lines a, a'.

The first claim is that there *exists* a motion carrying A into A', a into a', and α into α'.

The second claim is that there is *only one* such motion.

We could perhaps refer to the fact that both these statements have already been proved in Chapter III, §14, but we prefer to prove them here without getting involved, as was the case in Chapter III, with other more general problems.

Let us show that the first statement is true in our model (i.e., in the appropriate interpretation of the terms "half line," "half plane," "motion").

To begin with let us assume that the point A' is at the center of the circle. We take coordinate axes so that the origin is at the center of the circle and the Ox-axis passes through the point A (figure 14). In the preceding section we have investigated the transformation

$$x' = \frac{x+a}{1+ax}, \quad y' = \frac{y\sqrt{1-a^2}}{1+ax}. \tag{6}$$

We have proved there that it is a "motion" (i.e., that it carries the given circle into itself and straight lines into straight lines).

Let x_0 be the abscissa of A, its ordinate being $y_0 = 0$. Therefore, when we choose $a = -x_0$, then, by (6), A goes over into the point with the coordinates (0, 0), i.e., into A'.

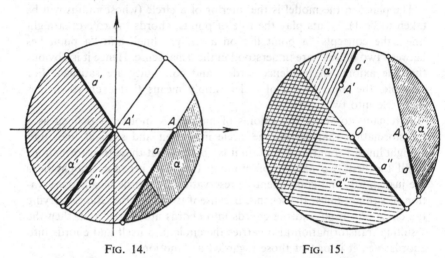

FIG. 14. FIG. 15.

Now since straight lines go over into straight lines, the "half line" (i.e., segment of a chord) a assumes some position a'' (figure 14). By a rotation around the center we can now carry a'' into a'. The "half plane" α is one of the segments bounded by the "straight line" (chord) a. If after our motion it coincides with a', then the transformation is complete; if not, then by a revolution (reflection in the diameter a') we carry it into the semicircle α'.

Thus, by combining the "shifting" process (6) with a rotation and, if necessary, a revolution, we have carried A, a, α into A', a', α'. But the result of all these "motions" is again a "motion"; this "motion" carries A, a, α into A', a', α'; i.e., the existence of the required motion is now proved.

So far we have restricted ourselves to the special case when the point A' is at the center. Let us now assume that it has some other position. Then by what we have already proved we can carry it into the center by a certain "motion" which we denote by D_1. Then the "half line" a' goes over into some "half line" a'' passing through the center, and the "half plane" α' into a certain "half plane" (semicircle) α'' (figure 15). As we have already proved, by means of a certain "motion" D_2 we can also carry A into the center, the "half line" a into a'', the "half plane" α

§5. AXIOMS OF GEOMETRY; THEIR VERIFICATION

into α''. Finally by the inverse "motion" of D_1 we carry A' into its former place and at the same time a'', α'' return to their original positions a', α' *

Thus, as a result of combining the "motion" D_2 and the "motion" inverse to D_1 we carry A, a, α into A', a', α'. But a combination of "motions" is again a "motion"; and so we have ascertained that there

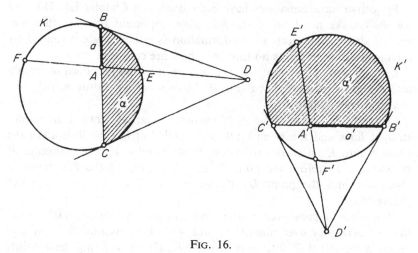

FIG. 16.

exists a motion carrying A, a, α into A', a', α' whatever the position of A and A' in the circle. So the first statement included in the third axiom of motion is proved in its full extent.

Let us now show that the second statement is also satisfied in our model. In the sense of this statement, we have to prove the following.

Let A, A' be two points inside the circle, a, a' segments of chords emanating from them, and α, α' parts of the circle bounded by these chords. For the sake of clarity we illustrate these points, chords, and parts of the circle in two different drawings (figure 16) although, of course, they lie in the one circle in question. It is claimed that the "motion" carrying A into A', a into a', α into α', respectively, is unique, i.e., is completely determined by these data.

For the proof we shall consider a transformation not only of the given circle, but of the whole plane.† A "motion," by definition, carries straight lines into straight lines. A transformation having these properties is

---

* The "motion" inverse to D_1 is expressed by the formulas (4) (§4) if D_1 is expressed by (3).

† It can be proved that a transformation of a circle into itself carrying straight lines into straight lines can be extended uniquely with preservation of this property to the whole projective plane, i.e., the plane to which points at infinity have been adjoined.

called *projective*. Consequently, we can say that for us a "motion" means a projective transformation carrying the given circle into itself. (In figure 16 we have illustrated this so that the circle K goes over into the circle K'. We only have to imagine that by a parallel shift the circle K' is superimposed on K.)

Projective transformations have been studied in Chapter III, §13, and we shall make use here of the following important theorem that was proved there: A projective transformation is completely determined by the images of four points no three of which are collinear.

Let us return to the "motion" in question. It carries the chord segment a into a' and therefore carries B into B'. Since it carries chords into chords, it also carries C into C'.

Furthermore, since a "motion" generally carries straight lines into straight lines and the given circle into itself but in our illustration the circle K into K', it carries the tangents at B and C into the tangents at B' and C'. Therefore the point D of intersection of the first tangents goes over into the point D' of intersection of the second tangents* (figure 16).

Now since A goes over into A' and straight lines into straight lines, the line AD goes over into $A'D'$. But AD intersects our circle in two points E, F, and $A'D'$ intersects it in E', F'. (In the drawing these points lie on the circumferences of K and K'.) Since the circle goes over into itself, the points E, F go over into E', F'. Let E lie on the arc bounding the part α and E' on that bounding α'. Then, since α by assumption goes over into α', E goes over precisely into E' and F into F'.

Thus we have found that under the "motion" in question the points B, C, E, F on the circumference go over into B', C', E', F'. But it is obvious that of the points B, C, E, F and B', C', E', F' no three can be collinear. Therefore, by the theorem quoted earlier, the projective transformation carrying B, C, E, F into B', C', E', F' is unique. But a "motion" is a projective transformation. Consequently our "motion" carrying A, a, α into A', a', α' is unique, and this is what we had to prove.

Thus we have shown that in the model in question all the axioms of Euclidean geometry are in fact satisfied, except the axiom of parallelism; or in other words, that all the axioms of Lobačevskiĭ geometry are satisfied in the model. The model therefore is a realization of Lobačevskiĭ geometry. This geometry is, as it were, reduced to Euclid's geometry inside the circle, presented in a special manner with the agreed interpretation of the terms "straight line" and "motion" in the model.

---

* If, for example, the tangents at B and C are parallel, then the point D is "at infinity."

§6. SEPARATION OF GEOMETRIC THEORIES

Incidentally, this enables us to develop the Lobačevskiĭ geometry on the given concrete model, a method that turns out to be more convenient in many problems.

From the point of view of a logical analysis of the foundations of geometry, the proof we have given shows, first, that the Lobačevskiĭ geometry is noncontradictory and, second, that the parallel postulate cannot possibly be deduced from the remaining axioms enumerated previously.

§6. Separation of Independent Geometric Theories from Euclidean Geometry

1. Projective geometry. A fundamental development of geometry parallel with the creation of Lobačevskiĭ geometry came about in yet another way. Within the wealth of all the geometric properties of space, separate groups of properties, distinguished by a peculiar interrelatedness and stability, were singled out and subjected to an independent study. These investigations, with their separate methods, gave rise to new chapters of geometry, i.e., to the science of spatial forms, just as for example anatomy or physiology form distinct chapters in the science of the human organism.

Initially, geometry was not divided up at all. It studied mainly the metrical properties of space connected with the measurement of the dimensions of figures. Circumstances not connected with the measurement but the qualitative character of the natural location of figures were considered in passing only, although it was noted long ago that a part of these properties is distinguished by a peculiar stability, in that they are

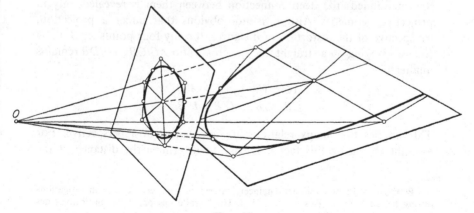

FIG. 17.

preserved under rather substantial distortion of the form and displacement of the location of figures.

Let us consider, for example, the projection of a figure from one plane into another (figure 17). The lengths of segments are changed in the process and so are the angles, the outlines of objects are visibly distorted. However, for example, the property of a number of points of lying on one straight line is preserved and so is the property of a straight line of being a tangent to a given curve, etc.

Projections and projective transformations have already been treated in Chapter III, where we mentioned their obvious connection with perspective, i.e., the drawing of spatial figures on a plane. The study of properties of perspective goes back in antiquity right to Euclid, to the work of the ancient architects; artists concerned themselves with perspective: Dürer, Leonardo da Vinci, and the engineer and mathematician Desargues (17th century). Finally, at the beginning of the 19th century Poncelet was the first to separate out and study systematically the geometrical properties that are preserved under arbitrary projective transformations of the plane (or of space) and so to create an independent science, namely *projective geometry.**

It might seem that there are only a few, very primitive properties that are preserved under arbitrary projective transformations, but this is by no means so. For example, we do not notice immediately that the theorem stating that the points of intersection of opposite sides (produced) of a hexagon inscribed in a circle lie on a straight line also holds for an ellipse, parabola, and hyperbola. The theorem only speaks of projective properties, and these curves can be obtained from the circle by projection. It is even less obvious that the theorem to the effect that the diagonals of a circumscribed hexagon meet in a point is a peculiar analogue of the theorem just mentioned; the deep connection between them is revealed only in projective geometry. Also it is not obvious that under a projection, irrespective of the distortion of distances, for any four points A, B, C, D (figure 18) lying on a straight line the cross ratio $AC/CB : AD/DB$ remains unaltered

$$\frac{AC}{CB} : \frac{AD}{DB} = \frac{A'C'}{C'B'} : \frac{A'D'}{D'B'}.$$

This implies that many relations are maintained in a perspective. For example, by using this fact it is easy to determine the distance of the

---

* Poncelet, a French military engineer, carried out his geometrical investigations during his captivity in Russia after 1812. His "Traité des propriétés projectives des figures" appeared in 1822.

§6. SEPARATION OF GEOMETRIC THEORIES

telegraph poles A, B, C from the point D (figure 18) on a photograph of the road leading into the distance, when their spacing is known.

With projective geometry and its application to aerial photography, we have dealt in Chapter III. It stands to reason that its laws are used in architecture, in the construction of panoramas, in decorating, etc.

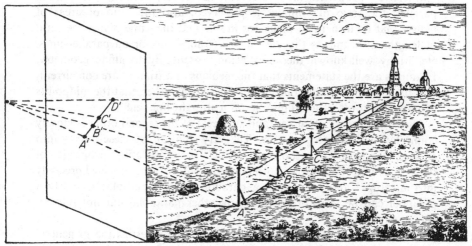

Fig. 18.

The separation of projective geometry played an important role in the development of geometry itself.

2. Affine geometry. As another example of an independent geometry we can take *affine geometry*. Here one studies the properties of figures that are not changed by arbitrary transformations in which the Cartesian coordinates of the original (x, y, z) and the new (x', y', z') position of each point are connected by linear equations:

$$x' = a_1 x + b_1 y + c_1 z + d_1,$$
$$y' = a_2 x + b_2 y + c_2 z + d_2,$$
$$z' = a_3 x + b_3 y + c_3 z + d_3$$

(where it is assumed that the determinant

$$\begin{vmatrix} a_1 & b_1 & c_1 \\ a_2 & b_2 & c_2 \\ a_3 & b_3 & c_3 \end{vmatrix}$$

is different from zero).

It turns out that every affine transformation reduces to a motion, possibly a reflection, in a plane and then to a contraction or extension of space in three mutually perpendicular directions.

Quite a number of properties of figures are preserved under each of these transformations. Straight lines remain straight lines (in fact all "projective" properties are preserved); moreover, parallel lines remain parallel; the ratio of volumes is preserved, also the ratio of areas of figures that lie in parallel planes or in one and the same plane, the ratio of lengths of segments that lie on one straight line or on parallel lines, etc. Many well-known theorems belong essentially to affine geometry. Examples are the statements that the medians of a triangle are concurrent, that the diagonals of a parallelogram bisect each other, that the midpoints of parallel chords of an ellipse lie on a straight line, etc.

The whole theory of curves (and surfaces) of the second order is closely connected with affine geometry. The very division of these curves into ellipses, parabolas, hyperbolas is, in fact, based on affine properties of the figures: Under affine transformations an ellipse is transformed precisely into an ellipse and never into a parabola or a hyperbola; similarly a parabola can be transformed into any other parabola, but not into an ellipse, etc.

The importance of the separation and detailed investigation of general affine properties of figures is emphasized by the fact that incomparably more complicated transformations turn out to be essentially linear, i.e., affine in the infinitely small, and the application of the methods of the differential calculus is linked exactly with the consideration of infinitely small regions of space.

3. Klein's Erlanger Programm. In 1872 Klein, in a lecture at the University of Erlangen which is now known as the "Erlanger Programm," in summing up the results of the developments of projective, affine, and other "geometries" gave a clear formulation of the general principle of their formation: We can consider an arbitrary group of single-valued transformations of space and investigate the properties of figures that are preserved under the transformations of this group.*

From this point of view the properties of space are stratified, as it were, with respect to their depth and stability. The ordinary Euclidean

---

* The word "group" is used here not merely in the sense of a collection. When we speak of a group of transformations (see Chapter XX) we have in mind a set of transformations which must contain the identity transformation (leaving all points in place), which contains together with every nonidentical transformation the one inverse to it (restoring all points to their previous place), and which contains together with any two transformations of the set also the transformation that is equivalent to the two carried out in succession.

§6. SEPARATION OF GEOMETRIC THEORIES

geometry was created by disregarding all properties of real bodies other than the geometrical; here, in the special branches of geometry, we perform yet another abstraction within geometry, by disregarding all geometrical properties except the ones that interest us in the given branch of geometry.

In accordance with this principle of Klein, we can construct many geometries. For example, we can consider the transformations that preserve the angle between arbitrary lines (conformal transformations of space), and when studying properties of figures preserved under such transformations we talk of the corresponding conformal geometry. We can consider transformations of not necessarily the whole space. Thus, by considering the points and chords of a circle under all its transformations into itself that carry chords into chords and by singling out the properties that are preserved under such transformations, we obtain a geometry which as we have shown in §§4 and 5 coincides with Lobačevskiĭ geometry.

4. Abstract spaces. The further development of the theories thus separated, even from the theoretical point of view (to say nothing of their factual content), did not stop at what we have said here.

If we are interested, for example, only in affine properties of figures we can, by abstracting from all other properties, imagine a space and geometrical figures in it that have *only* properties of interest to us and, as it were, no other properties at all. In this "space," figures do not have any properties except affine ones. It is natural to try and give also an axiomatic account of the geometry of such an abstract space, i.e., to assume that we are dealing with some abstract objects: "points," "straight lines," and "planes" whose properties are expressed in certain axioms (there are, of course, fewer of these properties than in the case of Euclidean geometry) so that consequences derived from these axioms correspond to affine properties of figures of the ordinary space.

This can indeed be done; and such a collection of abstract "points," "straight lines" and "planes" with the system of their properties is called an *affine space*.

In exactly the same way we can imagine an abstract system of objects having only that range of properties that correspond to projective properties of figures of Euclidean space. (This time the difference of the axiom system from the axioms of ordinary geometry turns out to be even more substantial.)

5. Abstract topological space. When we go deeper into the nature of geometric forms, we may note that in quite a number of problems we are concerned with properties, even deeper than projective ones, which are so firmly connected with the given figure that they are preserved under arbitrary distortions, provided only that these distortions do not cause the figure to break up or parts of it to become "pasted together." To make the notion of such a continuous distortion more precise than the intuitive description does, we refer to the definition of a continuous function known to us from analysis and say that we are dealing with an arbitrary transformation of all the points of the figure into a new position under which the Cartesian coordinates of the points in the new position are expressed as continuous functions of the old coordinates, while the old coordinates in turn can be expressed as continuous functions of the new ones.

Properties of figures that are preserved under arbitrary transformations of this kind are called topological and the science which investigates them is *topology* (see Chapter XVIII).

Topological properties connected with figures are in the simplest cases distinguished by their exceptionally intuitive character. For example, it is almost obvious that every line in a plane that can be obtained by a continuous deformation of the circumference of a circle divides the plane into two parts, the interior and exterior, no matter how much the contour winds; therefore

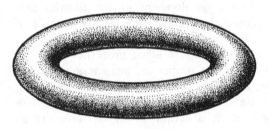

Fig. 19.

the property of a circumference of dividing the plane is topological. Also it is visibly obvious, no doubt, that the torus surface (figure 19) cannot possibly be turned into a sphere by a continuous transformation, so that the property of an arbitrary surface of admitting a continuous transformation into, say, a torus surface is a topological property that distinguishes it from many other surfaces.

Arguments in connection with continuity are of an intuitive character and often clarify the essence of the matter so well that it is very tempting to try to turn them into rigorous proofs and, even more, to extend such methods to other, incomparably more complicated problems.

For example, take the argument that establishes the truth of the fundamental theorem of algebra to the effect that every equation

$$z^n + a_1 z^{n-1} + a_2 z^{n-2} + \cdots + a_{n-1} z + a_n = 0 \qquad (7)$$

has at least one real or complex root.

§6. SEPARATION OF GEOMETRIC THEORIES

Let z be a point of one complex plane and $w = f(z)$ the corresponding point of another complex plane w, where $f(z)$ denotes the left-hand side of (7). For very large absolute values of z, the function $f(z)$ differs relatively little from z^n; but the function z^n is very simple. In particular, it is easy to verify that if z by a continuous motion describes in the complex plane a circumference with center at the origin, then the point $w_1 = z^n$ goes precisely n times around a similar circumference of radius $|z|^n$ in the w-plane.* So the point $w = f(z)$ describes n loops forming some contour Γ comparatively near to the line described by the point $w_1 = z^n$ (figure 20).

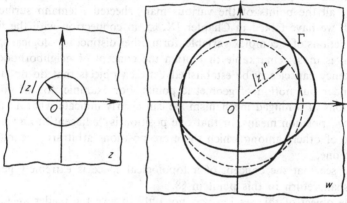

Fig. 20.

Now when the circumference described by z is continuously contracted to one point, then the n times looped-contour Γ described by $w = f(z)$ is continuously deformed and also contracts to a point. But it is rather obvious that it cannot contract to a point without passing through the origin O which this contour surrounds initially. Hence it goes through O at least once, and for such a z we have $w = f(z) = 0$. This z is, then, a root of the equation (7). Strictly speaking it is also clear that in a certain sense there must be exactly n roots, since each of the n loops of the contour Γ on contraction passes through O.

Our argument requires, of course, a rigorous establishment of just those topological properties of the contour and its deformations that we have used here.

---

* For if $z = \rho(\cos \phi + i \sin \phi)$ then [see Chapter IV, §3] we know that
$$z^n = \rho^n(\cos \phi n + i \sin \phi n);$$
therefore, when the argument ϕ of z changes from 0 to 2π, then the argument of z^n changes from 0 to $2\pi n$, i.e., the radius vector leading to z^n makes n complete revolutions under this motion.

We could give many examples of the use of topological properties in various branches of mathematics often very far removed from geometry.

With the study of topological properties before us, we can again imagine an abstract set of objects having only properties of this kind (see §7, Chapter XX). Such a set is called an *abstract topological space*.

This point of view is already incomparably wider than the study of topological properties of geometric figures only. Topological spaces can be extremely varied; for example, all the points of a torus surface with their specific properties of adjacency to one another form one topological space, all the points of the plane another, the whole Euclidean space a third; all the points of the various many-sheeted Riemann surfaces of which we have talked in Chapter IX, §5, in connection with the theory of functions of a complex variable, form other distinct topological spaces. But it is most remarkable that often the concept of neighborhood and adjacency can clearly be established between objects that do not fall at all under our notion of geometric points. For example, for all possible positions of a hinged mechanism we can clearly indicate what a "neighboring" position means, or that one position is "adjacent" to an infinite range of others among which there are positions arbitrarily near to the given one.

We see that the concept of a topological space is extremely general. We shall return to this point in §8.

The object of this section was not only to give the reader an idea of the various geometries, but also to show that certain concrete problems lead to the isolation and investigation of separate groups of geometric properties; that these investigations entail the creation of the idea of abstract geometric objects having *only* these properties, i.e., that the isolation of these properties in their pure form leads us to the idea of the corresponding abstract space.

Other developments, leading to the construction of a different kind of abstract spaces, will be discussed in the following section.

§7. Many-Dimensional Spaces

1. The geometry of n-dimensional space. An important step in the development of new geometric ideas was the creation of the geometry of many-dimensional spaces to which we have already referred in the preceding chapter. One of the moving forces was the tendency to use geometric arguments for the solution of problems in algebra and analysis. The geometric approach to the solution of analytical problems is based on the method of coordinates. Let us give a simple example.

We want to know how many integral solutions the inequality $x^2 + y^2 < N$

§7. MANY-DIMENSIONAL SPACES

has. By regarding x and y as Cartesian coordinates in a plane, we see that the problem reduces to the following: How many points with integer coordinates are contained in a circle of radius $\sqrt{N}$. The points with integer coordinates are the vertices of squares with sides of unit length covering the plane (figure 21). The number of such points inside the circle is approximately equal to the number of squares lying in the circle, i.e., to the area of a circle of radius $\sqrt{N}$. Thus, the number of solutions of the inequality we are interested in is approximately equal to πN. Furthermore, it is not difficult to prove that the relative error occurring here tends to zero for $N \to \infty$. A more accurate study of this error is a very difficult problem in the theory of numbers and has become the object of deep investigations in comparatively recent years.

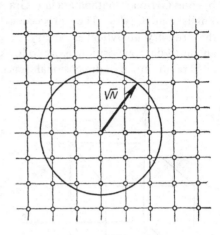

FIG. 21.

In our example it was sufficient to translate the problem into geometric language in order to obtain at once a result which is by no means obvious from the point of view of "pure algebra." The corresponding problem for an inequality with three unknowns can be solved in exactly the same way. However, when there are more than three unknowns, this method is not applicable because our space is three-dimensional, i.e., the position of a point in it is determined by a triple of coordinates. To preserve the convenient geometrical analogy in similar cases we introduce the idea of an abstract "n-dimensional space" whose points are determined by n coordinates $x_1, x_2, \cdots, x_n$. The fundamental concepts of geometry are then generalized in such a way that the geometric arguments turn out to be applicable to the solution of problems with n variables; this makes it much easier to obtain results. The possibility of such a generalization is based on the uniformity of the algebraic laws thanks to which many problems can be solved simultaneously for an arbitrary number of variables. This enables us to apply geometric arguments that are valid for three dimensions to an arbitrary number of dimensions.

2. Coordinates in n-dimensional geometry. Rudiments of the concepts of a four-dimensional space can already be found in Lagrange who, in his papers on mechanics, formally regarded the time as a "fourth coordinate" beside the three spatial coordinates. But the first systematic account of the elements of many-dimensional geometry was given in 1844 by the German mathematician Grassmann and independently by the Englishman Cayley. They proceeded by way of a formal analogy with the ordinary analytical geometry. A general outline of this analogy in an up-to-date exposition is as follows.

A point in an n-dimensional space is determined by n coordinates $x_1, x_2, \cdots, x_n$. A figure in n-dimensional space is a geometric locus or set of points satisfying certain conditions. For example, an n-dimensional "cube" is defined as the geometric locus of points whose coordinates are subject to the inequalities $a \leqslant x_i \leqslant b$ ($i = 1, 2, \cdots, n$). The analogy with the ordinary cube is here completely evident: When $n = 3$, i.e., when the space is three-dimensional, our inequalities in fact define the cube whose edges are parallel to the coordinate axes and of length $b - a$ (figure 22 illustrates the case $a = 0$, $b = 1$).

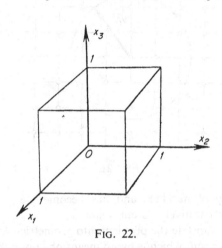

FIG. 22.

The distance between two points can be defined as the square root of the sum of the squares of the differences of the coordinates

$$d = \sqrt{(x_1' - x_1'')^2 + (x_2' - x_2'')^2 + \cdots + (x_n' - x_n'')^2}.$$

This is a direct generalization of the well-known formulas for the distance in a plane or in three-dimensional space, i.e., for $n = 2$ or 3.

It is now possible to define equality of figures in n-dimensional space. Two figures are regarded as equal if a correspondence can be established between their points under which the distances between pairs of corresponding points are equal. A transformation preserving distances can be called a generalized motion.* Then we can say by analogy with the usual Euclidean geometry that the objects of study in n-dimensional geometry are the properties of figures that are preserved under generalized motions.

---

* The generalization consists not only in the transition to n variables but also in the fact that a reflection in a plane is incorporated among motions, because it also does not alter distances between points.

§7. MANY-DIMENSIONAL SPACES

This definition of the content of n-dimensional geometry was set up in the 1870's and provided a precise foundation for its development. Since then n-dimensional geometry has been the object of numerous investigations in all directions similar to those of Euclidean geometry (elementary geometry, general theory of curves, etc.).

The concept of distance between points also enables us to transfer to n-dimensional space other concepts of geometry such as segment, sphere, length, angle, volume, etc. For example, an n-dimensional sphere is defined as the set of points whose distance from a given point is not more than a given R. Therefore a sphere is given analytically by an inequality

$$(x_1 - a_1)^2 + \cdots + (x_n - a_n)^2 \leqslant R^2,$$

where $a_1, \cdots, a_n$ are the coordinates of its center. The surface of the sphere is given by the equation

$$(x_1 - a_1)^2 + \cdots + (x_n - a_n)^2 = R^2.$$

The segment AB can be defined as the set of points X such that the sum of the distances from X to A and B is equal to the distance from A to B. (The length of a segment is the distance between its end points.)

3. Hyperplanes. Let us dwell in some detail on planes of various dimensions.

In three-dimensional space there are the one-dimensional "planes" (namely, the straight lines), and the ordinary (two-dimensional) planes. In n-dimensional space for $n > 3$ we also have to take many-dimensional planes into account, of dimensions 3 to $n - 1$.

In three-dimensional space a plane is, of course, given by one linear equation, and a straight line by two such equations.

By a direct generalization we come to the following definition: A *k-dimensional plane* (usually called a *hyperplane*) in an n-dimensional space is the geometric locus of points whose coordinates satisfy a system of $n - k$ linear equations

$$\begin{aligned} a_{11} x_1 + a_{12} x_2 + \cdots + a_{1n} x_n + b_1 &= 0, \\ a_{21} x_1 + a_{22} x_2 + \cdots + a_{2n} x_n + b_2 &= 0, \\ &\cdots \\ a_{n-k,1} x_1 + a_{n-k,2} x_2 + \cdots + a_{n-k,n} x_n + b_{n} &= 0, \end{aligned} \qquad (8)$$

provided the equations are compatible and independent (i.e., none of them is a consequence of the others). Each of these equations represents an $(n - 1)$-dimensional hyperplane and together they determine the common points of $n - k$ such hyperplanes. The fact that the equations (8) are compatible means that there are some points that satisfy them, i.e., that the $n - k$ given $(n - 1)$-dimensional hyperplanes intersect. The fact that none of the equations is a consequence of the others means that none of them can be omitted. Otherwise the system would reduce to a smaller number of equations and would define a hyperplane of a larger number of dimensions. Thus, to speak geometrically, a k-dimensional hyperplane is determined as the intersection of $n - k$ $(n - 1)$-dimensional hyperplanes represented by independent equations. In particular, when $k = 1$, we have $n - 1$ equations which determine a "one-dimensional hyperplane," i.e., a straight line. Thus, this definition of a k-dimensional hyperplane is a natural formal generalization of well-known results of analytical geometry. The advantage of this generalization becomes apparent in the fact that conclusions concerning systems of linear equations receive a geometric interpretation which makes them more lucid. This geometric approach to problems of linear algebra was also discussed in Chapter XVI.

An important property of a k-dimensional hyperplane is the fact that it can be regarded as a k-dimensional space. For example, a three-dimensional hyperplane is itself an ordinary three-dimensional space. This enables us to transfer to a space of higher dimension many conclusions obtained for spaces of lower dimension, by means of the usual argument from n to $n + 1$.

If the equations (8) are compatible and independent, then it is proved in algebra that we can choose k of the n variables x_i at will and then the remaining $n - k$ variables can be expressed in terms of them.* For example:

$$x_{k+1} = c_{11} x_1 + c_{12} x_2 + \cdots + c_{1k} x_k + d_1,$$
$$x_{k+2} = c_{21} x_1 + c_{22} x_2 + \cdots + c_{2k} x_k + d_2,$$
$$\cdots\cdots\cdots\cdots\cdots\cdots\cdots\cdots\cdots\cdots\cdots\cdots\cdots\cdots\cdots\cdots\cdots\cdots$$
$$x_n = c_{n-k,1} x_1 + c_{n-k,2} x_2 + \cdots + c_{n-k,k} x_k + d_k.$$

Here arbitrary values can be given to $x_1, x_2, \cdots, x_k$, and the remaining x_i are determined by them. This means that the position of a point in a k-dimensional hyperplane is determined by k coordinates that can assume arbitrary values. It is in this sense that the hyperplane has k dimensions.

---

* These k variables cannot, in general, be chosen arbitrarily from the x_i. For example, in the system $x_1 + x_2 + x_3 = 0$, $x_1 - x_2 - x_3 = 0$ the value of x_1 is uniquely determined: $x_1 = 0$, and obviously neither x_2 nor x_3 can be expressed in terms of it. All that is stated, however, is that the necessary k of the x_i can always be found.

§7. MANY-DIMENSIONAL SPACES

From the definition of the hyperplanes of various dimensions, we can derive in a purely algebraic way the following fundamental theorems.

1. One and only one k-dimensional hyperplane passes through $k+1$ points that do not lie in a $(k-1)$-dimensional hyperplane.

The complete analogy with known facts of elementary geometry is obvious here. The proof of this theorem is based on the theory of systems of linear equations and is somewhat complicated, so that we shall not write it out.

2. If an l-dimensional and a k-dimensional hyperplane in an n-dimensional space have at least one point in common and $l+k \geqslant n$, then they intersect in a hyperplane of dimension not less than $l+k-n$.

Hence it follows as a special case that two two-dimensional planes in a three-dimensional space, if they do not coincide and are not parallel, intersect on a straight line ($n=3$, $l=2$, $l+k-n=1$). But in a four-dimensional space two two-dimensional planes may well have a single point in common. For example, the planes given by the system of equations:

$$\left.\begin{matrix} x_1 = 0 \\ x_2 = 0 \end{matrix}\right\}, \quad \left.\begin{matrix} x_3 = 0 \\ x_4 = 0 \end{matrix}\right\},$$

obviously intersect only in the point with the coordinates $x_1 = 0$, $x_2 = 0$, $x_3 = 0$, $x_4 = 0$.

The proof of the theorem is extremely simple: An l-dimensional hyperplane is given by $n-l$ equations, a k-dimensional one by $n-k$; the coordinates of the points of intersection must satisfy simultaneously all these $(n-l)+(n-k) = n-(l+k-n)$ equations. If none of the equations is a consequence of the others, then by the very definition of a hyperplane we have as intersection an $(l+k-n)$-dimensional hyperplane; otherwise we have a hyperplane with a larger number of dimensions.

To these two theorems we can add another two.

3. In each k-dimensional hyperplane there are at least $k+1$ points that do not lie in a hyperplane of smaller dimension. In an n-dimensional space there are at least $n+1$ points that do not lie in any hyperplane.

4. If a straight line has two points in common with a hyperplane (of an arbitrary number of dimensions), then it lies entirely in that hyperplane. Generally, if an l-dimensional hyperplane has $l+1$ points in common with a k-dimensional hyperplane that do not lie in an $(l-1)$-dimensional hyperplane, then it lies entirely in this k-dimensional hyperplane.

Note that n-dimensional geometry can be built up starting from axioms that generalize the axioms listed in §5. In this approach the four theorems mentioned here assume the role of axioms of incidence. This shows, by the way, that the concept of axiom is relative: One and the same statement can emerge as a theorem in one buildup of a theory, and as an axiom in another.

4. Various examples of an *n*-dimensional space. We have now obtained a general idea of the mathematical concept of a many-dimensional space. In order to clarify the actual physical meaning of this concept, let us turn again to the problem of graphical illustration. Suppose, for example, we wish to illustrate the dependence of the pressure of a gas on its volume. We take coordinate axes in a plane and plot on one axis the volume v, on the other the pressure p. The dependence of the pressure on the volume under the given conditions is then illustrated by a certain curve (by the well-known Boyle-Mariotte law this would be a hyperbola for an ideal gas with a fixed temperature). But when we have a more complicated physical system, whose state is given not by two data (like volume and pressure in the case of the gas) but by say five, then the graphical illustration of its behavior leads to the notion of a five-dimensional space.

Suppose, for example, that we are concerned with an alloy of three metals or a mixture of three gases. The state of the mixture is determined by four data: the temperature T, the pressure p, and the percentage contents c_1, c_2 of two gases (the percentage content of the third gas is then determined by the fact that the sum total of the percentage contents is 100% so that $c_3 = 100 - c_1 - c_2$). The state of such a mixture is, therefore, determined by four data. A graphical illustration of it either requires a combination of several diagrams, or else we have to represent the state in the form of a point of a four-dimensional space with four coordinates T, p, c_1, c_2. Such a representation is, in fact, used in chemistry; the application of the methods of many-dimensional geometry to chemistry was developed by the American scientist Gibbs and the school of Soviet physicochemists of Academician Kurnakov. Here the introduction of a many-dimensional space is dictated by the endeavor to preserve the useful geometrical analogies and arguments arising from the simple device of graphical illustration.

Let us give an example from the realm of geometry. A sphere is given by four data: the three coordinates of its center and its radius.* The special geometry of spheres which was built up about a century ago by several mathematicians can, therefore, be regarded as a four-dimensional geometry.

---

* Translator's note: Therefore a sphere can be represented as a point in four-dimensional space.

§7. MANY-DIMENSIONAL SPACES

From all we have said the real basis for the introduction of the concept of a many-dimensional space will be clear. If some figure or the state of some system, etc., is given by n data, then this figure, this state, etc., can be conceived as a point of some n-dimensional space. The advantage of this representation is approximately the same as that of ordinary graphs: It consists in the possibility of applying well-known geometric analogies and methods to the study of the phenomena in question.

There is, therefore, no mysticism in the mathematical concept of a many-dimensional space. It is not more than a certain abstract concept developed by the mathematicians for the purpose of describing in geometric language those things that do not admit a simple geometric illustration in the usual sense. This abstract concept has an entirely real basis, it reflects actuality and was created by the demands of science, not by idle play of the imagination. It reflects the fact that there exist such things as a sphere or a mixture of three gases that are characterized by several data so that the collection of all these things is many-dimensional. The number of variables in a given case is just the number of these data; just as a point moving in space changes its three coordinates, so a sphere moving, expanding, and contracting changes its four "coordinates," i.e., the four quantities that determine it.

In the subsequent sections, we shall dwell upon many-dimensional geometry. It is important here to understand that this is a method of mathematical description of real things and phenomena. The idea that there exists some sort of four-dimensional space in which our real space is embedded, an idea that has been used by certain litterateurs and spiritualists, has no relation to the mathematical concept of a four-dimensional space. If one can speak here of a relationship to science at all, it is perhaps possible only in the sense of an imaginative distortion of scientific concepts.

5. Polyhedra in n-dimensional space. As we have already said, the geometry of a many-dimensional space was built up at first by way of a formal generalization of the usual analytic geometry to an arbitrary number of variables. However, such an approach to the matter could not satisfy the mathematicians completely. As a matter of fact, the purpose was not so much a generalization of the geometric concepts as of the geometric method of investigation itself. It was, therefore, important to give a purely geometric exposition of n-dimensional geometry, independently of the analytical apparatus. This was first done by the Swiss mathematician Schlaefli in 1852, one of whose articles deals with the problem of regular polyhedra of a many-dimensional space. True, Schlaefli's article was not appreciated by his contemporaries, because to understand it one has to rise, to a certain extent, to an abstract view of geometry. Only subsequent developments of mathematics have brought complete clarity into this problem, by an exhaustive elucidation of the mutual relationship of the analytic and geometric approach. Since we cannot go deeper into this problem, we confine ourselves to examples of a geometric exposition of n-dimensional geometry.

Let us consider the geometric definition of an n-dimensional cube. When we move a segment in a plane perpendicular to itself by a distance equal to its length, we sweep out a square, i.e., a two-dimensional cube (figure 23a). Similarly, when we move a square in the direction perpen-

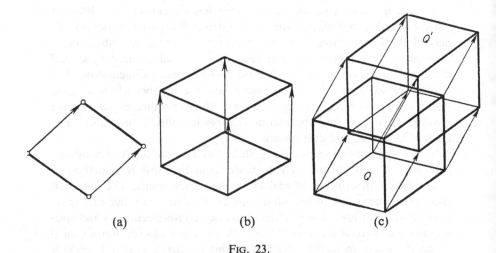

(a)   (b)   (c)

FIG. 23.

dicular to its plane by a distance equal to its side, we sweep out a three-dimensional cube (figure 23b). In order to obtain a four-dimensional cube, we use the same construction: We take a three-dimensional hyperplane in a four-dimensional space and in it a three-dimensional cube and we move it in the direction perpendicular to this three-dimensional hyperplane by a distance equal to its edge (by definition a straight line is perpendicular to a k-dimensional hyperplane if it is perpendicular to every straight line lying in that hyperplane). This construction is symbolically represented in figure 23c. Two three-dimensional cubes Q and Q' are drawn here, namely the given cube in its initial and its final position. The lines joining the vertices of these cubes illustrate the segments traced out by the vertices in the movement of the cube. We see that a four-dimensional cube has 16 vertices in all: 8 for the cube Q and 8 for Q'. Further, it has 32 edges: 12 edges of the moving three-dimensional cube in its initial position Q,

§7. MANY-DIMENSIONAL SPACES

12 edges in the final position Q', and 8 "lateral" edges. It has 8 three-dimensional faces which are themselves cubes. In the motion of the three-dimensional cube each of its faces sweeps out a three-dimensional cube so that we obtain 6 cubes as the lateral faces of the four-dimensional cube, and in addition there are two faces: "front" and "back" or the initial and final position of the moving cube. Finally, a four-dimensional cube also has two-dimensional square faces, 24 in number: 6 each in the cubes Q and Q', and another 12 squares that are swept out by the edges of Q in its motion.

So a four-dimensional cube has 8 three-dimensional faces, 24 two-dimensional faces, 32 one-dimensional faces (edges), and finally 16 vertices; every face is a "cube" of the appropriate number of dimensions: a three-dimensional cube, a square, a segment, and a vertex (which we can regard as a zero-dimensional cube).

Similarly, by shifting a four-dimensional cube "into the fifth dimension" we obtain a five-dimensional cube and so, by repeating the construction, we can build up a cube of an arbitrary number of dimensions. All the faces of an n-dimensional cube are themselves cubes of a smaller number of dimensions: $(n - 1)$-dimensional, $(n - 2)$-dimensional, etc., finally one-dimensional, i.e., edges. For the inquisitive reader it is not a difficult task to find out how many faces of each number of dimensions an n-dimensional cube has. It is easy to see that the number of $(n - 1)$-dimensional faces is $2n$ and that there are 2^n vertices. But how many edges are there, for example?

Let us look at another polyhedron in an n-dimensional space. In the plane the simplest polygon is a triangle; it has the least possible number of vertices. In order to obtain a polyhedron with the least number of vertices it is sufficient to take a point not in the plane of a triangle and join it by segments to each point of the triangle. The segments so obtained fill a three-sided pyramid, i.e., a tetrahedron (figure 24). In order to obtain

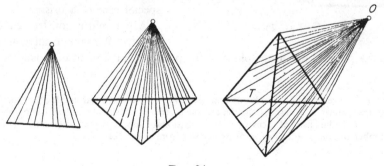

FIG. 24.

the simplest polyhedron in a four-dimensional space we argue as follows. We take an arbitrary three-dimensional hyperplane and in it a certain tetrahedron T. Next we take a point not in the given three-dimensional hyperplane and join it by segments to all the points of the tetrahedron T. On the right of figure 24 we have illustrated this construction symbolically. Each of the segments joining the point O to a point of the tetrahedron T has no other points in common with the tetrahedron, because otherwise it would lie entirely in the three-dimensional space containing T. All these segments, as it were, "go into the fourth dimension." They form the simplest four-dimensional polyhedron, the so-called *four-dimensional simplex*. Its three-dimensional faces are tetrahedra: one at the base and 4 lateral faces resting on the two-dimensional faces of the basis; altogether 5 faces. Its two-dimensional faces are triangles; there are 10 of them: 4 in the basis and 6 lateral. Finally, it has 10 edges and 5 vertices.

By repeating the same construction for an arbitrary number n of dimensions we obtain the simplest n-dimensional polyhedron, the so-called *n-dimensional simplex*. As is clear from the construction, it has $n + 1$ vertices. One can see that all its faces are also simplexes of a smaller number of dimensions: $(n - 1)$-dimensional, $(n - 2)$-dimensional, etc.*

It is also easy to generalize the concepts of a prism and a pyramid. If we give a parallel shift to a plane polygon into the third dimension, then it sweeps out a prism. Similarly, by shifting a three-dimensional polyhedron into the fourth dimension, we obtain a four-dimensional prism (illustrated symbolically in figure 25). A four-dimensional cube is, of course, a special case of a prism.

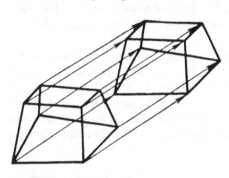

Fig. 25.

A pyramid is constructed as follows. We take a polygon Q and a point O not in the plane of the polygon. Each point of Q is joined

---

* Any m vertices of a simplex determine the $(m - 1)$-dimensional simplex "spanned" by them: an $(m - 1)$-dimensional face of the given n-dimensional simplex. The number of $(m - 1)$-dimensional faces of an n-dimensional simplex is therefore equal to the number of combinations of m of its vertices from $n + 1$, i.e.,

$$C_m^{n+1} = \frac{(n + 1)!}{m!\,(n - m + 1)!}.$$

§7. MANY-DIMENSIONAL SPACES

by a segment to O and these segments fill a pyramid with the base Q (figure 26). Similarly, if a three-dimensional polyhedron Q is given in a four-dimensional space and a point O not in the same three-dimensional plane, then the segments joining the points of Q to O form a four-dimensional pyramid with the base Q. A four-dimensional simplex is nothing but a pyramid with a tetrahedron as base.

In exactly the same way, by starting from an $(n - 1)$-dimensional polyhedron Q, we can define an n-dimensional prism and an n-dimensional pyramid.

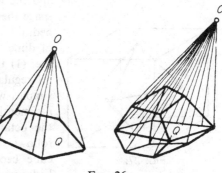

FIG. 26.

Generally, an n-dimensional polyhedron is a part of an n-dimensional space bounded by a finite number of portions of $(n - 1)$-dimensional hyperplanes; a k-dimensional polyhedron is a part of a k-dimensional hyperplane bounded by a finite number of portions of $(k - 1)$-dimensional hyperplanes. The faces of a polyhedron are themselves polyhedra of a smaller number of dimensions.

The theory of n-dimensional polyhedra is a generalization of the theory of ordinary three-dimensional polyhedra and is full of concrete results. In a number of cases theorems on three-dimensional polyhedra generalize to an arbitrary number of dimensions without special difficulties, but there are also problems whose solution for n-dimensional polyhedra runs into enormous difficulties. We might mention here the deep investigations of G. F. Voronoĭ (1868-1908) which arose, by the way, in connection with problems of the theory of numbers; they were continued by Soviet geometers. One of the problems, the so-called "Voronoĭ problem," is still not completely solved.*

As an example for the essential difference that can hold between spaces of different dimensions, we can take the regular polyhedra. In the plane a regular polyhedron can have an arbitrary number of sides. In other words, there are infinitely many distinct forms of regular "two-dimensional

---

* It concerns the search for those convex polyhedra by which the space can be filled by joining them one to another parallel and along whole faces. In the case of three-dimensional space this problem was raised and solved by Fedorov in connection with the needs of crystallography; Voronoĭ and his successors made some progress with the same problem for the n-dimensional space, but the final solution is known only for the spaces of two, three, and four dimensions.

polyhedra." There are altogether five forms of three-dimensional regular polyhedra: the tetrahedron, the cube, the octahedron, the dodecahedron, and the icosahedron. In four-dimensional space there are six forms of regular polyhedra, but in any space of a larger number of dimensions there are only three. They are: (1) the analogue to the tetrahedron, the regular n-dimensional simplex, i.e., the simplex whose edges are all equal; (2) the n-dimensional cube; (3) the analogue to the octahedron which is constructed as follows: The centers of the faces of the cube become the vertices of this polyhedron so that it is spanned by them, as it were. In the case of a three-dimensional space this construction is carried out in figure 27. So we see that as far as regular polyhedra are concerned, spaces of two, three and four dimensions occupy a special position.

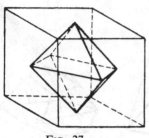

FIG. 27.

6. Calculation of volumes. Now let us discuss the problem of the volume of a body in n-dimensional space. The volume of an n-dimensional body is defined similarly to the way it is done in ordinary geometry. Volume is a numerical characteristic attached to a figure, and of the volume it is postulated that equal bodies have equal volumes, i.e., that the volume does not change when the figure moves as a rigid body, and that in case one body is composed of two others, its volume is equal to the sum of their volumes. As unit of volume one takes the cube with edges of unit length. It is then shown that the volume of a cube with edge a is a^n. This is done exactly as in the plane and in three-dimensional space, by filling up the cube with layers of cubes (figures 28). Since the cubes are packed in n directions, this gives us a^n.

In order to define the volume of an arbitrary n-dimensional body, we

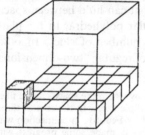

FIG. 28.

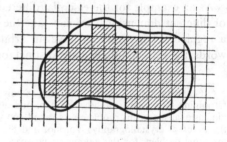

FIG. 29.

§7. MANY-DIMENSIONAL SPACES

replace it by an approximate body composed of very many n-dimensional cubes similarly to the way in which in figure 29 a plane figure is replaced by a figure of squares. The volume of the body is defined as the limit of such step-shaped bodies when the cubes that make it up decrease in size *ad infinitum*.

The k-dimensional volume of a k-dimensional figure lying in some k-dimensional hyperplane is defined in exactly the same way. From the definition of the volume it is easy to deduce an important property of it: Under a similarity magnification of a body, i.e., when all its linear dimensions are increased λ fold, the k-dimensional volume is increased λ^k fold.

If a body is divided into parallel layers, then its volume is the sum of the volumes of these layers

$$V = \Sigma V_i.$$

The volume of each layer can be represented approximately as the product of its height Δh_i with the $(n-1)$-dimensional volume ("area") of the corresponding section S_i. As a result, the total volume of the body is represented approximately by the sum

$$V \approx \Sigma S_i \Delta h_i.$$

Passing to the limit for all $\Delta h_i \to 0$ we obtain a representation of the volume in the form of an integral

$$V = \int_0^H S(h)\, dh, \qquad (9)$$

where H is the length of the body in the direction perpendicular to the section.

All this is completely analogous to the calculation of volumes in three-dimensional space. For example, for a prism all the sections are equal and, therefore, their "area" does not depend on h. Hence for the prism $V = SH$; i.e., the volume of a prism is equal to the product of the "area" of the base and the height. Let us determine the volume of an n-dimensional pyramid. Suppose that a pyramid is given with height H and area of the base S. We cut it by a plane parallel to the base at a distance h from the vertex. Then a pyramid of height h is cut off. We denote the area of its base by $s(h)$. This smaller pyramid is, obviously, similar to the original one: All its dimensions are smaller in the ratio of h to H, i.e., they are multiplied by h/H. Therefore the $(n-1)$-dimensional volume (i.e., the "area") of its base is

$$s(h) = \left(\frac{h}{H}\right)^{n-1} S,$$

because when the linear dimensions of an $(n-1)$-dimensional figure are changed by a factor λ, then the volume is multiplied by λ^{n-1}.

By formula (9) the volume of the whole pyramid is equal to

$$V = \int_0^H s(h)\, dh,$$

hence

$$V = \int_0^H S\left(\frac{h}{H}\right)^{n-1} dh = \frac{S}{H^{n-1}} \int_0^H h^{n-1}\, dh = \frac{S}{H^{n-1}} \cdot \frac{H^n}{n} = \frac{1}{n} SH;$$

i.e., the volume of an n-dimensional pyramid is equal to $1/n$th of the product of the "area" [$(n-1)$-dimensional volume] of the base and the height. For n equal to 2 or 3 we obtain as special cases the well-known results: The area (two-dimensional volume) of a triangle is equal to half the product of the base and the height, and the volume of a three-dimensional pyramid is one third of the product of the area of the base and the height.

A sphere can be represented approximately as composed of very narrow pyramids with a common vertex at the center of the sphere. The heights of these pyramids are equal to the radius R and the areas of their bases σ_i cover the whole surface S of the sphere approximately. Since the volume of each pyramid is equal to $1/n\, R\sigma_i$, we obtain by adding up these volumes that the volume of the sphere is

$$V \approx \frac{1}{n} R \sum \sigma_i \approx \frac{1}{n} RS.$$

In the limit this gives us the exact formula: $V = 1/n\, RS$; i.e., the volume of a sphere is equal to $1/n$th of the product of its radius and its surface. For n equal to 2 or 3 this relation is widely known.*

We mention one important property of a sphere which can be proved, generally speaking, for an n-dimensional space in exactly the same way as for a three-dimensional space: Among all bodies of a given volume the sphere alone has the least surface area.

7. "Higher" n-dimensional geometry. So far we have confined ourselves to the elementary geometry of the n-dimensional space; but we can also develop the "higher" geometry, for example the general theory

---

* The calculation of the volume of a sphere can also be effected by applying formula (9); a section of an n-dimensional sphere is an $(n-1)$-dimensional sphere, and therefore the volume of an n-dimensional sphere can be calculated by going from $n-1$ to n.

§8. GENERALIZATION OF THE SCOPE OF GEOMETRY

of curves and surfaces. In n-dimensional space surfaces may have various numbers of dimensions: one-dimensional "surfaces," i.e., curves, two-dimensional surfaces, three-dimensional, $\cdots$, and finally $(n-1)$-dimensional surfaces. A curve can be defined as the geometric locus of points whose coordinates depend continuously on a variable or parameter t

$$x_1 = x_1(t), \quad x_2 = x_2(t), \cdots, \quad x_n = x_n(t).$$

A curve is so to speak the trajectory of the motion of a point in n-dimensional space with varying t. If our space serves as an illustration of the state of some physical system such as we have discussed in subsection 4, then a curve illustrates a continuous sequence of states or the course of the change of state in dependence on the parameter t (for example the time). This generalizes the usual graphical illustration of the process of change of state by means of curves.

With every point of a curve in n-dimensional space, we connect not only a tangent ("one-dimensional tangential plane") but also tangent hyperplanes of all dimensions from 2 to $n-1$. The rate of rotation of each of these hyperplanes with respect to the rate of increase of the arc length of the curve gives the corresponding curvature. Thus, a curve has $n-1$ tangent hyperplanes, from the one-dimensional to the $(n-1)$-dimensional, and accordingly $n-1$ curvatures. The differential geometry in an n-dimensional space turns out to be far more complicated than in three-dimensional space.

So far we have only talked of the n-dimensional geometry which is an immediate generalization of the ordinary Euclidean geometry. But we know already that, apart from Euclidean geometry, there exist also Lobačevskiĭ geometry, projective geometry and others. These geometries are just as easily generalized to an arbitrary number of dimensions.

§8. Generalization of the Scope of Geometry

1. The space of colors. When we spoke in the preceding section of the real meaning of n-dimensional space, we came up against the problem of generalizing the scope of geometry, the problem of the general concept of space in mathematics. But before giving the corresponding general definition, let us consider a number of examples.

Experience shows, and this was already mentioned by M. V. Lomonosov, that the normal human vision is three-colored, i.e., every chromatic perception, of a color C, is a combination of three fundamental perceptions: red R, green G and blue B, with specific intensities.* When we

---

* We are concerned here with the perception of color, not of light. Perception of color is also an objective phenomenon—a reaction to light. One and the same per-

denote these intensities in certain units by x, y, z, we can write down that $C = xR + yG + zB$. Just as a point can be shifted in space up and down, right and left, back and forth, so a perception of color, of a color C, can be changed continuously in three directions by changing its constituent parts red, green, and blue. By analogy we can say, therefore, that the set of all possible colors is the "three-dimensional color space." The intensities x, y, z play the role of coordinates of a point, of a color C. (An important difference from the ordinary coordinates consists in the fact that the intensities cannot be negative. When $x = y = z = 0$, we obtain a perfectly black color corresponding to complete absence of light.)

A continuous change of color can be represented as a line in the "color space"; the colors of the rainbow, for example, from such a line; a color line is also formed by a number of perceptions produced on an object of homogeneous coloration by a continuous change of the brightness of illumination. In this case only the intensity of the perception changes, its "coloredness" remains unchanged.

Further, when two colors are given, say red R and white W, then by mixing them in varying proportions* we obtain a continuous sequence of colors from R to W which we can call the segment RW. The conception that a rose color lies between red and white has a clear meaning.

In this way there arises the concept of the simplest geometric figures and relations in the "color space." A "point" is a color, the "segment" AB is the set obtained by mixing the colors A and B; the statement that "the point D lies on the segment AB" means that D is a mixture of A and B. The mixture of three colors gives a piece of a plane—a "color triangle." All this can also be described analytically by using the color coordinates x, y, z, and the formulas giving color lines and planes are entirely analogous to the formulas of ordinary analytic geometry.[†]

In the color space the relations of Euclidean geometry concerning the disposition of points and segments are satisfied. The system of these relations forms an affine geometry, and we can say that the set of all

---

ception may be produced by different light waves. For example, a green color may be obtained not only from spectrally pure green light, but also from a mixture of red and blue. On the other hand, people suffering from "color blindness" (Daltonism) have only two fundamental perceptions; cases of "complete color blindness," when there is only one fundamental perception of color, are extremely rare.

* Such a mixture can be obtained by mixing in varying proportions very fine colored powders provided the illumination remains unchanged.

† For example, if the colors C_0 and C_1 are determined by the intensities, namely the coordinates x_0, y_0, z_0 and x_1, y_1, z_1, then a color C between C_0 and C_1 has the coordinates $x = (1 - t)x_0 + tx_1$, $y = (1 - t)y_0 + ty_1$, $z = (1 - t)z_0 + tz_1$, where t is the portion of C_1 and $1 - t$ the portion of C_0 in the mixture that makes up C.

§8. GENERALIZATION OF THE SCOPE OF GEOMETRY

possible color perceptions realizes an affine geometry. (This is not quite accurate, because, as we have already said, the color coordinates x, y, z cannot be negative. Therefore the color space corresponds only to that part of the space where in the given coordinate system all the coordinates of points are positive or zero.)

Further, we have a natural idea of the degree of distinctness of colors. For example, it is clear that pale pink is nearer to white than deep pink, and crimson nearer to red than to blue, etc. Thus, we have a qualitative concept of distance between colors as the degree of their distinctness. This qualitative concept can be made into a quantitative measure. However, to define the distance between colors as in Euclidean geometry by the formula $r = \sqrt{(x_0 - x_1)^2 + (y_0 - y_1)^2 + (z_0 - z_1)^2}$ turns out to be unnatural. A distance so defined does not correspond to real perception; with this definition it would happen in a number of cases that two colors that differ from a given one in varying degree would have one and the same distance from it. The definition of distance must reflect the real relations between color perceptions.

Guided by this principle we introduce a peculiar measure of distance in the space of colors. This is done as follows. When a color is altered continuously, a human being does not perceive this change at once, but only when it reaches a certain extent exceeding the so-called threshold of distinction. In this connection it is assumed that all colors that are exactly on the threshold of distinction from a given one are equidistant from it. We are then led automatically to the idea that the distance between any two colors must be measured by the smallest number of thresholds of distinction that can be laid between them. The length of a color line is measured by the number of such thresholds covering it. The distance between two colors is defined as the length of the shortest line joining them. This is similar to the fact that distances between two points in a plane are measured by the length of the shortest line joining them.

Thus, measurement of length and distance in the color space proceeds in very small, as it were infinitely small, steps.

As a result, a certain peculiar non-Euclidean geometry is defined in the color space. This geometry has a perfectly real meaning: It describes in geometrical language properties of the set of all possible colors, i.e., properties of the reaction of the eye to a light stimulus.

The concept of color space arose about a century ago. Many physicists have studied the geometry of this space; for example, we may mention Helmholtz and Maxwell. These investigations continue; they have not only theoretical but also practical value. They give an accurate mathematical foundation for the solution of problems on the difference of color signals, on dyes in the textile industry, and others.

2. Phase spaces in physics and chemistry.

Now let us turn to another example already mentioned in the preceding section.

Suppose that we study some physicochemical system such as a mixture of gases, an alloy, etc. Suppose that the state of this system is determined by n values (as the state of a gaseous mixture is determined by pressure, temperature, and the concentrations of its constituent components). Then one says that the system has n degrees of freedom, meaning that its state can be changed in n independent directions under a change of each of the values that determine this state. These values that determine the state of the system play as it were the role of its coordinates. Therefore the set of all its states can be regarded as an n-dimensional space, the so-called phase space of the system.

Continuous changes of state, i.e., processes occurring in the system, are presented by lines in this space. Separate domains of states, distinguished by one feature or another, are domains of the phase space. The states bordering two such domains form a surface in this space.

In physical chemistry it is particularly important to study the form and the mutual contiguity of those domains of the phase space of a system that correspond to qualitatively distinct states. The surfaces dividing these domains correspond to such qualitative transitions as melting, evaporation, precipitation of a sediment, etc. A state of a system with two degrees of freedom is illustrated by a point in a plane. As an example we can take a homogeneous substance whose state is determined by the pressure p and temperature T; they are the coordinate points describing the state. Then the question reduces to studying the lines of division between domains corresponding to qualitatively distinct states. In the case of water, for example, these domains are ice, liquid water, and steam (figure 30). Their division lines correspond to melting (freezing), evaporation (condensation), sublimation of ice (precipitation of ice crystals from steam).

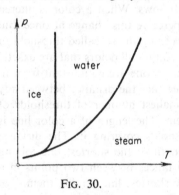

FIG. 30.

For an investigation of systems with many degrees of freedom, the methods of many-dimensional geometry are required.

The concept of phase space applies not only to physicochemical but also to mechanical systems, and generally it can be applied to any system if its possible states form a certain continuous collection. In the kinetic theory of gases one considers, for example, the phase spaces of systems

§8. GENERALIZATION OF THE SCOPE OF GEOMETRY 155

of material particles, the molecules of the gas. The state of motion of one particle at each moment is determined by its position and velocity, which gives altogether six values: three coordinates and three velocity components (with respect to the three coordinate axes). The state of N particles is given by $6N$ values and since there are very many molecules, $6N$ is an enormous number. This does not disturb the physicists in the least, who speak of a $6N$-dimensional phase space of a system of molecules.

A point in this space describes the state of the whole mass of molecules with coordinates and velocities. A motion of a point describes a change of state. This abstract presentation turns out to be very useful in many deep theoretical developments. In a word, the concept of phase space has a secure place in the arsenal of the exact natural sciences and is applicable in diverse problems.

3. The generalization of geometry. The examples we have given enable us to reach a conclusion on how the scope of geometry is to be generalized.

Suppose that we wish to study some continuous collection of objects, events, or states of one kind or another, for example the set of all possible colors or of the states of a group of molecules. The relations holding in such a collection may happen to be similar to the ordinary spatial relations; for example "distance" between colors or the "mutual position" of domains of a phase space. In that case, by abstracting from the qualitative peculiarities of the objects in question and by taking into account only the aforementioned relations between them, we can regard the given collection as a space of its own kind. The "points" of this "space" are the objects, events or states themselves. A "figure" in this space is an arbitrary aggregate of its points, as for example the "line" of rainbow colors or the "domain" of steam in the "space" of states of water. The "geometry" of such a space is determined by those spacelike relations that hold between the given objects, phenomena or states. Thus, the "geometry" of the color space is determined by the laws of color mixing and the distances between colors.

The real significance of this point of view is that it makes it possible to use the concepts and methods of abstract geometry for the investigation of diverse phenomena. The realm of applicability of geometric concepts and methods is extended immensely in this way. As a result of the generalization of the concept of space the term "space" assumes two meanings in science: On the one hand it is the ordinary real space (the universal form of existence of matter), on the other hand it is the "abstract space," a collection of homogeneous objects (events, states, etc.) in which spacelike relationships hold.

It is worth noting that the ordinary space as we visualize it in a somewhat simplified manner can also be regarded as a collection of homogeneous states. Namely, as the collection of all possible positions of an infinitely small body, a "material point." This remark does not pretend to give a definition of space but aims at making the connection between the two concepts of space clearer. The concept of an abstract space will be further expounded in the next subsection, and the relation of abstract geometry to the ordinary real space will be treated in the last section of this chapter.

4. Generalized spaces in mathematics. The widest application of the concept of an abstract space occurs in mathematics itself. In geometry one considers the "space" of certain figures as, for example, the "space of spheres" of which we have spoken, "the space of straight lines," and so forth.

This method turns out to be particularly fruitful in the theory of polyhedra. For example, in §5 of Chapter VII a theorem was mentioned on the existence of a convex polyhedron with a given development. The proof of this theorem is based on the discussion of two "spaces": the "space of polyhedra" and the "space of developments." The set of convex polyhedra having a given number of vertices is regarded as a space of its own whose points represent polyhedra; similarly, the set of admissible developments is also treated as a certain space whose points represent developments. The process of fitting together polyhedra from developments establishes a correspondence between polyhedra and developments, i.e., a correspondence between the points of the "space of polyhedra" and the "space of developments." The problem consists in showing that to every development there corresponds a polyhedron; i.e., to every point of one space there corresponds a point of the other. And precisely this can be proved by means of an application of topology.

A whole series of other theorems on polyhedra can be proved similarly, and this "method of abstract spaces" turns out in a number of cases (like the theorem on the existence of a polyhedron with a given development) to be the simplest of the known methods of proving such theorems. Unfortunately, however, the method itself is still rather complicated, and we cannot give a more accurate account of it here.

Extensive applications of the generalized concept of space also occur in analysis, algebra, and number theory. This stems from the usual representation of functions by means of curves. The values of one variable x are usually plotted as points on a line. Similarly, the values of two variables are plotted as points in a plane, the values of n variables as points in an n-dimensional space; we represent the set of values of the

§8. GENERALIZATION OF THE SCOPE OF GEOMETRY 157

variables $x_1, x_2, \cdots, x_n$ by the point with the coordinates $x_1, x_2, \cdots, x_n$. We speak of the "domain of variation of the variables" or of the "domain of values" of a function $f(x_1, x_2, \cdots, x_n)$ of these variables; we speak of points, lines, or surfaces of discontinuity of the function, etc. This geometric language is in constant use and is not only a mode of expression; the geometric representation makes many facts of analysis "intuitive" by analogy with the ordinary space and permits the use of geometric methods of proof, generalized to n-dimensional space.

The same takes place in algebra, when equations with n unknown or algebraic functions of n variables are under discussion. In the preceding section it was mentioned that a linear equation with n unknowns determines a hyperplane in n-dimensional space, that m such equations determine m hyperplanes and every solution of them represents a point that is common to all these hyperplanes. The hyperplanes need not intersect at all, or intersect in a single point or in a whole straight line in a two-dimensional or, generally, a k-dimensional hyperplane. All in all, the problem of solubility of systems of linear equations is expressed as a problem on the intersection of hyperplanes. This geometric approach has a number of advantages. Quite generally, "linear algebra," which comprises the study of linear equations and linear transformations is usually set forth to a large extent in geometrical form, as it was done in Chapter XVI.

5. Infinite-dimensional space; definition of a "space." In all our examples we were concerned with a continuous collection of objects of one sort or another being treated as a space of its own particular kind. These objects were colors, states of one system or another, figures, or aggregates of values of variables. In all cases one object was given by a finite number of data so that the corresponding space had a finite number of dimensions, namely the number of these data.

However, at the beginning of the present century mathematicians began to discuss also "infinite-dimensional spaces," namely collections of objects each of which cannot be given by a finite number of data. This is so, above all, in a "functional space."

The idea of treating the collection of functions of one type or another as a space of its own is one of the basic ideas of a new branch of analysis, namely functional analysis, and turns out to be extremely fruitful in the solution of many problems. The reader will find an account of it in Chapter XIX, which is devoted specially to functional analysis.

One can discuss the spaces of continuous functions of one or several variables. One also regards as "spaces" various classes of discontinuous functions, for which one defines a distance between functions by one method or another, depending on the character of the problems awaiting solution. In a word, the number of possible "function spaces" is unlimited, and in fact many such spaces are used in mathematics.

In just the same way one can discuss the "space of curves," the "space of convex bodies," the "space of possible motions of a mechanical system," etc. For example, in §5 of Chapter VII the theorem was mentioned that on every closed convex surface there exist at least three closed geodesics and that any two points can be joined by an infinite number of geodesics. For the proofs of these theorems one uses the space of curves on the surface: in the first the space of closed curves, in the second the space of curves joining two given points. We introduce in the set of all possible curves joining two given points a kind of distance and so turn this set into a space. The proof of the theorem is based on an application of certain deep results of topology to this space.

Let us now formulate a general conclusion.

By a "space" we understand in mathematics quite generally an arbitrary collection of homogeneous objects (events, states, functions, figures, values of variables, etc.) between which there are relationships similar to the usual spatial relations (continuity, distance, etc.). Moreover, in regarding a given collection of objects as a space we abstract from all properties of these objects except those that are determined by these spacelike relationships in question. These relations determine what we can call the structure or the "geometry" of the space. The objects themselves play the role of "points" of such a space; "figures" are sets of its "points."

The scope of the geometry of a given abstract space consists in those properties of the space and the figures in it that are determined by the spacelike relationships taken into account. For example, in discussing the space of continuous functions the properties of an individual function on its own are completely ignored. The function here plays the role of a point and consequently "has no parts," has no structure at all in this sense, no properties unconnected with other points; more accurately, all this is neglected. In a function space, properties of functions are determined only by their relations to one another, by their distance and by other relations that can be derived from distance.

To the variety of possible sets of objects and diverse relations between them, there corresponds the unlimited variety of spaces studied in mathematics. Spaces can be classified with respect to the types of those spacelike relations that underlie their definition. Without aiming at a full account of all the various types of abstract spaces, let us mention, above all, two very important types: topological and metric spaces.

§8. GENERALIZATION OF THE SCOPE OF GEOMETRY

6. Topological spaces. A *topological space* (see Chapter XVIII) is any collection of points (an arbitrary set of elements), in which a relation of neighborhood of one point to a set of points is defined and, consequently, a relation of neighborhood or adherence of two sets (figures) to one another. This is a generalization of the intuitive intelligible relation of neighborhood or adherence of figures in the ordinary space.

Already Lobačevskiĭ with remarkable foresight pointed out that of all the relations of figures, the most fundamental is the relation of neighborhood. "Neighborhood forms a distinctive appurtenance of bodies and gives them the same *geometric* when we retain in them this property and do not take into consideration all others whether they be essential or accidental."* For example, every point on the circumference is adherent to the set of all interior points of a circle; two parts of a connected body are adherent to one another. As the subsequent development of topology has shown, it is precisely the property of neighborhood that underlies all other topological properties.

The concept of neighborhood expresses the notion of a point being infinitely near to a set. Therefore every collection of objects in which there is a natural concept of continuity, of being infinitely near, turns out to be a topological space.

The concept of a topological space is extremely general and the study of such spaces, abstract topology, represents the most general mathematical study of continuity.

A rigorous mathematical definition of a general topological space can be given in the following way.

A set R of arbitrary elements "points" is called a general topological space if for every set M contained in it neighborhood points are defined such that the following conditions, namely the axioms of the space, are satisfied:

1. Every point of M is counted among its neighborhood points. (It is perfectly natural to assume that a set is adherent to each of its points.)

2. If a set M_1 contains a set M_2, then the neighborhood points of M_1 must contain all the neighborhood points of M_2. (To put it briefly, but less accurately, the larger set does not have fewer neighborhood points.)

Usually other axioms are added to these, so that various types of topological spaces are thereby defined.

With the help of the concept of neighborhood, it is easy to define a number of very important topological concepts. These are at the same time the most fundamental and general concepts of geometry and their definitions are intuitively altogether clear. Let us give some examples.

---
* N. I. Lobačevskiĭ, "Collected works," Vol. II, Gostehizdat, 1949, page 168.

1. *Adherent*. We shall say that sets M_1 and M_2 are adherent to one another if one of them contains at least one neighborhood point of the other. (In this sense, for example, the circumference of a circle is adherent to the interior.)

2. Continuity or, as the mathematicians say, *connectedness* of a figure. A figure, i.e., a set of points M, is connected if it cannot be split into parts that are not adherent to one another. (For example, a segment is connected, but a segment without its midpoint is disconnected.)

3. *Boundary*. The boundary of a set M in a space R is the set of all points that are adherent both to M and to its complement $R - M$, i.e., to the remaining part of the space R. (This is, obviously, a perfectly natural concept of boundary.)

4. *Interior point*. A point of a set M is called interior if it does not lie on its boundary, i.e., if it is not adherent to the complement $R - M$.

5. *Continuous mapping or transformation*. A transformation of a set M is called continuous if it does not disrupt neighborhoods. (One could hardly give a more natural definition of a continuous transformation.)

Other important definitions could be added to this list, such as for example a definition of the concept of convergence of a sequence of figures to a given figure or the concept of the number of dimensions of a space.

We see that the most fundamental geometric concepts can be defined in terms of neighborhoods. The significance of topology, in particular, lies in the fact that it gives rigorous general definitions for these concepts, thereby providing a basis for the strict application of arguments connected with the intuitive conception of continuity.

Topology is the study of those properties of spaces, of figures in them, and of their transformations that are defined by the relation of neighborhood.

The generality and fundamental nature of this relation makes topology into a very general geometrical theory that penetrates the diverse branches of mathematics, wherever continuity only is under discussion. But precisely because of its generality topology in its most abstract parts goes beyond the framework of geometry proper. All the same, at its basis lies a generalization of the properties of real space and the most fruitful and powerful of its results are connected with the application of methods that spring from intuitive geometric ideas. An example is the method of approximating general figures by polyhedra which was developed by P. S. Aleksandrov and was extended by him, though in an abstract form, to extremely general types of topological spaces.

Nowadays every specialist, no matter what his subject of study,

§8. GENERALIZATION OF THE SCOPE OF GEOMETRY

investigates when he discovers that there is a natural way of introducing into it the concept of nearness or adherence, and immediately has at his disposal the ready-made, widely ramified apparatus of topology, which enables him to draw conclusions that are far from trivial even in their application to his special field.

7. Metric spaces. A *metric space* is a set of arbitrary elements, to be called points, between which a distance is defined; i.e., with each pair of points X, Y a number $r(X, Y)$ is associated so that the following conditions, namely the axioms of a metric space, are satisfied:

1. $r(X, Y) = 0$ if and only if the points X, Y coincide.
2. For any three points X, Y, Z

$$r(X, Y) + r(Y, Z) \geqslant r(Z, X).$$

This condition is called the "triangle inequality," since it is quite analogous to the well-known property of the ordinary distance between points A_1, A_2, A_3 of Euclidean space (figure 31):

$$r(A_1, A_2) + r(A_2, A_3) \geqslant r(A_3, A_1).$$

As examples of metric spaces we may take
 1. the Euclidean space of an arbitrary number n of dimensions,
 2. the Lobačevskiĭ space,
 3. any surface in its intrinsic metric (Chapter VII, §4),
 4. the space C of continuous functions with distance defined by the formula $r(f_1, f_2) = \max |f_1(x) - f_2(x)|$,
 5. the Hilbert space to be described in Chapter XIX, which is an "infinite-dimensional Euclidean" space.

The Hilbert space is the most important of the spaces used in functional analysis; it is closely connected with the theory of Fourier series and more generally with the expansion of functions in series by orthogonal functions (the coordinates $x_1, x_2, x_3, \cdots$ are then the coefficients of such series). This space also plays an important role in mathematical physics and has acquired much significance in quantum mechanics. It turns out that the set of all possible states (not only stationary) of an atomic system, for example a hydrogen atom, can be regarded from an abstract point of view as a Hilbert space.

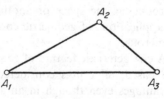

Fig. 31.

The number of examples of metric spaces that are actually discussed in mathematics could be increased considerably; in the next section we shall become acquainted with one important class of metric spaces, the so-called Riemannian spaces, but the examples we have given so far will be sufficient to show how widely the general concept of a metric space extends.

In a metric space it is always possible to define all topological concepts and, moreover, to introduce other "metric" concepts; for example, the concept of the length of a curve. Length is defined in any metric space precisely in the same way as usual, and the basic properties of length are then preserved. Indeed, by the length of a curve we understand the limit of the sum of the distances between points $X_1, X_2, \cdots, X_n$ that are arranged in a sequence on the curve, subject to the condition that the points come to lie closer and closer on the curve.

8. Advantages of the geometric method. Many types of spaces besides the general topological or the metric spaces are discussed in mathematics. In fact, we have already become acquainted in §6 with a whole class of such spaces. These are the spaces in which some group of transformations is given (for example, the projective or affine spaces). In such spaces we can define "equality" of figures. Figures are "equal" if they can be carried into one another by a transformation of the given group.

We shall not go deeper into the definitions of possible types of spaces; they are rather diverse and the reader can turn to the special literature on the various branches of present-day geometry.

But what is the sense of extending the range of geometric concepts so much? For what purpose, for example, does one have to introduce the concept of the space of continuous functions? Is it not sufficient to solve problems of analysis by the usual means without resorting to such abstract spaces?

The general answer to these questions consists briefly in this: that by introducing one space or another into the discussion we open the way to applications of geometric concepts and methods, which are extremely numerous.

A characteristic feature of geometric concepts and methods is that they are based, all things considered, on intuitive ideas and preserve their advantages even though in an abstract form. What the analyst achieves by long calculations, the geometer can occasionally grasp at once. An elementary example of this can be seen in a graph that gives a completely clear picture of one dependence or another between quantities. The geometrical method can be characterized as an all-embracing synthetic method, in contrast to the analytical method. Of course, in abstract

§8. GENERALIZATION OF THE SCOPE OF GEOMETRY

geometrical theories the immediate intuitiveness fades away, but the intuitive arguments by analogy remain and so does the synthetic character of the geometrical method.

The reader is already familiar with the application of geometric pictures in analysis, with a geometric representation of complex numbers and functions of a complex variable, with geometric arguments in a proof of the fundamental theorem of algebra, and with other applications of geometric concepts and methods. Everywhere he may observe what we have described here in general terms. We recall the examples, in the beginning of §7, and also here §8, under subsection 5, of theorems that can be proved by applying many-dimensional geometry. Let us give one further example of a problem in analysis that can be solved by an application of the concept of a function space.

It is proved in topology that if we take in the ordinary plane any domain that has the form of a distorted circle and then deform it in a continuous manner as we please, but so that in the end it becomes embedded inside its original contour, then at least one point of the domain comes to lie after the transformation where it was before. This is a purely topological fact.

Now let us consider a problem altogether remote from geometry: A function $y(x)$ is to be found that satisfies the differential equation,*

$$y' = f(x, y) \tag{10}$$

and assumes for $x = 0$ the value $y = 0$.

Obviously, instead of this equation we can look for a solution of the equation

$$y = \int_0^x f[t, y(t)]\, dt. \tag{11}$$

Naturally the problem arises: Does there exist, in general, a function $y(x)$ satisfying this condition?

Let us look at the problem in another way. We represent every continuous function $y(x)$ by a point of some abstract space. The result of computing the integral

$$\int_0^x f[t, y(t)]\, dt = z(x)$$

is again a continuous function of x, i.e., a "point" of our abstract space. By taking various "points" y, i.e., various functions $y(x)$, we obtain, generally speaking, various points z. In this way, the set of points of

---

* Here, $f(x, y)$ is assumed to be a continuous function of the variables x and y.

our space is again mapped into points of the same space. The problem of finding a solution of the equation (11) has been reduced to the question: Can a "point" of our space be found such that after this transformation it coincides with its previous "place"?

A natural problem in the theory of differential equations has become a problem concerning a property of an abstract function space. The analogy with the aforementioned theorem tells us that we are evidently dealing here with a topological property of the corresponding space.

In this way we obtain, by means of the requisite topological investigations, what are probably the shortest proofs of many theorems on the existence of solutions of differential equations; in particular we can make it clear that the equation (10) does in fact have a solution for every continuous function $f(x, y)$.

§9. Riemannian Geometry

1. History of Riemannian geometry. The ideas explained previously that every continuous collection of homogeneous phenomena can be treated as a space of its own was first expressed by Riemann in his lecture "On the hypotheses that underlie geometry," given at the University of Göttingen in 1854. This was a sort of test lecture, somewhat like a report or a dissertation that a lecturer or professor had to make to the faculty before taking up his post. In his lecture Riemann set out in general lines, without calculations or mathematical proofs, the original idea of a vast geometric theory that is now called Riemannian geometry. It is said that nobody in the audience understood it except the aged Gauss. Riemann provided the formal apparatus of his theory in another paper, with an application to the problem of heat conduction, so that the abstract Riemannian geometry was born in close connection with mathematical physics. In the development of geometry, Riemann's ideas came next after Lobačevskiĭ's decisive step. However, Riemann's papers were not at once duly appreciated. His lectures and papers on heat conduction were published only posthumously in 1868. It is worth while mentioning that in 1868 the first interpretation of Lobačevskiĭ's geometry also appeared, by Beltrami, and in 1870 the second one, by Klein. In 1872 Klein expounded his general view of the various geometries: Euclidean, Lobačevskiĭ, projective, affine, etc., as the study of properties of figures that remain unchanged under the transformations of one group or another. In the same year many-dimensional geometry was finally consolidated in mathematics. Thus, the seventies of the 19th century were that critical period in the history of geometry when the new geometrical ideas, accumulated in the course of the preceding fifty years, were finally

§9. RIEMANNIAN GEOMETRY

understood by a wide circle of mathematicians and assumed a secure place in the science.

Riemann's work was then continued and at the end of the 19th century Riemannian geometry had reached a considerable development and had found applications in mechanics and physics. When, in 1915, Einstein in his general theory of relativity applied Riemannian geometry to the theory of universal gravitation, this event drew particular attention to Riemannian geometry and resulted in its brisk development and in various generalizations.

2. The basic ideas of Riemannian geometry. Riemann's ideas, which had such a brilliant success, are really rather simple if one sets aside the mathematical details and concentrates on the basic essentials. Such an intrinsic simplicity is a feature of all great ideas. Was not Lobačevskiĭ's idea simple: to regard the consequences of the negation of the Fifth Postulate as a possible geometry? Was not the idea of evolution of organisms simple, or the idea of the atomic structure of matter? All of these are simple and at the same time very complicated, because new ideas must, first of all, work their way over a wide field and must not be pressed into a rigid framework, and second, their foundation, development, and application is a many-sided task, requiring an immense amount of labor and ingenuity, and impossible without the specialized apparatus of science. For Riemannian geometry this apparatus consists in its formulas; they are complicated and therefore accessible to a specialist only. But we shall not deal with complicated formulas and turn now to the essence of Riemann's ideas. As we have already said, Riemann began by considering an arbitrary continuous collection of phenomena as a space of its own. In this space the coordinates of points are quantities that determine the corresponding phenomenon among others, as for example the intensities x, y, z that determine the color $C = xR + yG + zB$. If there are n such values, say $x_1, x_2, \cdots, x_n$, then we speak of an n-dimensional space. In this space we may consider lines and introduce a measurement of their length in small (infinitely small) steps, similar to the measurement of the length of a curve in ordinary space.

In order to measure lengths in infinitely small steps, it is sufficient to give a rule that determines the distance of any given point from another infinitely near to it. This rule of determining (measuring) distance is called a *metric*. The simplest case is when this rule happens to be the same as in Euclidean space. Such a space is Euclidean in the infinitely small. In other words, the geometrical relations of Euclidean geometry are satisfied in it, but only in infinitely small domains; it is more accurate to say that they are satisfied in any sufficiently small domain, though not exactly, but with an accuracy that is the greater, the smaller the domain. A space in which distance is measured by such a rule is called *Riemannian*; and the geometry of such spaces is also called *Riemannian*. A Riemannian space is, therefore, a space that is Euclidean "in the infinitely small."

The simplest example of a Riemannian space is an arbitrary smooth surface in its intrinsic geometry. The intrinsic geometry of a surface is a Riemannian geometry of two dimensions. For in the neighborhood of each of its points a smooth surface differs only a little from its tangent plane, and this difference is the smaller, the smaller the domain of the surface that we consider. Therefore the geometry in a small domain of the surface also differs little from the geometry in a plane; the smaller the domain, the smaller this difference. However, in large domains the geometry of a curved surface turns out to be different from the Euclidean, as was explained in §4 of Chapter VII and is easy to see in the examples of the sphere or pseudosphere. Riemannian geometry is nothing but a natural generalization of the intrinsic geometry of a surface with two dimensions to an arbitrary number n. Like a surface, considered only from the point of view of its intrinsic geometry, a three-dimensional Riemannian space, although Euclidean in small domains, may differ from the Euclidean in large domains. For example, the length of a circle may not be proportional to the radius; it will be proportional to the radius with a good approximation for small circumferences only. The sum of the angles of a triangle may not be two right angles; here the role of rectilinear segments in the construction of a triangle is played by the lines of shortest distance, i.e., the lines having the smallest length among all the lines joining the given points.

One can speculate that the real space is Euclidean only in domains that are small in comparison with the astronomical scale. The smaller a domain is, the more accurately Euclidean geometry holds, but we can imagine (and, in fact, it turns out to be so) that on a very large scale the geometry differs somewhat from the Euclidean. This idea, as we know, was already put forward by Lobačevskiĭ. Riemann generalized it so that it applied to an arbitrary geometry and not only to Lobačevskiĭ geometry, which now appears as a special case of Riemannian geometry.

From what we have said it is clear that Riemannian geometry has grown by a synthesis and generalization of three ideas that contributed to the successful development of geometry. First came the idea of the possibility of a geometry other than the Euclidean, second was the concept of the intrinsic geometry of a surface, and third the concept of a space of an arbitrary number of dimensions.

3. Measurement of distance. In order to make it clear how a

§9. RIEMANNIAN GEOMETRY

Riemannian space is defined mathematically, we recall first of all the rule for measuring distances in a Euclidean space.

If rectangular coordinates x, y are introduced in a plane, then by Pythagoras' theorem the distance between two points whose coordinates differ by Δx and Δy is expressed by the formula

$$s = \sqrt{\Delta x^2 + \Delta y^2}.$$

Similarly in a three-dimensional space

$$s = \sqrt{\Delta x^2 + \Delta y^2 + \Delta z^2}.$$

In a n-dimensional Euclidean space the distance is defined by the general formula

$$s = \sqrt{\Delta x_1^2 + \cdots + \Delta x_n^2}.$$

Hence it is easy to conclude how the rule for measuring distance in a Riemannian space ought to be given. The rule must coincide with the Euclidean, but only for an infinitely small domain in the neighborhood of each point. This leads to the following statement of the rule.

A Riemannian n-dimensional space is characterized by the fact that in the neighborhood of each of its points A coordinates x_1, x_2, $\cdots$, x_n can be introduced such that the distance from A of an infinitely near point X is expressed by the formula

$$ds = \sqrt{dx_1^2 + \cdots + dx_n^2}, \tag{12}$$

where dx_1, $\cdots$, dx_n are the infinitely small differences of the coordinates of A and X. This can also be expressed more accurately in another way: The distance from A to an arbitrarily near point X is expressed by the same formula as in Euclidean geometry, but only with a certain accuracy which is the greater the nearer the point X is to A, i.e.,

$$s(AX) = \sqrt{\Delta x_1^2 + \cdots + \Delta x_n^2} + \epsilon,$$

where ϵ is a small quantity in comparison with the first term and is smaller, the smaller the coordinate differences Δx_1, $\cdots$, Δx_n are.*

---

* Usually the precise meaning of the formula (12) is expressed as follows. Suppose that a curve starts out from A so that the coordinates of the point X on it are given as functions $x_1(t), x_2(t), \cdots, x_n(t)$ of some variable t. Then the differential ds of the arc length of this curve at A is expressed by (12).

Now this is the exact mathematical definition of a Riemannian metric and a Riemannian space. The difference of Riemannian metric, i.e., the rule for measuring distances, from Euclidean consists in that this rule holds only in the neighborhood of each given point. Moreover, the coordinates in which it is expressed so simply have to be taken differently for different points.* The difference between the general Riemannian metric and the Euclidean will be further specified later on.

The fact that a Riemannian space coincides with a Euclidean in the infinitely small enables us to define in it the fundamental geometric quantities similarly to the way this was done for the intrinsic geometry of a surface by approximating an infinitely small portion of the surface by a plane (Chapter VII, §4). For example, an infinitely small volume is expressed just as in Euclidean space. The volume of a finite domain is obtained by summing infinitely small volumes, i.e., by integrating the differential of the volume. The length of a curve is determined by summing infinitely small distances between infinitely near points on it, i.e., by integrating the differential of the length ds along the curve. And this is a rigorous analytic expression for the fact that the length is determined by laying off a small (infinitely small) measuring rod along the curve. The angle between curves at a common point is defined exactly as in a space. Further, in an n-dimensional Riemannian space we can define surfaces of various numbers of dimensions from 2 to $n-1$. Moreover, it is easy to prove that each such surface in its turn represents a Riemannian space of the corresponding number of dimensions, just as a surface in the ordinary Euclidean space turns out to be a two-dimensional Riemannian space.

It has also been proved that a Riemannian space can always be represented as a surface in a Euclidean space of a sufficiently large number of dimensions, namely: for every n-dimensional Riemannian space one can find in an $n(n+1)/2$-dimensional Euclidean space an n-dimensional surface which from the point of view of its intrinsic geometry does not differ from this Riemannian space (at least in a given limited part of it).

4. The fundamental quadratic form. In order to obtain the actual analytical expression for various geometric quantities in a Riemannian geometry, we have to define, first of all, a general expression for the rule of measuring lengths in a Riemannian space independent of the specific coordinates at each point. True, the formula (12) holds at every point A for a special choice of coordinates at that point, so that on transition

---

* If in the whole space coordinates could be introduced so that for any pair of neighboring points this rule for measuring distance would hold, then the space would be Euclidean.

§9. RIEMANNIAN GEOMETRY

from one point to another the coordinates themselves must be changed, and this is of course inconvenient. But this can easily be avoided, for we can prove the following.

Suppose that in some domain of a Riemannian space coordinates $y_1, y_2, \cdots, y_n$ are introduced arbitrarily. Then the "infinitely small distance" or, as one says, the "element of length" from the point A with the coordinates $y_1, y_2, \cdots, y_n$ to the point X with the coordinates $y_1 + dy_1, y_2 + dy_2, \cdots, y_n + dy_n$ is expressed by the formula

$$ds = \sqrt{\sum_{i,k=1}^{n} g_{ik} \, dy_i \, dy_k}, \quad \text{or} \quad ds^2 = \sum g_{ik} \, dy_i \, dy_k, \tag{13}$$

where the coefficients g_{ik} are functions of the coordinates $y_1, y_2, \cdots, y_n$ of A.

The expression on the right of the last formula is called a quadratic form* in the differentials of the coordinates $dy_1, \cdots, dy_n$. In expanded form it can be written as follows:

$$\sum g_{ik} \, dy_i \, dy_k = g_{11} \, dy_1^2 + g_{12} \, dy_1 \, dy_2 + g_{21} \, dy_2 \, dy_1 + g_{22} \, dy_2^2 + \cdots.$$

Since $dy_1 \, dy_2 = dy_2 \, dy_1$, it is convenient to take the second and third term as equal: $g_{12} = g_{21}$ and generally $g_{ik} = g_{ki}$; this is possible, because only their sum $(g_{ik} + g_{ki})dy_i \, dy_k$ is important.

The quadratic form is positive definite, since obviously $ds^2 > 0$, except when all the differentials dy_i are equal to zero.

The converse also holds. Namely, if in an n-dimensional space, where coordinates $y_1, y_2, \cdots, y_n$ have been introduced, the element of length is given by the formula (13) with the condition that the quadratic form is positive definite (i.e., always greater than zero except when all the $dy_i = 0$), then the space is Riemannian. In other words, in the neighborhood of each point A one can introduce new special coordinates $x_1, x_2, \cdots, x_n$ so that in the new coordinates the element of length at this point is expressed in the simple form (12)

$$ds^2 = dx_1^2 + dx_2^2 + \cdots + dx_n^2.$$

Thus, Riemannian metric (i.e., a definition of length that is Euclidean in the infinitely small) can be given by any positive definite quadratic form (13) with coefficients g_{ik} that are functions of the coordinates y_i. This is the general method of giving a Riemannian metric.

---

* A quadratic form of several quantities is an algebraic expression that is a homogeneous polynomial of degree 2 in these quantities.

A curve in a Riemannian space is given by the fact that all n coordinates of a point vary in dependence on a single parameter t which ranges over a certain interval

$$y_1 = y_1(t), y_2 = y_2(t), \cdots, y_n = y_n(t) \quad (a \leqslant t \leqslant b). \tag{14}$$

The length of the curve is expressed by the integral

$$s = \int ds = \int \sqrt{\sum g_{ik}\, dy_i\, dy_k}.$$

In the case of the curve given by the equations (14) we have

$$dy_1 = y_1'\, dt, \cdots, dy_n = y_n'\, dt,$$

therefore

$$s = \int_a^b \sqrt{\sum g_{ik} y_i' y_k'}\, dt. \tag{15}$$

Since the g_{ik} are known functions of the coordinates $y_1, \cdots, y_n$ and the latter depend in a known manner on t in accordance with the formulas (14), the function of t under the integral sign in (15) is completely determined for the given curve. Consequently its integral has a definite value and so the curve has a definite length.

The length of the shortest curve joining two given points A, B is taken to be the distance between these points. This curve itself, called a geodesic, plays the role of an analogue to the rectilinear segment AB. One can show that in a small domain any two points are joined by a unique shortest line. The problem of finding the geodesic (shortest) lines is that of minimizing the integral (15). This is a problem in the calculus of variations which was discussed in Chapter VIII. A standard application of the methods of the calculus of variations permits us to derive a differential equation that determines the geodesic lines and to establish their general properties for every Riemannian space.

Let us prove the principal statement made previously, namely, that a Riemannian metric is given in arbitrary coordinates by the general formula (13).

Suppose that in some domain of a Riemannian space certain coordinates $y_1, y_2, \cdots, y_n$ are introduced. We take an arbitrary point A in this domain and assume that $x_1, x_2, \cdots, x_n$ are the special coordinates in which the element of length at A is expressed by the formula (12) or, what is the same,

$$ds^2 = dx_1^2 + dx_2^2 + \cdots + dx_n^2. \tag{16}$$

§9. RIEMANNIAN GEOMETRY

The coordinates x_i are expressed in terms of the y_j ($i, j = 1, \cdots, n$) by certain formulas,

$$x_1 = f_1(y_1, y_2, \cdots, y_n),$$
$$x_2 = f_2(y_1, y_2, \cdots, y_n),$$
$$\cdots\cdots\cdots\cdots\cdots\cdots\cdots\cdots\cdots$$
$$x_n = f_n(y_1, y_2, \cdots, y_n).$$

Then

$$dx_1 = \frac{\partial f_1}{\partial y_1} dy_1 + \frac{\partial f_2}{\partial y_2} dy_2 + \cdots + \frac{\partial f_n}{\partial y_n} dy_n$$

and similarly for $dx_2, \cdots, dx_n$. We substitute these expressions in (16). When we then square the right-hand side and combine the terms with dy_1^2, $dy_1 dy_2$, dy_2^2, etc., we obtain an expression of the form

$$ds^2 = g_{11} dy_1^2 + 2g_{12} dy_1 dy_2 + g_{22} dy_2^2 + \cdots + g_{nn} dy_n^2,$$

where the coefficients $g_{11}, g_{12}, \cdots, g_{nn}$ are expressed in terms of the partial derivatives $\partial f_i / \partial y_j$ (the form of these expressions is of no interest to us). But this is simply formula (13) written in expanded form, and so our statement is proved.

Let us now show that, conversely, the formula (13) defines a Riemannian metric, i.e., that at each point by a special choice of the coordinates x_i it can be transformed into the simple form (16). Suppose that

$$ds^2 = \Sigma g_{ik} dy_i dy_k,$$

where the g_{ik} are functions of the coordinates $y_1, \cdots, y_n$ and the quadratic form on the right-hand side is positive definite. Then the coefficients g_{ik} are given numbers and the variables on which the form depends are $dy_1, \cdots, dy_n$. From algebra it is known that every positive-definite quadratic form (with arbitrary numerical coefficients) can be reduced to a sum of squares by a linear transformation of the variables (see Chapter XVI),* i.e., that there exists a transformation

$$dy_1 = a_{11} dx_1 + \cdots + a_{1n} dx_n,$$
$$\cdots\cdots\cdots\cdots\cdots\cdots\cdots\cdots\cdots\cdots\cdots\cdots\cdots\cdots \quad (17)$$
$$dy_n = a_{n1} dx_1 + \cdots + a_{nn} dx_n,$$

---

* It does not matter that in our case the variables of the form are differentials; we can regard them simply as certain independent variables.

such that when these expressions are substituted in (13), we obtain

$$ds^2 = dx_1^2 + \cdots + dx_n^2.$$

If we make the change of the coordinates $y_1, \cdots, y_n$ to $x_1, \cdots, x_n$ by

$$y_1 = a_{11}x_1 + \cdots + a_{1n}x_n,$$
$$\cdots\cdots\cdots\cdots\cdots\cdots\cdots\cdots\cdots\cdots$$
$$y_n = a_{n1}x_1 + \cdots + a_{nn}x_n,$$

then the differentials dy_j are expressed in terms of the differentials dx_i precisely by the formulas (17). Consequently this change of coordinates solves our problem: in the coordinates $x_1, \cdots, x_n$ at the point we have chosen the square of the differential ds^2 is expressed in the simple "Euclidean" form (16). So we have proved that the general formula (13) does in fact give a Riemannian metric.

5. The curvature tensor. A Euclidean space is the simplest special case of a Riemannian space.* It is an important task of Riemannian geometry to give an analytical expression for the difference of a general Riemannian space from a Euclidean by defining a measure, so to speak, for the non-Euclideanness of a Riemannian space. This measure is the so-called curvature of the space.

We must emphasize right away that the concept of curvature of a space is not at all connected with the idea that the space is situated in some higher enveloping space in which it is somehow curved. Curvature is defined within the given space and expresses its difference from a Euclidean space in the sense of its intrinsic geometric properties. This must be clearly understood in order to avoid linking the concept of a curved space with something extraneous. When it is said that our real space has curvature, this only means that its geometric properties differ from the properties of a Euclidean space. But it does not mean at all that our space lies within some higher space in which it is somehow curved. Such an idea has no relation whatsoever with an application of Riemannian geometry to the real space and belongs in the realm of speculative phantasy.

---

* In a Euclidean space the element of length in rectangular coordinates is expressed by the formula (16): $ds^2 = \Sigma \, dx_i^2$. If we go over to other coordinates, then by what we have deduced under subsection 4 earlier ds^2 is expressed by some quadratic form (13). Consequently the same general formula (13) for the element of length holds in *arbitrary* coordinates in a Euclidean space. The Euclidean space, however, differs from any other by the fact that in it coordinates can be introduced (and these will be rectangular coordinates) such that the formula (16) holds everywhere with one and the same coordinate system and not only near one point or another, as is the case in a general Riemannian space.

§9. RIEMANNIAN GEOMETRY

The concept of curvature of a Riemannian space generalizes to n dimensions that of the Gaussian curvature of a surface. As was explained in §4 of Chapter VII, the Gaussian curvature is a measure of the deviation of the intrinsic geometry of the surface from the geometry in a plane and can be treated purely from an internal-geometric point of view. It is nothing other than the curvature of that two-dimensional Riemannian space that represents the given surface.

Let us recall, for example, two formulas of the intrinsic geometry in which the Gaussian curvature occurs. Suppose that there is a small triangle on the surface near a certain point O whose sides are geodesic lines; let its angles be α, β, γ and its area σ. The quantity $\alpha + \beta + \gamma - \pi$ expresses the difference of the sum of its angles from the sum of the angles of a triangle in the plane.

When the triangle is contracted towards the point O, then the ratio of $\alpha + \beta + \gamma - \pi$ to its area σ tends to the Gaussian curvature K at O. In other words, for a small triangle

$$\frac{\alpha + \beta + \gamma - \pi}{\sigma} = K + \epsilon,$$

where $\epsilon \to 0$ as the triangle is contracted to O. This shows exactly that the Gaussian curvature K is a measure for the difference of the sum of the angles of a triangle in a plane and the sum of the angles of a triangle on the surface.

Now let us consider a small circle on the surface with its center at O (i.e., the geometric locus of points equidistant from O in the sense of the distance on the surface). If r is the radius of the circle and l its length, then

$$l = 2\pi r - \frac{\pi}{3} K r^3 + \epsilon,$$

where K is again the Gaussian curvature at O and ϵ denotes a quantity that is small compared with r^3.

Here the Gaussian curvature emerges as a measure of the deviation of the length of a small circle from the value $2\pi r$ to which it is equal in Euclidean geometry.

Now the curvature of a Riemannian space plays a similar role. It can be defined, for example, in the following manner. In the given Riemannian space we construct a smooth surface F formed from geodesic lines passing through a given point O. The Gaussian curvature of this surface is taken to be the curvature of the space at O in the direction of the surface F. Generally speaking, this curvature will differ not only at different points

O, but also for various geodesic surfaces G passing through one and the same point O. The curvature of a space at a given point is, therefore, not characterized by a single number. Already Riemann introduced a general rule connecting the curvatures of the various surfaces F at one and the same point. Owing to these connections, the curvature at a point is completely characterized by a certain system of numbers, the so-called *curvature tensor*.

However, we cannot dwell here on an explanation of this situation, since it would require extensive mathematical apparatus. The only important thing is to grasp that the curvature is a measure of the non-Euclideanness of a Riemannian space; it is defined intrinsically as a measure of the deviation of its metric from the metric of Euclidean space. It determines, for example, the difference of the sum of the angles of a triangle from π or the difference of the length of a circle from $2\pi r$. At different points it has, in general, distinct values and at one and the same point it is given not by one number but by a certain system of numbers.

A Riemannian space need not be homogeneous in its properties, and in that case free mobility of figures without altering the distances between their points is impossible. So there arises the question in what Riemannian spaces free motion for figures is possible with the same number of degrees of freedom as in a Euclidean space. These are the most homogeneous Riemannian spaces.

It turns out that a Euclidean space is homogeneous without curvature (a space of zero curvature). Another type of homogeneous space is the Lobačevskiĭ space, so that Lobačevskiĭ geometry, just like Euclid's geometry, is a special case of the general Riemannian geometry.

Generally, a Riemannian space in which free motion of figures is possible is a space of constant curvature: In it the curvature has one and the same value at all points and for all geodesic surfaces. (Instead of the "curvature tensor" which changes from point to point it is given this time by a single number common to all points.) A space of zero curvature is Euclidean; a space of negative curvature is a Lobačevskiĭ space; a space of positive curvature has the same geometry as an n-dimensional sphere in an $(n + 1)$-dimensional Euclidean space.

6. Applications of Riemannian geometry. Riemannian geometry did not have long to wait for applications. Riemann himself, as we have already said, applied its formal apparatus to the solution of a problem of heat conduction, but this was merely an application of its formulas, not of the idea of an abstract space with a Euclidean measure of distance in infinitely small domains. Such an application was made to the color

§9. RIEMANNIAN GEOMETRY

space, where the distance between neighboring colors can be expressed by using a Riemannian metric; the color space has been treated as a special three-dimensional Riemannian space.

Another important application of Riemannian geometry emerged in mechanics. In order to understand its essence let us consider, to begin with, the motion of a point on a surface. We imagine a material point, for example a particle that can move freely on a certain smooth surface without leaving the surface. The point moves, as it were, in the surface itself. We can introduce arbitrary coordinates x_1, x_2 on the surface; then the motion of the point is completely determined by the dependence of these coordinates on the time, and its velocity on the velocities of the change of coordinates, i.e., on their derivatives with respect to the time $\dot{x}_1$, $\dot{x}_2$. So we see that the point moves, as it were, in two-dimensional space; but this space is not Euclidean and has its own geometry, the intrinsic geometry of the surface. The laws of motion can be transformed so as to contain only the coordinates x_1, x_2 of a point on the surface and their first and second derivatives.

If a force acts on the point, then its component perpendicular to the surface is annihilated by the resistance (by the reaction of the surface) and there remain only the components tangential to the surface; so the force acts only along the surface. In this manner the forces acting on a point can also be regarded as acting in the surface itself. The intrinsic geometry of the surface is a special case of a Riemannian geometry. Therefore the motion of the point on the surface is a motion in a two-dimensional Riemannian space. The laws of this motion have the same character as the usual laws of motion, with the difference only that the intrinsic geometry of the surface is taken into account. This becomes perfectly clear from the following fact, mentioned before, in §4 of Chapter VII: A point moving on the surface under inertia and without friction moves on a geodesic line with constant velocity. Since the geodesic lines play the role of straight lines on the surface, this fact is analogous to the law of inertia: it is the same law of inertia, but for the motion on a surface or, abstractly, in a two-dimensional Riemannian space.

Of course, so far there is no advantage visible in this abstract presentation, because we are concerned only with the motion on an ordinary surface.

The benefit of the abstract point of view makes itself felt at once when we go over to mechanical systems whose state is given by more than two quantities. Then a representation of the motion as the motion of a point on a surface becomes impossible. We have encountered this circumstance in §7 when we talked of how graphical methods fail in an abstract presentation of a many-dimensional space.

Suppose, then, that there is a mechanical system whose configuration, i.e., the distribution of whose parts, is given by n quantities $x_1, x_2, \cdots, x_n$. If we are concerned with a system of several material points, then their distribution is determined by giving all their coordinates, three for each point. As another example we can take a gyroscope (a wheel spinning on an axis that itself can turn around a stationary point). The rotation of the gyroscope around the axis is given by the angle of deflection and the inclination of its axis by the two angles it forms with two given directions. Altogether there are three quantities that determine the position of such a gyroscope (figure 32).

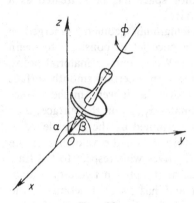

FIG. 32.

Every configuration (every position of the parts of the system) can be thought of as a "point" in the space of all possible configurations. This is the so-called configuration space of the system.* The number of its dimensions is equal to the number of quantities $x_1, x_2, \cdots, x_n$ that determine the configuration. These quantities serve as coordinates of a "point" in the configuration space. For a system of say three material points, we obtain three coordinates each for three points, or nine coordinates altogether. For the case of a gyroscope, we have three coordinates, namely the three angles, so that the configuration space of a gyroscope is three-dimensional.

The motion of the system is represented as the motion of a point in a configuration space. The velocity of the motion is determined by the velocities of change of the coordinates $x_1, x_2, \cdots, x_n$.

Such spaces will be discussed again in Chapter XVIII in connection with their topological structure. Here we wish only to emphasize that we can introduce in a configuration space a special rule of measuring distances which is closely connected with the mechanical properties of the system. Indeed, if the kinetic energy of the system is expressed by the formula

$$T = \frac{1}{2} \sum_{i,k=1}^{n} a_{ik} \dot{x}_i \dot{x}_k,$$

---

* This must not be confused with the "phase space" mentioned in §8 under subsection 2. In the phase space a point determines not only the position but also the velocity of motion of the points of the system at every moment.

§9. RIEMANNIAN GEOMETRY

where the $\dot{x}_i$ are the velocities of change of the corresponding coordinates, then the square of an infinitely small distance is given by the formula

$$ds^2 = \sum_{i,k=1}^{n} a_{ik}\, dx_i\, dx_k .$$

Thus, the configuration space becomes a Riemannian space. Moreover, not only is the motion of the system represented as the motion of a point in the configuration space but the very equations that describe the motion of the system coincide with the equations of motion of this point; in a word, the mechanics of the system are represented as the mechanics of a point in the configuration space. In particular, the motion of the system under inertia, i.e., without the action of forces (like the free rotation of a gyroscope) becomes a uniform motion of a point on a geodesic line in that space.

This representation is expedient in a number of cases and is used, along with certain generalizations and modifications, in theoretical mechanics.

Thus, Riemannian geometry has its applications as a method of abstract-geometrical description of physical phenomena. This description is not at all arbitrary and is not an idle play of the mathematical mind; it reflects the real mechanism of the phenomena in question but reflects it in an abstract form. But this is the nature of every mathematical description of physical phenomena. This is also the nature of every application of abstract geometry; the difference consists only in that more powerful, more delicate abstractions are applied, but the essence remains the same.

The most brilliant application of Riemannian geometry came in the theory of relativity. Of this we shall speak in the following section, where we will be concerned with the important and difficult problem of the relationship between abstract geometry and properties of the real space.

7. Generalizations. In the last thirty years the geometry of various non-Euclidean spaces has been the subject of remarkable developments and generalizations in several directions. New theories have arisen in which Riemannian geometry is included as a special case. The first of these was the so-called Finsler geometry, the idea of which goes back right to Riemann;* then came a general theory of spaces of the eminent French geometer E. Cartan, which combines Riemannian geometry with Klein's Erlanger Programm, and other theories. We cannot possibly give

---

* Finsler was the German mathematician who in 1916 initiated a detailed treatment of the geometry mentioned here.

an account of these new directions of geometry and shall mention only that they are worked out essentially by means of a special adaptation of their analytical apparatus. A group of Soviet geometers has been taking part in the development of these new directions; we could perhaps name here the new "polymetric" geometry created by P. K. Raševskiĭ and the investigations of V. V. Vagner which extend from the most general problems of the theory of curved spaces to the applications of non-Euclidean geometry in mechanics.

§10. Abstract Geometry and the Real Space

1. Difficulties of visualizing our actual space as non-Euclidean. In the course of the preceding account of the development of geometrical ideas beginning with Lobačevskiĭ, we have gone deep into abstract spaces and quite far away from the original object of geometry, that real space in which all phenomena take place. We shall now return to this space in the usual sense and shall set ourselves the task of explaining what the development of abstract geometry has contributed to our knowledge of its properties.

We know that geometry has grown from experiment, from an experimental investigation of spatial forms and relations of bodies: from the measurement of lots of land, of volumes of containers, etc. So in origin it is a physical theory such as, say, mechanics. The axioms of Euclidean geometry were conclusions clearly formulated on the basis of protracted experiments; they express laws of nature, and they can be called laws of geometry, just as the fundamental laws of mechanics are now often called axioms of mechanics.* But it is wrong to assert that these laws are absolutely exact and never require modification or generalization in connection with new experimental data; the real properties of space may differ more or less from what Euclidean geometry states.

We have already brought forward these arguments and now they must appear, we would think, quite obvious. But this was not so a hundred years ago, when Lobačevskiĭ's ideas failed to achieve general recognition. Before Lobačevskiĭ and Gauss, it had not entered into anybody's head that the Euclidean geometry could turn out to be not entirely accurate,

---

* An abstract conception of axioms that dissociates the axioms from their original content has arisen during the last fifty years; this changes nothing in the fact that the axioms of Euclidean geometry express laws of nature. In speaking of axioms and not of laws of geometry or mechanics we wish to place the logical deductive construction of these sciences in the forefront, but they do not lose their experimental foundation on that account. Any statement of a theorem is called an axiom when it is taken as the basis for the deductive construction of the theory, and other statements of the theory (theorems) are deduced from the basic ones (the axioms) by logical reasoning.

§10. ABSTRACT GEOMETRY AND THE REAL SPACE

that the real properties of space could be somewhat different. Lobačevskiĭ developed his geometry as a theory of possible properties of the real space. Later Riemann and some other scientists also raised the problem of possible properties of space, of possible laws of measuring lengths that could be discovered by more accurate measurements. Quite generally, even abstract geometry in some of its parts can be regarded as a theory of possible properties of space. All this remained, however, in the realm of hypothesis until in 1915 Einstein in his general theory of relativity corroborated the ideas of Lobačevskiĭ and Riemann. This theory claims that the geometry of the real space in fact differs somewhat from Euclidean geometry, and this was discovered on just that astronomical scale that Lobačevskiĭ had anticipated.

In what we have just said about space at least three difficulties are involved. The problem of the relation of abstract geometry to physical geometry, i.e., to the properties of real space, reduces in the ultimate analysis to a clarification of these difficulties.

The first difficulty consists in visualizing at all how and in what sense the properties of real space can possibly differ from the statements of Euclidean geometry. We are so accustomed to it that we cannot easily imagine anything else, and explanations are obviously necessary here.

The second difficulty lies in the very expression "properties of the real space." Space by itself is conceived of as empty and homogeneous. It would seem that in the concept of space itself the idea of its homogeneity is already included. How then can the empty space, i.e., the "emptiness," have any properties? We speak of "properties of space" without thinking of these problems, but they are worth thinking about, as is apparent from the difficulty just mentioned.

The third difficulty lies in the concept of the truth of one geometry or another. The question may appear very simple: That geometry is true which corresponds to reality. This is so, of course. But on the other hand we have seen, for example, that the geometry inside a circle can be regarded as a Lobačevskiĭ geometry, because every geometrical fact inside the circle can be presented as a theorem of Lobačevskiĭ geometry. Consequently it turns out that the same geometrical facts can be presented both as theorems of Euclid's geometry and as theorems of Lobačevskiĭ's geometry. Hence both geometries correspond to reality. So which of them is true and in what sense, and why do we assume all the same that, in fact, Euclidean geometry is satisfied in the circle and that Lobačevskiĭ geometry is only illustrated or interpreted in it?

Clearly, in these problems lies a considerable difficulty which has baffled at times even some eminent mathematicians.

Our explanation must begin with the second of the difficulties mentioned, because an understanding of what these "properties of space" are will lead us to a solution of the other difficulties.

2. Space and matter. The subject matter of geometry, namely the "properties of space," is made up of the properties of real bodies, their material relationship and forms. In real space a "place," a "point," a "direction," etc., is determined by material bodies. "Here" and "there," "hither" and "thither" have a meaning only in connection with one material object or another. "Here" can mean "on the earth," "in this room" or something else of the sort; in a word, "here" always denotes a place determined by material criteria of one kind or another. Similarly, for example, a straight line does not exist by itself, but only as a taut thread or the edge of a ruler or a ray of light. A straight line, "a line like this," is altogether an abstraction which reflects the common properties of these material lines, just as, say, the "house in itself" is an abstraction reflecting the common properties of houses; the "house in itself" does not exist outside of or independently of the various real houses.

This objective character of the properties of space is expressed by the well-known statement of dialectic materialism: *Space is the form of existence of matter.* The form of an object is determined by the connections and relations of its parts. The structure of space is the common regularity of a number of relations of material bodies and events. There are the spatial relationships, the spatial order of objects, their mutual positions, distances, etc. But as every form is inseparable from its content, except in abstractions and in certain contexts, so is space inseparable from matter. The idea of a space "in itself," of a space without matter is an abstraction that must not be abused. Real spatial relationships and forms: "here," "between," "inside," "straight," "sphere," etc., these are always relations and forms of material bodies. Geometry, however, considers them abstractly. This abstraction is necessary, because otherwise it would not be possible to perceive generality in the diverse concrete relations of objects. But this abstraction must not be made absolute by substituting abstract concepts for the objective reality itself.

In an absolutely empty space, void of all traces of matter, nothing would distinguish a place, a direction, consequently there are no places, no directions, so that the absolutely empty space reduces to nothing. Even in the abstract idea of an empty space we imply tacitly that various places and directions are distinguishable in it. In other words, in the abstract idea of space we retain the properties of distinctness of places, directions, distances which exist in real space precisely because this space is inseparably connected with material bodies.

Thus, space is the form of existence of matter; "properties of space"

§10. ABSTRACT GEOMETRY AND THE REAL SPACE

are, therefore, properties of matter, properties of certain relations of material bodies, their mutual positions, dimensions, etc.

Furthermore, if the theorems of geometry are to have a physical meaning, we must know what we shall understand by "straight line," "distance," and other geometrical concepts in them. In §4 we have seen that one and the same geometrical theory admits different interpretations.

Consequently, for comparing geometry with experiment it is necessary to define as accurately as possible the physical meaning of geometrical concepts, because geometry describes the properties of real space only on condition that the corresponding physical meaning is attributed to its concepts. Without this physical meaning the theorems of geometry are of an abstract mathematical, formal character. Herein lies the solution to the second difficulty specified earlier. This difficulty arises because, instead of the real space, which is inseparably connected with matter, one wants to think of a "pure" space, a space "as such," which is, however, nothing more than an abstraction.

3. Intuition and understanding. Now it is easy to understand how the other two difficulties are resolved.

First of all, how can one imagine that the real space in its properties is other than Euclidean? Suppose we wish to verify some statement of Euclidean geometry, for example, that the sum of the angles of a triangle is 180° or that the length of a circle is equal to $2\pi R$. To verify the first we have to determine what physically defined triangles are to be considered and how their angles are to be defined. Suppose a side of a triangle is a light ray in empty space. In that case there is nothing inconceivable in the fact that very accurate experiments may show that sums of the angles of a triangle are different from 180°. In the same way one can imagine that measurement of a radius and circumference on one and the same scale leads to results that do not satisfy the relation $l = 2\pi R$ accurately. In fact, this is so on the surface of the earth, where the length of a circle is not proportional to the radius but grows slowly and reaches a maximum when the radius is made equal to half a meridian. But it may be objected that on the surface of the earth the role of straight lines is played by arcs of great circles so that the radius is understood here in another sense and our result, therefore, does not contradict Euclidean geometry. However, according to the theory of relativity, near a body of large mass the ratio of the length of a circle to the "genuine" radius is all the same somewhat different from 2π and, in fact, the following approximate formula holds for the ratio of the length of the equator of a homogeneous sphere-shaped body and its radius:

$$\frac{\text{length of circumference}}{\text{radius}} = 2\pi \left(1 - \frac{kM}{3Rc^2}\right),$$

where M is the mass of the body (in metric tons), R the radius of the body (in kilometers), $c = 300{,}000$ km/sec the velocity of light and k the constant of gravitation which for this choice of units of measurement is equal to $66.6 \cdot 10^{-18}$.

So we see that the ratio of the length of a circle to its radius is not equal to 2π, but a little less. Computations show that on the surface of the sun this ratio differs from 2π by approximately .000004, and on the surface of the companion of Sirius, whose average density is 50,000 times that of water, the deviation reaches .00014.

It may be objected, of course, that all this is nevertheless impossible to imagine, that in an intuitive picture space is always Euclidean. This objection need not disturb us, first of all, because the task of science does not consist in giving intuitive pictures of the phenomena but in arriving at an understanding of them. An intuitive picture is restricted and conditioned by the customary forms transmitted by our sense organs. Therefore we are not in a position to have an intuitive picture of ultraviolet rays, of the propagation of radio waves, of the motion of an electron in an atom, or of many other phenomena, except by substituting models in place of them. But this does not mean at all that these phenomena are incomprehensible for us. On the contrary, the successes of radio technology, for example, show clearly that we have mastered radio waves completely and therefore understand them quite well. Second, the solution of the problem of what we can and what we cannot imagine depends on habit and training of the imagination. Can we imagine the antipodes, where from our point of view people walk with their heads hanging downwards? Nowadays we are able to imagine this, but there was a time when the "unimaginability" of the antipodes served as an argument against the spherical shape of the earth.

4. Geometry and truth. Now let us turn to the last difficulty, i.e., to the problem as to which geometry can be regarded as true. When we posed this problem, we indicated that the geometrical facts inside a circle can be interpreted as theorems of Euclid's geometry and of Lobačevskiĭ's geometry. Therefore, both these geometries correspond to reality; i.e., both are true. And after all, there is nothing astonishing here. One and the same phenomenon can always be described by various methods; one and the same quantity can be measured in various units; one and the same curve can be given by various equations depending on the choice of the coordinate system. Similarly a given isolated collection of geometrical relations (in the example in question, the relations inside a circle) can be described by various methods. But we raise the problem not of some isolated collection of geometrical facts but of the spatial

§10. ABSTRACT GEOMETRY AND THE REAL SPACE

relationships in their entire generality. Space is the universal form of the existence of matter and, consequently, when the problem of the properties of space is raised, no domain of facts can be separated artificially.

When the problem is posed in this way, then geometrical quantities or geometrical facts cannot be considered isolated from other phenomena with which they are necessarily connected. For example, the length of a segment is determined by laying off a rigid rule so that the measure of length is necessarily connected with the motion of rigid bodies, i.e., with mechanics. Geometry is inseparable from mechanics. However, the measurement of length inside a circle in the interpretation of Lobačevskiĭ geometry proceeds quite differently, as we have explained in §4; a chord here becomes infinitely long. It is clear that the so-defined measure does not correspond to the original idea of measurement which has grown on the basis of mechanical transportation of the real bodies to be compared. In general, figures that are equal in the sense of Euclidean geometry are, by the very origin of Euclidean geometry, figures that can be superimposed on one another by means of a mechanical motion. In the interpretation of Lobačevskiĭ geometry equality is defined differently, the role of motions is taken here by other transformations. Therefore, when the geometrical facts inside a circle are taken in conjunction with their necessary link with mechanics, then we must admit that it is in fact the Euclidean geometry which holds inside a circle (with great accuracy).

Euclidean geometry is the one in which the role of motion is played by the ordinary mechanical motion of rigid bodies. It was precisely for this reason that Euclidean geometry, and not any other geometry, was the first to be discovered. But the development of physics has now led to the conclusion that the laws of Newtonian mechanics and with them the laws of Euclidean geometry are only approximations to more accurate and general laws. In this change of the laws of geometry, in the transition from Euclidean to Riemannian geometry which was accomplished in the theory of relativity, mechanics was not the only branch of physics to play a role: Of equal, if not greater, importance were the theories of electromagnetic phenomena and optics. Geometry as the science of the properties of space is connected with physics, depends on it, and can be separated from it only in abstraction and only in certain contexts.

The dependence of geometry on physics, or in other words the dependence of the properties of space on matter, was clearly indicated by Lobačevskiĭ, who foresaw the possibility of a change of the laws of geometry on transition to a new domain of physical phenomena. In contrast to this materialistic point of view, the famous mathematician Poincaré stated rather recently that the choice of one geometry or another is dictated solely by considerations of simplicity or "economy of thought" in the phrase of the well-known idealist Mach. On this basis Poincaré predicted further that science would more readily give up the law of rectilinear propagation of light than the Euclidean geometry, because it is the "simplest." However, Poincaré died three years before the general theory of relativity was finally set up, in which just the opposite is done: Euclidean geometry is abandoned, but the law of propagation of light is preserved, though in a generalized form; light is propagated on a geodesic line.

So we have reached the following conclusions. One and the same isolated collection of facts can, generally speaking, be described in a variety of ways and all these descriptions are true provided they reflect reality. However, it is wrong to consider the geometrical facts in their entire generality severed from other phenomena. Only so can the properties of space be established, because it is the universal form of existence of matter. But by taking geometry in conjunction with physics we must necessarily adapt them to one another, and then we see the essential difference between the various "geometries" which, when dissociated from physics, can only be distinguished by their greater or lesser simplicity. Euclidean geometry appeared not because it was simpler than the others, but because it corresponded to mechanics. In fact, in connection with the development of physics in the theory of relativity we now go over to a more complicated geometry, namely Riemannian.

To sum up, in reference to the properties of real space that geometry is true which reflects the properties of spatial relationships in their entire generality with sufficient accuracy and which consequently corresponds not only to the purely geometrical facts but also to mechanics and to the whole of physics.

5. The space-time of relativity. What little we have said above on the theory of relativity does not touch on its main contents. In the understanding of the problem of space, it went substantially further than Lobačevskiĭ and Riemann had thought of.

The most essential and basic proposition of the theory of relativity is: Space is completely inseparable from time and that together they form a single form of existence of matter, the four-dimensional manifold of space-time. An event in the world is characterized by its place and time and consequently by four coordinates: three spatial and the fourth temporal, the time of the event. The events form in this sense a four-dimensional collection. The theory of relativity is concerned above all with this four-dimensional collection from the point of view of its structure, disregarding the properties of the individual phenomena. It is not fundamentally a theory of fast motions, nor of cosmology, nor a new

§10. ABSTRACT GEOMETRY AND THE REAL SPACE

theory of space or time, but just a theory of space-time as a single form of existence of matter.

Of course, in Newtonian mechanics also we can combine space and time in a single four-dimensional manifold. As we have had occasion to recall, the idea itself of a many-dimensional space was born in Lagrange's work, when in considering the motion of a material point he added to the spatial coordinates x, y, z the time coordinate t. The motion of a point is then represented as a line in the four-dimensional space with the coordinates x, y, z, t; under a motion of the point all four coordinates change: the position (x, y, z) and the time t. However, this unification of space and time has a purely formal character. No internal necessary connection between space and time is set up here. Of course, in the law of motion of each given body there is its dependence on spatial position and on time. But this concerns only each given motion, no universal internal connection between space and time was established before the theory of relativity, neither in mechanics nor in physics generally. The spatial relationships, the spatial order of objects and phenomena were always carefully distinguished from their relationships and order in time. The temporal sequence of events, the duration of time intervals were regarded as absolute, as definite irrespective of what happened. In short, the concept of absolute time ruled supreme.

Einstein's greatest discovery, which not only laid the corner stone of the theory of relativity but revolutionized the whole physical and philosophical understanding of the problem of space and time, was the discovery that absolute time does not really exist. Shortly after Einstein developed his theory in 1905,* Minkowski showed that its essence consists not as much in the rejection of absolute time as in the institution of a mutual link of space and time, in virtue of which there exists a single absolute form of existence of matter: space-time. The separation of space (the spatial coordinates) from time (from the time coordinate t) is to a certain degree relative, depending on the material system (the "system of reference") in relation to which the spatial and temporal order of the phenomena is determined. Events that are simultaneous with reference to one system need not be so with reference to another system.

The definition of the order of phenomena cannot be, of course, completely dependent on the system of reference. The order of events that are connected by direct interaction, it stands to reason, remains one and the same with reference to all systems, so that an action always precedes its result. But for events that are not connected by interaction the order

---

*The theory that Einstein developed in 1905 is called the special theory of relativity in contradiction to Einstein's "general" theory of 1915.

of time turns out to be relative. Since the spatial order (in its pure form) refers to simultaneous events and simultaneity is relative, the separation of purely spatial relationships from the general aggregate of space-time relationships turns out to be relative, depending on the system of reference. Space in the abstract sense is, as it were, a "section" of the four-dimensional manifold of space-time that is laid through simultaneous events (in reference to a given system).

It cannot be our task to explain the foundations of the theory of relativity and so we shall try only to characterize in a few words its basic features in the form in which it is most natural to consider them in connection with the ideas of abstract geometry. This interpretation, incidentally, is quite different from that which comes from Einstein himself.

The world, the universe, can be regarded as a set of diverse events. By an event we understand here not an arbitrary phenomenon extending in space and lasting in time but as it were an instantaneous, pointlike phenomenon such as a momentary flash of a point source of light. To use geometrical language, events are points in the four-dimensional manifold of the universe.

Space-time is the form of existence of matter, the form of this world manifold. The structure of space-time, its "geometry," is nothing but a certain general world structure, i.e., in accordance with our analysis, the "geometry" of the set of events. This structure is determined by certain universal material connections and relations of events.

First, as we have explained previously for the spatial relationships, we must, in fact, be dealing here with material relationships and connections. This is also true for the relationships of phenomena in time. Spatial and temporal relationships "as such," in a pure form, are only abstractions.

Second, the relationships of events that determine the structure of space-time must have universal character in accordance with the universal character of space-time.

Such a universal material relation of events is their cause-effect connection. Every event acts in one way or another, directly on certain other events and, in turn, experiences the action of other events. This relation of the action of some events on others determines the structure of space-time.

Thus, the theory of relativity allows us to make the following definition. Space-time is the set of all events, irrespective of their concrete properties and relations except for the general relation of action of some events on others. This relation too must be understood here in the general sense, irrespective of its various concrete forms.

In the special theory of relativity, space-time is regarded as maximally

§10. ABSTRACT GEOMETRY AND THE REAL SPACE 187

homogeneous. This means that the manifold of events admits transformations that do not disturb the relations of action between events, and the group of these transformations is in a certain sense as large as possible. Suppose, for example, that there are two pairs of events A, B and A', B' and that neither A acts on B nor B on A and similarly for A' and B'. Then there exists a transformation between the events under which A corresponds to A' and B to B' and such that for any pairs of events the relation of action (or inaction) is not infringed; i.e., if X acts on Y, then the corresponding event X' also acts on Y' and if X does not act on Y, then X' does not act on Y'.

In accordance with these explanations, it turns out that from the point of view of the special theory of relativity, space-time is a four-dimensional space of its own kind whose geometry is determined by a certain group of transformations. These transformations are nothing other than the famous Lorentz transformations. The laws of geometry and physics do not change under these transformations. This outlook corresponds to the view of geometry put forward by Klein in his Erlanger Programm of which we have spoken in §6.

6. Gravitation and curvature. The general theory of relativity goes further and abandons the idea of homogeneity of space-time. It assumes that space-time is homogeneous only to a certain approximation in sufficiently small domains but is on the whole inhomogeneous. The inhomogeneity of space-time is determined according to Einstein's theory by the distribution and motion of matter. In its turn the structure of space-time determines the laws of motion of bodies, and this appears in the phenomenon of universal gravitation. The general theory of relativity is, properly speaking, a theory of gravitation that explains the gravitational link of the structure of space-time with the motion of matter.

The idea of space-time as homogeneous only to a certain approximation in small domains is similar to the idea of Riemannian space which is Euclidean only "in the infinitely small." The mathematical space-time of the general theory of relativity is treated as a kind of Riemann space, though in a substantially altered sense.

In fact, in a four-dimensional Riemannian space we can introduce coordinates in the neighborhood of every point such that the square of the line elements is expressed by the formula $ds^2 = dx_1^2 + dx_2^2 + dx_3^2 + dx_4^2$.

In space-time we can introduce coordinates x, y, z, t in the neighborhood of every event such that the line element is represented by the formula $ds^2 = dx^2 + dy^2 + dz^2 - c^2 dt^2$, where c is the velocity of light, which for a suitable choice of the units of measurement can be taken to be 1. Here x, y, z are the spatial coordinates and t is the time. The minus sign for dt^2 gives a formal expression to the essential radical difference of the time coordinate from the spatial ones, of time from space.

In the theory of gravitation the concept of curvature of space-time plays a very important role. The fundamental equations of the theory given by Einstein at once connect the quantities that characterize the curvature of space-time with the quantities that characterize the distribution and motion of matter. These equations are at the same time the equations of the gravitational field and thus, as Einstein has proved in collaboration with V. A. Fok, the laws of motion of bodies in a gravitational field can be derived from them.

The structure of space-time according to the general theory of relativity is complicated, and space cannot be separated from time even to the extent permitted by the special theory of relativity. However, with a certain approximation and under certain assumptions this can be done. Space turns out to be Euclidean with a sufficient accuracy in domains that are small in comparison with the cosmic scale, but in large domains the deviation from Euclidean geometry becomes apparent. This deviation depends on the distribution and motion of masses of matter and reaches appreciable, though still very small, values near a star of large mass, or in general when the magnitudes involved are on a cosmic scale. In a number of hypotheses on the structure of the universe as a whole it is assumed that on the average the distribution of mass is approximately uniform. In one of these hypothetical theories proposed by the Soviet physicist and mathematician Fridman the geometry of space on the whole coincides with Lobačevskiĭ geometry.

In the theory of relativity abstract geometry finds an application not only as a mathematical apparatus; the very ideas of an abstract space provide the means for a deeper formulation of the foundations of this theory. Possibilities contemplated in abstract geometry are discovered in reality, and theoretical thinking celebrates here its most brilliant triumph. Abstract geometry, which itself has grown from an experimental study of the spatial relationships and forms of bodies, now faces, as a well-developed mathematical method, the study of real space. Such is the general path of science: From what is immediately given by experiment it rises to theoretical generalizations and abstractions, and then turns again to the experiment as the instrument for deeper understanding of the essence of the phenomena; by thus giving explanations of known phenomena and predictions of new ones, it guides the practical activities of its investigators and in return finds herein its own justification and the source of its future development.

Suggested Reading

H. S. M. Coxeter, *Introduction to geometry*, Wiley, New York, 1961.
L. P. Eisenhart, *An introduction to differential geometry with use of the tensor calculus*, Princeton University Press, Princeton, N. J., 1940.
——, *Riemannian geometry*, Princeton University Press, Princeton, N. J., 1949.
W. C. Graustein, *Elementary differential geometry*, Macmillan, New York, 1935.
D. Hilbert, *Foundations of geometry*, Open Court, LaSalle, Ill., 1959.
A. V. Pogorelov, *Differential geometry*, Noordhoff, Groningen, 1959.
G. Y. Rainich, *Mathematics of relativity*, Wiley, New York, 1950.
D. J. Struik, *Lectures on classical differential geometry*, Addison-Wesley, Cambridge, Mass., 1950.
J. L. Synge, *Relativity: the special theory*, Interscience, New York, 1956.

PART 6

CHAPTER **XVIII**

TOPOLOGY

§1. The Object of Topology

"*Adjacency* is the distinguishing appurtenance of bodies and permits us to call them *geometric*, when we retain in them this property and abstract from all others, whether they be essential or accidental." With these words I. N. Lobačevskiĭ begins the first chapter of his work "New Elements of Geometry".*

Explaining by a diagram (figure 1) the words just quoted Lobačevskiĭ continues: "Two bodies A, B that touch each other form a single geometric body $C \cdots$. Conversely, every body C can be split by an arbitrary section S into two parts A, B."

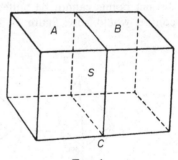

Fig. 1.

These concepts of adjacency, neighborhood, infinite proximity, and also the concept of a dissection of a body, which in a certain sense is dual to them, these are the concepts that Lobačevskiĭ places at the foundation of the whole structure of geometry and they are also in essence the fundamental, primordial concepts of topology in the full extent in which we now understand this discipline. Therefore contemporary commentators on the great geometer are right in saying† that "Lobačevskiĭ makes the first attempt in the history of the mathematical sciences to start, in the construction of

---

* I. N. Lobačevskiĭ, "Collected works," Vol. II, Gostehizdat, 1949, page 168.
† Comments to "New Elements of Geometry," loc. cit., page 465.

geometry, from topological properties of bodies .... The concepts of surface, line, point are defined by Lobačevskiĭ in terms of dissections and adjacencies of bodies." Some idea of the diversity of concrete geometric content reflected in the concepts of adjacency and dissection of bodies, as Lobačevskiĭ imagined them, can be obtained from the following diagrams (figure 2) taken from his work.

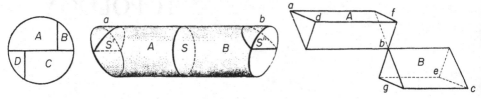

Fig. 2.

Every transformation of a geometric figure in which the relations of adjacency of various parts of the figure are not destroyed is called *continuous*; if the adjacencies are not only not destroyed, but also no new ones arise, then the transformation is called *topological*. Therefore, under a topological transformation of an arbitrary figure the parts of this figure that are in contact remain in contact, and the parts that are not in contact cannot come into contact; to put it briefly, in a topological transformation neither breaks nor fusions can arise. In particular, two distinct points cannot be united into a single point (in that case a new contact would arise: figure 3). Therefore a topological transformation of

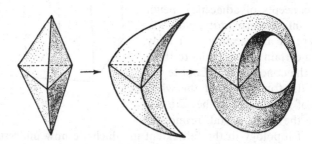

Fig. 3.

any geometrical figure, considered as the set of points forming it, is not only a continuous, but also a one-to-one transformation: Any two distinct points of the figure are transformed into two distinct points. Thus, topological transformations are single-valued and continuous both ways.

§1. THE OBJECT OF TOPOLOGY

Intuitively a topological transformation of an arbitrary geometric figure (a curve, a surface, etc.) can be represented in the following way.

Let us imagine that our figure is made of some flexible and stretchable material, for example of rubber. Then it can be subjected to all possible continuous deformations under which it will be extended in some of its

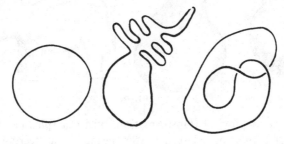

FIG. 4.

parts and contracted in others, and altogether will change its size and shape in every way. For example, when a closed rubber ring is given in the form of a circle, then we can stretch it into the shape of an extremely elongated ellipse, we can give it the form of a regular or an irregular polygon and even of very complicated closed curves, some of which are illustrated in figure 4. But we cannot, by a topological transformation, turn a circle into a figure of eight (for this would require the fusion of two distinct points of the circle; figure 5) or into an interval (for this

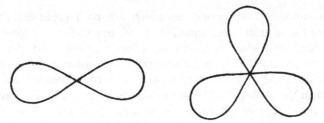

FIG. 5.

would require the fusion of one semicircle with another or else a break of the circle at an arbitrary point). The circle is the simplest closed curve, since it forms only one loop in contrast to the figure of eight, which forms two loops, or the trefoil curve (figure 5), which forms three loops. The property of a circle of being a simple closed curve is a property preserved under an arbitrary topological transformation or, as one says, it is a topological property.

If we take a spherical surface, which we can imagine in the form of a thin rubber sheet, then we can again make exceedingly great changes in its shape by means of a topological transformation (figure 6). But we

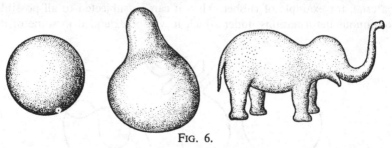

FIG. 6.

cannot, by a topological transformation, turn our spherical surface into a square or a ring-shaped surface (the surface of a steering wheel or a life belt), which is called a torus (figure 7). For the surface of a sphere

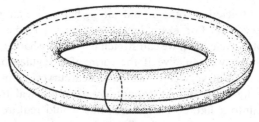

FIG. 7.

has the following two properties which are both preserved under an arbitrary topological transformation. First, our surface is closed: There are no edges on it (but the square has edges); second, every closed curve on a spherical surface is in Lobačevskiĭ's expression a dissection of it; if we make a cut along a given closed curve traced out on our rubber sheet, then the surface splits into two disconnected parts. The torus does not have this property: If a torus is cut along a meridian (figure 7), then it is not split into parts but is turned into a surface having the form of

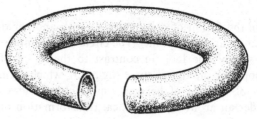

FIG. 8.

§2. SURFACES

a bent tube (figure 8) which we can then easily turn (straighten) into a cylinder by a topological transformation. Thus, in contrast to the sphere, not every closed curve on the torus is a dissection. Therefore the spherical surface cannot be turned into a torus by a topological transformation. We say that the sphere and the torus are topologically distinct surfaces or surfaces that belong to distinct topological types or, finally, that these surfaces are not homeomorphic to each other. Conversely, a sphere and an ellipsoid and quite generally any bounded convex surface belong to one and the same topological type; i.e., they are *homeomorphic*. This means that they can be carried into one another by a topological transformation.

§2. Surfaces

As mentioned earlier, every property of a geometrical figure that is preserved under an arbitrary topological transformation of it is called a topological property. Topology studies topological properties of figures; furthermore, it studies topological transformations and also arbitrary continuous transformations of geometrical figures.

We have just given some examples of topological properties. Such properties are: the property of a curve or a surface of being closed, the property of a closed curve of being simple (i.e., of forming only one loop), the property of a surface that every closed curve lying on it is a dissection of the surface (the spherical surface has this property, but the ring-shaped one has not), etc.

The largest number of closed curves that can be drawn on a given surface in such a way that these curves do not form dissections, i.e., that the surface does not split into parts when cuts are made along all these curves, is called the *order of connectivity* of the surface. This number gives us some important information on the topological layout of the surface. We have seen that for a spherical surface it is equal to zero (every closed curve on this surface is a dissection). On the torus we can find two closed curves that taken together do not form sections: One

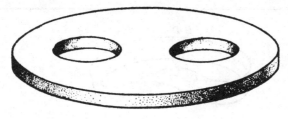

Fig. 9.

of them can be taken as an arbitrary meridian and the other as a parallel of the torus (figure 7). However, it is impossible to draw on the torus three closed curves that taken together do not form a dissection of it; the order of connectivity of the torus is 2. The order of connectivity of the pretzel surface (figure 9) is 4, etc. Quite generally, let us take a spherical surface and cut $2p$ spherical holes in it (in figure 10, the case $p = 3$ is

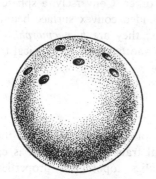
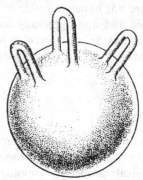

Fig. 10.

illustrated). We divide these holes into p pairs and attach to each pair of holes (at the edges) a cylindrical tube (a "handle"). We obtain a sphere with p "handles" or as it is called, a normal surface of genus p. The order of connectivity of this surface is $2p$.

All these surfaces, in Lobačevskiĭ's expression, are "dissections" of space: Each of them divides the space into two domains, an interior and an exterior, and they are the common boundary of these two domains. This fact is connected with another, namely that every one of our surfaces has two sides: an interior and an exterior (one side can be painted in one color, and the other in another).

However, apart from these there also exist the so-called one-sided surfaces on which there are not two distinct sides. The simplest of these

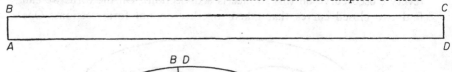

Fig. 11.

§2. SURFACES

is the well-known "Möbius band," which is obtained when we take a rectangular strip of paper $ABCD$ and paste together the two opposite short sides AB and CD so that the vertex A coincides with C, and B with D. The surface so obtained is illustrated in figure 11; it is called the Möbius band or strip. It is easy to verify that there are not two sides on it that could be painted in different colors: When we go along the middle line of the surface beginning our path at the point E, say, then we arrive again at the point E by proceeding on the surface, but on the other side of it, although we have not crossed an edge on our path. Incidentally, the edge of the Möbius surface consists of a single closed line.

The problem now arises: Do there exist closed one-sided surfaces, i.e., one-sided surfaces that do not have edges? It turns out that they exist, but that such surfaces, no matter how we arrange them in three-dimensional space, always have self-intersections. A typical example of a closed one-sided surface is illustrated in figure 12; this is the so-called "one-sided

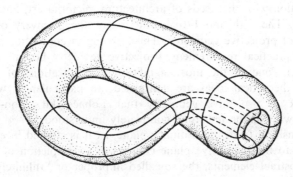

FIG. 12.

torus" or the Klein bottle. If without worrying about self-intersections we imagine two copies of a Möbius strip pasted together along their edges, (the edge of a Möbius strip, as was mentioned earlier, consists of a single contour), then we obtain a Klein bottle.

Now we can formulate the fundamental theorem of the topology of surfaces as applied to two-sided surfaces: Every closed two-sided surface is homeomorphic to some *normal surface* of genus p, i.e., to a "sphere with p handles"; two closed two-sided surfaces are homeomorphic if and only if they are of one and the same *genus p* (the same order of connectivity $2p$), i.e., when they are homeomorphic to a sphere with one and the same number of handles p.

For one-sided surfaces there also exist "normal forms," similar to the normal forms of two-dimensional surfaces of genus p, but they are complicated to describe. For this purpose we have to take a sphere, cut p circular holes in it, and attach to each of them a Möbius surface by pasting the edge of this surface to the edge of the corresponding hole. The complication that arises in an attempt to imagine such a pasting comes from the fact that there is no physical realization of it: Self-intersections of the surface arise at once in such a pasting, and they are unavoidable in every realization of a one-sided closed surface in the form of a spatial model.

One must not think that closed one-sided surfaces belong to the domain of mathematical curiosities, unconnected with serious problems of science. To see how wrong such an opinion is we need only recall that one of the fundamental achievements of geometric thinking was the creation of the so-called projective geometry, the elements of which occur nowadays in geometry courses of universities and teacher training colleges. Practical sources of projective geometry lie in the theory of perspective which dates back to the Renaissance (Leonardo da Vinci) in connection with the needs of architecture, pictorial art, and technical projection. The 16th and 17th centuries saw the discovery of the first theorems of projective geometry. Thus, arising in connection with quite a definite practical requirement, projective geometry became in its full development one of the most significant generalizations of geometry, as far as theoretical ideas are concerned. In particular, it was in the framework of projective geometry that Lobačevskiĭ's non-Euclidean geometry was for the first time completely understood.*

The transition from the ordinary plane, as it is studied in elementary geometry, to the projective plane consists in a completion of the plane by new abstract elements, the so-called improper or "infinitely distant" points. Only after such a completion does the operation of projecting one plane onto another (for example, the projection onto a screen by means of a projection lantern) become a one-to-one transformation of the one plane onto the other. The completion of the plane by improper points, which in coordinate geometry corresponds to a transition from the ordinary Cartesian coordinates to homogeneous coordinates, proceeds in the following way. Every straight line is completed by a single improper point ("at infinity"), and two straight lines have the same improper point if and only if they are parallel. A straight line completed by a single point at infinity becomes a closed line and the set of all points at infinity of all possible straight lines forms by definition an improper line or a line at infinity.

---

* See, for example, Chapter XVII, §6, or the book by P. S. Aleksandrov "What is non-Euclidean geometry," Moscow, 1951.

§2. SURFACES

Since parallel lines have a common point at infinity, in the representation of the whole process of completion of the plane by improper points it is sufficient to consider the lines passing through an arbitrary point of the plane, for example through the origin of coordinates O (figure 13). The improper points of these lines already exhaust all the improper points of the whole projective plane (since every line has the same improper point as the line through O parallel to it). Therefore we obtain a "model" of the projective plane by regarding it as a circle of "infinitely large" radius with center at O and assuming that every pair of diametrically opposite points A, A' of the circumference of this circle are united into the single "infinitely distant" point of the line AA'.

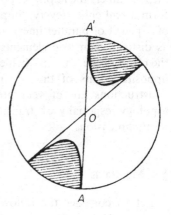

Fig. 13.

The circumference of our circle then becomes the line at infinity, but we must strictly keep in mind that any two diametrically opposite points of this circumference must be thought of as identified with one another. Hence it is at once clear that the projective plane is a closed surface, that there are no edges on it.

If we take a curve of the second order in a projective plane, which we can illustrate in the form of a hyperbola (see figure 13), then it is obvious that this hyperbola in the projective plane is a closed curve (only cut into two branches by the line at infinity). Bearing in mind that diametrically opposite points of the circumference of our fundamental circle are identified with one another, we can see without difficulty that the interior of our hyperbola which is shaded in figure 13 is homeomorphic to the interior of an ordinary circle, and the complement, the unshaded part of the projective plane, is homeomorphic to a Möbius band. Thus, from the point of view of topology the projective plane is the result of pasting together a circle (in our case the interior of a hyperbola) with a Möbius band along their edge. Hence it follows that the projective plane, i.e., the basic object of study of plane projective geometry, is a closed one-sided surface.

The example of the projective plane, apart from its great intrinsic geometrical value, is also interesting because it throws into relief one peculiarity of contemporary geometric thinking as it is moulded on the basis of Lobačevskiĭ's discovery. Geometric thinking has always been abstract, by the very character of the concept of a geometrical figure.

Now it rises to a new degree of abstraction, which in our case becomes apparent in the completion of the ordinary plane by new abstract elements, namely the improper points. Of course, even these abstract elements form a real entity (every "improper point" is nothing but an abstraction of a pencil of parallel lines), but they are introduced into our discussion as distinct geometric elements which we can only indirectly imagine as the result of the "fusion" (which does not exist physically) of diametrically opposite points of the circumference of some circle. Similar abstract constructions are of very great value in the whole of contemporary topology, especially on transition from planes to manifolds of three or more dimensions.

§3. Manifolds

Let us consider the following simple apparatus, sometimes called a compound plane pendulum (figure 14). It consists of two rods OA and AB, hinged together at A; the point O remains immovable, the rod OA turns freely in a fixed plane around O, and the rod AB turns freely in the same plane around A. Every possible position of our system is completely determined by the magnitude of the angles ϕ and

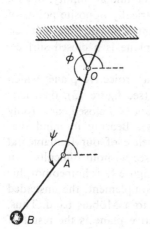
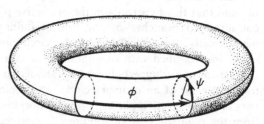

Fig. 14.   Fig. 15.

ψ that the rods OA and AB form with an arbitrary fixed direction in the plane, for example with the positive direction of the abscissa axis. We can regard these two angles, which change from O to 2π, as "geographical coordinates" of a point on a torus, counting from the "equator" of the torus and one of its "meridians," respectively, (figure 15).

§3. MANIFOLDS

Thus, we can say that the manifold of all possible states of our mechanical system is a manifold of two dimensions, namely a torus. When we replace each of the two angles ϕ, ψ by a corresponding point on the circumference of a circle on which an initial point and a direction are given (figure 16), then we can also say that every possible state of our mechanical system is completely characterized by giving one point on each of two circles (one of these is taken as the latitude ϕ and the other as the longitude ψ). In other words, just as in analytic geometry we identify a point of the plane with a pair of numbers, namely its coordinates, so in our case we can identify a point of the torus (and hence an arbitrary position of our pendulum) with the pair of its geographic coordinates, i.e., with a pair of points one of which lies on one circle and the other on another. The essence of the situation is expressed by saying that the manifold of all possible states of our compound plane pendulum, i.e., the torus, is the topological product of two circles.

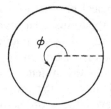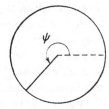

FIG. 16.

Now we modify our apparatus in the following way. Suppose that, as before, it consists of two rods OA and AB and that OA can turn freely in a definite plane around the point O but that AB is now hinged to OA by a spherical hinge at A, so that for a given position of this point it can freely turn around it in space, keeping parallel to an arbitrary original ray through A. Now the position of our system is given by three parameters of which the first is the previous angle ϕ formed by the rod OA with the positive direction of the abscissa axis, and the other two determine the direction of the rod AB in space. The latter direction can be determined, for example, by giving the point B' on the unit sphere with the centre at the origin of coordinates O at which the radius OB' parallel to AB intersects the sphere, or by giving on the sphere the two geographical coordinates of B'. Thus, the manifold of all positions of our new hinged system is a certain three-dimensional manifold, and the reader will easily realize that it can be treated as the topological product of a circle and a sphere. This manifold is closed, i.e., it has no edges, therefore it cannot be realized in the form of a figure lying in three-dimensional space. If nevertheless we wish to get a somewhat intuitive idea of it, we can consider the part of space lying between two concentric spheres. Each ray emanating from the common center of these spheres pierces them in two points. If we regard each pair of such points as identified (fused into a single point), then we obtain a three-dimensional manifold which is the topological product of a sphere with a circle.

We can make our hinged apparatus even more complicated if we not only connect the rods OA and AB by a spherical hinge at A, but also assume that the rod OA can turn freely in space around the point O. The set of possible positions of the system so obtained is then a four-dimensional closed manifold, namely the product of two spheres.

Thus we have seen that even the simplest mechanical (kinematical) considerations lead us to topological manifolds of three and more dimensions. Of even greater value in the practical, more detailed discussion of mechanical problems are certain manifolds (in general, many-dimensional), the so-called *phase spaces* of dynamical systems. Here we take into account not only the configurations that the given mechanical system can have, but also the velocities with which its various constituent points move. Let us confine ourselves to one of the simplest examples. Suppose we have a point that can move freely on a circle with an arbitrary velocity. Every state of this system is determined by two data: the position of the point on the circle and the velocity at the given instant. The manifold of states (the phase space) of this mechanical system is, of course, an infinite cylinder (a product of a circle with a straight line).

The number of dimensions of the phase space increases as we increase the number of degrees of freedom of the given system. Many of the dynamical characteristics of a mechanical system can be expressed in terms of the topological properties of its phase space. For example, to every periodic motion of the given system there corresponds a closed curve in its phase space.

The study of the phase spaces of dynamical systems occurring in various problems of mechanics, physics, and astronomy (celestial mechanics, cosmogony) drew the attention of mathematicians to the topology of many-dimensional manifolds. It was precisely in connection with these problems that the famous French mathematician Poincaré in the 1890's inaugurated the systematic construction of the topology of manifolds, by applying the so-called combinatorial method, which up to the present day is one of the fundamental methods of topology.

§4. The Combinatorial Method

Historically the first theorem in topology is the theorem or formula of Euler (which was apparently known even to Descartes). It consists in the following. Let us take the surface of an arbitrary convex polyhedron. We denote by α_0 the number of its vertices, by α_1 the number of its edges,

§4. THE COMBINATORIAL METHOD

and by α_2 the number of its faces; then the following relation is known as *Euler's formula*

$$\alpha_0 - \alpha_1 + \alpha_2 = 2. \tag{1}$$

This geometrical theorem belongs to topology, because our formula obviously remains true when we subject the convex polyhedron in question to an arbitrary topological transformation. Under such a transformation the edges will, in general, cease to be rectilinear, the faces cease to be plane, the surface of the polyhedron goes over into a curved surface, but the relation (1) between the number of vertices and the numbers of edges and faces, now curved, remains valid. The most important case is when all the faces are triangles and then we have a so-called *triangulation* (a division of our surface into triangles, rectilinear or curvilinear). It is easy to reduce the general case of arbitrary polygonal faces to this case: It is sufficient to divide these faces into triangles (for example by drawing diagonals from an arbitrary vertex of the given face). Thus, we can restrict our attention to the case of a triangulation. The combinatorial method in the topology of surfaces consists in replacing the study of such a surface by the study of one of its triangulations, and of course we are only interested in properties of the triangulation that are independent of the accidental choice of one triangulation or another and so, being common to all triangulations of the given surface, express some property of the surface itself.

Euler's formula leads us to one of such properties, and we shall now consider it in more detail. The left-hand side of Euler's formula, i.e., the expression $\alpha_0 - \alpha_1 + \alpha_2$, where α_0 is the number of vertices, α_1 the number of edges, and α_2 the number of triangles of the given triangulation, is called the *Euler characteristic* of this triangulation. Euler's theorem states that for all triangulations of a surface homeomorphic to a sphere the Euler characteristic is equal to two. Now it turns out that for every surface (and not only for a surface homeomorphic to a sphere) all triangulations of the surface have one and the same Euler characteristic.

It is easy to figure out the value of the Euler characteristic for various surfaces. First of all, for the cylindrical surface it is equal to zero. For when we remove from an arbitrary triangulation of the sphere two nonadjacent triangles but preserve the boundaries of these triangles, then we obviously obtain a triangulation of a surface homeomorphic to the curved surface of a cylinder. Here the number of vertices and of edges remains as before, but the number of triangles is decreased by two, therefore the Euler characteristic of the triangulation so obtained is zero. Now let us take the surface obtained from a triangulation of a sphere after removal of $2p$ triangles of this triangulation that are pairwise not adjacent (i.e., do not have any common vertices nor common sides).* Here the Euler characteristic is decreased by $2p$ units. It is easy to see that the Euler characteristic does not change when cylindrical tubes are attached to each pair of holes made in the surface of the sphere. This comes from the fact that the characteristic of the tube to be pasted in is, as we have seen, zero and on the rim of the tube the number of vertices is equal to the number of edges. Thus, a closed two-sided surface of genus p has the Euler characteristic $2 - 2p$ (a fact that was first proved by the French admiral de Jonquières).

We now give an important property of triangulations which satisfies the so-called condition of topological invariance (i.e., every triangulation of the given surface has the property if at least one of them does). This is the property of *orientability*. Before we formulate it, let us observe that every triangle can be oriented, i.e., that a definite direction of traversing its boundary can be furnished. Each of the two possible orientations of a triangle is given by a definite order of the sequence of its vertices.† Now let us suppose that on an arbitrary surface we are given two triangles which have a common side and no other common points (figure 17). Two orientations of these triangles are called compatible if they generate opposite directions on the common side of the triangle. (In the plane or on any other two-sided surface this means that the two triangles, when they are regarded as lying on one side of the surface, are traversed in the same direction, i.e., either both counterclockwise or both clockwise.) A triangulation of a given closed surface is called orientable if the orientations of all the triangles occurring in it can be so chosen that any two triangles, adjoining in a common side, turn out to be compatibly oriented. Then the following fact holds: Every triangulation of a two-sided surface is orientable, every triangulation of a one-sided surface is nonorientable. Therefore two-sided surfaces are also called orientable, and one-sided nonorientable. Choosing an arbitrary triangulation of a Möbius band the reader can easily convince himself of its nonorientability. In order to obtain the simplest triangulation of the projective plane, we have only to draw in it any three straight lines that do not pass through one and the same point (figure 18). They divide the projective plane into four triangles one of which lies in the finite part of the plane, while each of the

---

* In order to do this we have only to make the triangulation in question sufficiently "fine." This can always be achieved by a suitable subdivision of an arbitrary triangulation.

† Moreover, it is easy to see that two orders of the vertices determine one and the same orientation (one and the same direction of circuit) if and only if they go over into one another by an "even" permutation. Thus (ABC), (BCA), (CAB) determine one orientation of the triangle, and (BAC), (ACB), (CBA) the other. (About even and odd permutations see, e.g., Chapter XX, §3.)

§4. THE COMBINATORIAL METHOD

other three is cut by the line at infinity into two parts. In figure 18 one of these triangles, extending to infinity, is shaded. From the same figure it is clear that when we attempt to give all four triangles compatible orientations we are inevitably doomed to failure. In particular, in the

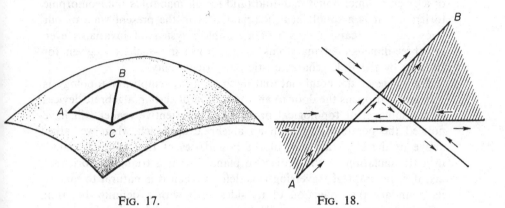

Fig. 17.　　　　　Fig. 18.

choice of orientations of our four triangles as it is made in figure 18, we obtain for the algebraic sum of its boundaries not zero, as it should be for compatible orientations, but the line AB taken twice.

Euler's characteristic and the property of orientability or nonorientability of closed surfaces give us, so to speak, a complete system of topological invariants of closed surfaces. The meaning of this statement is that two surfaces are homeomorphic if and only if, first, their triangulations have one and the same Euler characteristic, and second, both are orientable or not orientable.

The combinatorial method is applicable not only to the study of surfaces (two-dimensional manifolds) but also to manifolds of an arbitrary number of dimensions. But in the case of three-dimensional manifolds, for example, the role of the ordinary triangulations is now taken by decompositions into tetrahedra. They are called three-dimensional triangulations or simplicial divisions of the manifold. The Euler characteristic of a three-dimensional triangulation is defined as the number $\alpha_0 - \alpha_1 + \alpha_2 - \alpha_3$, where α_i, $i = 0, 1, 2, 3$, is the number of i-dimensional elements of this triangulation (i.e., α_0 the number of vertices, α_1 the number of edges, α_2 the number of two-dimensional boundaries, α_3 the number of tetrahedra). For a number of dimensions $n > 3$ the manifolds are divided into n-dimensional simplexes, i.e., the simplest convex n-dimensional polyhedra analogous to triangles ($n = 2$) and tetrahedra ($n = 3$). The simplexes into which an n-dimensional manifold is divided and their boundaries form an n-dimensional triangulation of this manifold. As before we can speak of an Euler characteristic, interpreting it as the sum $\sum_{i=0} (-1)^i \alpha_i$, where α_i is the number of i-dimensional elements occurring in this triangulation ($i = 0, 1, 2, \cdots, n$), and as before the Euler characteristic has one and the same value for all triangulations of a given n-dimensional manifold (and for all manifolds homeomorphic to it); i.e., it is a topological invariant. But in the present state of our knowledge, we cannot dream of a complete system of invariants even for three-dimensional manifolds (in the sense in which it is given for surfaces by the Euler characteristic and orientability).

The value of the combinatorial method in contemporary topology is very great. It opens the door to an application of certain algebraic devices in the solution of topological problems. The attentive reader will have noticed the possibility of such an algebraic approach when we talked above of the algebraic sum of the boundaries of the oriented triangle in a triangulation of the projective plane. For if a triangle is oriented, i.e., if a direction of traversing it is defined, then it is natural to take as its boundary the collection of its sides each with a definite direction, namely the one that continues the existing circuit of the triangles.

Now let us consider all the triangles T_i^2, $i = 1, 2, \cdots, \alpha_2$, that occur in a given triangulation of a surface. To each of them we can give two orientations; let us denote the triangle T_i^2 with one of the two possible orientations by t_i^2, and the same triangle with the other (opposite) orientation by $-t_i^2$. In exactly the same way we can orient each of the one-dimensional elements (edges) T_k^1 ($k = 1, 2, \cdots, \alpha_1$) that occur in the given triangulation, i.e., provide it with one of the two possible directions. We denote the segment T_k^1 with one of these orientations by t_k^1, and with the other by $-t_k^1$. Now if the sides of the triangle T_i^2 are T_1^1, T_2^1, T_3^1, then the boundary of the oriented triangle t_i^2 is the set of the same sides, but taken with a definite direction, i.e., the boundary consists of the directed segments $\epsilon_1 t_1^1$, $\epsilon_2 t_2^1$, $\epsilon_3 t_3^1$; here $\epsilon_i = 1$ if this direction for the edge T_i^1 coincides with its appropriate direction t_i^1, and $\epsilon_i = -1$ in the opposite case. The boundary of t_i^2 is denoted by Δt_i^2. As we have seen, $\Delta t_i^2 = \sum \epsilon_k t_k^1$ and this sum can be imagined as extending over all edges of our triangulation when we consider the coefficients ϵ_k for segments not in the boundary of t_i^2 as being equal to zero.

It now becomes natural to consider more general sums of the form $x^1 = \sum a_k t_k^1$ extended over all edges of the given triangulation.* The geometric meaning of such sums is very simple: Every summand of the sum is a certain segment that occurs in our triangulation, taken with a

---

* The coefficients a_k are assumed to be integers.

§4. THE COMBINATORIAL METHOD

definite direction and a definite coefficient (a definite "multiplicity"). The whole algebraic sum so described expresses a path composed of segments in which every segment is assumed to occur as often as its coefficient indicates. For example, if we begin by running around the polygon $ABCDEF$ (figure 19) in the direction of the arrow, then go along AA' to the polygon $A'B'C'D'E'F'$ and traverse it in the indicated direction, and then return along $A'A$ and go around $ABCDEF$ again in the same direction as before, we obtain a sum in which the segments AB, BC, CD, DE, EF, FA, occur with the coefficient 2, the segments $A'B', B'C', C'D', D'E', E'F', F'A'$, with the coefficient 1, and the segment AA' does not occur at all (it has the coefficient zero because it is traversed twice in opposite directions).

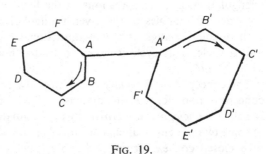

FIG. 19.

Sums of the form $x^1 = \sum a_k t_k^1$ are called *one-dimensional chains* of the given triangulation. From the algebraic point of view they represent linear forms (homogeneous polynomials of the first degree); they can be added and subtracted and also multiplied by an arbitrary integer according to the usual rules of algebra. Of particular importance among the one-dimensional chains are the so-called one-dimensional cycles. Geometrically they correspond to closed paths (and in figure 19 we have just dealt with such a path).

For a purely algebraic definition of a cycle, let us make the convention that of the two vertices of a segment AB the end point B occurs in the boundary of $\overrightarrow{AB}$ with the plus sign (with the coefficient $+1$) and the initial point A with the minus sign (with the coefficient -1). Then the *boundary* of the segment $\overrightarrow{AB}$ can be written in the form $\Delta(\overrightarrow{AB}) = B - A$.

When we accept this convention, we observe immediately that the sum of the boundaries of the segments $\overrightarrow{AB}, \overrightarrow{BC}, \overrightarrow{CD}, \cdots, \overrightarrow{FA}$, which form a closed path (in the usual sense of the word), is zero. This leads us naturally to the general definition of a one-dimensional *cycle* as a one-dimensional chain $z^1 = \sum_k a_k t_k^1$ for which the sum of the boundaries of the terms, i.e., the sum $\sum_k a_k \Delta t_k^1$, is zero. It is easy to verify that the sum of two cycles is a cycle. When we multiply a cycle as an algebraic expression by an arbitrary integer, we obtain again a cycle. This enables us to speak of linear combinations of cycles $z_1^1, z_2^1, \cdots, z_s^1$, i.e., of cycles of the form $z = \sum_{\nu=1}^s c_\nu z_\nu^1$, where c_ν are integers.

In analogy to the concept of a one-dimensional chain of a given triangulation, we can also speak of two-dimensional chains of this triangulation, i.e., of expressions of the form $x^2 = \sum_i a_i t_i^2$, where the t_i^2 are oriented triangles of the given triangulation. Since the boundary of each oriented triangle is a one-dimensional cycle, the chain $\sum_i a_i \Delta t_i^2$ is also a cycle. This cycle is taken to be the boundary Δx^2 of the chain $x^2 = \sum_i a_i t_i^2$.

The concept of a *boundary of a chain* now enables us to formulate the concept of homology: A one-dimensional cycle z^1 of a given triangulation is called *homologous to zero* in this triangulation if it is the boundary of some two-dimensional chain of this triangulation. In every triangulation of a closed convex surface and more generally of any surface homeomorphic to a sphere, every one-dimensional cycle is homologous to zero; geometrically this is perfectly clear: Every closed polygon on a convex surface is the boundary of some piece of the surface. But this is not so on the torus: A meridian of the torus as well as its equator are not boundaries of any piece of the surface. If we take an arbitrary triangulation of the torus, then we can find in it cycles similar to the meridian or the equator of the torus and these cycles are not homologous to zero.

We observe an entirely new phenomenon in the triangulation of the projective plane we have constructed. If we regard a straight line, for example AB (see figure 18), as a cycle of this triangulation, then this cycle is not homologous to zero in it. However, the same line taken with the coefficient 2 does turn out to be homologous to zero. Thus, the introduction of coefficients other than $+1$ in the definition of chains, which appears in the first instance formal and unnecessary, enables us to discern important geometric properties of surfaces and more generally of manifolds. In the given case this is the so-called property of torsion which consists in the existence of cycles that in the given manifold are not homologous to zero (do not bound any piece of the surface) but become homologous to zero when they are provided with certain integer coefficients.

In connection with what we have said, let us finally introduce the extremely important concept of *homological independence* of cycles. The cycles $z_1, \cdots, z_s$ are called homologically independent in the given triangulation if no linear combination $\sum c_i z_i$ of them in which at least one coefficient c_i is different from zero is homologous to zero in this triangulation. As examples of homologically independent cycles on the torus, we can take an arbitrary meridian and equator considered as cycles of some triangulation of the torus.

The fundamental concepts of the whole of combinatorial topology (the concepts of boundary, cycle, homology) were defined by us for one-

§4. THE COMBINATORIAL METHOD

dimensional formations, but they can be extended verbatim to an arbitrary number of dimensions. For example, a two-dimensional chain $z^2 = \sum a_i t_i^2$ is called a cycle if its boundary $\Delta z^2 = \sum a_i \Delta t_i^2$ is equal to zero. A three-dimensional chain is an expression of the form $x^3 = \sum a_h t_h^3$, where the t_h^3 are oriented three-dimensional simplexes (tetrahedra).

As in the case of the triangle, the orientation of the three-dimensional simplex (tetrahedron) is given by a definite order of its vertices, where two orders of the vertices that can be carried into one another by an even permutation determine one and the same orientation. The boundary of a three-dimensional oriented simplex $t^3 = (ABCD)$ is the two-dimensional chain (cycle) $\Delta t^3 = (BCD) - (ACD) + (ABD) - (ABC)$ (figure 20). The boundary of a three-dimensional chain is defined as the sum of the boundaries of its simplexes taken with the same coefficient with which these simplexes occur in the given chain. The reader can easily verify that the boundary of an arbitrary three-dimensional chain is a two-dimensional cycle (it is sufficient to prove this for the boundary of a single three-dimensional simplex). We say that a two-dimensional cycle is homologous to zero in a given manifold if it is the boundary of some three-dimensional chain of this manifold, and so on. Observe that from

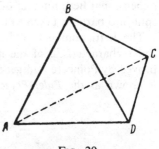

FIG. 20.

the definition of orientable and nonorientable triangulations* given previously it follows easily that in every orientable triangulation there occur cycles (two-dimensional in the case of surfaces) different from zero, and in nonorientable triangulations there are no such cycles; this result can also be generalized immediately to an arbitrary number of dimensions.

The concepts introduced enable us to define the order of the one-dimensional, two-dimensional, etc., connectivity of given manifolds of an arbitrary number of dimensions. The maximal number of homologically independent one-dimensional, two-dimensional, etc., cycles occurring in an arbitrary triangulation of a given manifold does not depend on the choice of the triangulation of this manifold and is called its *order of connectivity*, or its Betti number (of the corresponding dimension).

The one-dimensional Betti number of a closed orientable surface of genus p is equal to $2p$ (i.e., to the order of connectivity of the surface,

---

* This definition, given earlier for triangulations of surfaces, can be extended to the case of triangulations of manifolds of an arbitrary number of dimensions.

as defined in §2). The one-dimensional Betti number of the projective plane is equal to zero. (Here every cycle not extending to infinity bounds part of the plane, i.e., is homologous to zero, but a cycle extending to infinity, for example a projective line, turns out to be homologous to zero if it is taken twice.) The two-dimensional Betti number of every nonorientable surface is equal to zero (there is not a single two-dimensional cycle different from zero on such a surface).

The two-dimensional Betti number of every orientable surface is equal to one. For if we orient all the triangles of an arbitrary triangulation of an orientable surface in a suitable manner, then we obtain a cycle (the so-called fundamental cycle of the surface). It is not difficult to observe that every two-dimensional cycle is obtained from a fundamental cycle by multiplying it by an arbitrary integer. These results can be generalized immediately to n-dimensional manifolds. We remark that the zero-dimensional Betti number of a connected manifold (i.e., one that does not split into parts) is taken to be one.

The Betti numbers of the various dimensions are connected with the Euler characteristic of the manifold by a remarkable formula that was proved by Poincaré and generalizes Euler's theorem. This formula, which is known as the *Euler-Poincaré formula*, has the following simple form:

$$\sum_{r=0}^{n} (-1)^r \alpha_r = \sum_{r=0}^{n} (-1)^r p_r.$$

Here we have on the left-hand side the Euler characteristic of an arbitrary triangulation of the given manifold, and the numbers p_r on the right are the Betti numbers of the various dimensions r of this manifold. In particular, for orientable surfaces we have, as we have just seen, $p_0 = p_2 = 1, p_1 = 2p$, where p is the genus of the surface. This gives us Euler's theorem for orientable surfaces

$$\alpha_0 - \alpha_1 + \alpha_2 = 2 - 2p.$$

§5. Vector Fields

Let us consider the simplest differential equation

$$\frac{dy}{dx} = F(x, y), \qquad (2)$$

in a given plane domain G. Its geometrical meaning is that at every point (x, y) of G a direction is defined whose slope is equal to $F(x, y)$, where $F(x, y)$ is a certain continuous function of the point (x, y). We say

§5. VECTOR FIELDS

that in G a continuous field of directions is given; we can easily turn it into a continuous vector field by taking, for example, a vector of unit length in each of the given directions. The task of integrating the differential equation (2) consists in splitting, if this is possible, the given plane domain into pairwise nonintersecting curves (the "integral curves" of the equation) such that at each point of the domain the direction given in it is the direction of the tangent to the unique integral curve passing through this point.

Let us consider, for example, the equation

$$\frac{dy}{dx} = \frac{y}{x}.$$

At every point $M(x, y)$ of the plane the direction corresponding to it is obviously that of the ray $\overrightarrow{OM}$ (where O is the origin of coordinates). The integral curves are the straight lines passing through O. Through every point of the plane other than O there passes a unique integral curve. As regards the origin of coordinates, this is a singular point of the given differential equation (a so-called "node") through which all the integral curves pass.

If we take the differential equation

$$\frac{dy}{dx} = -\frac{x}{y},$$

then we see that it associates with every point $M(x, y)$ other than O the direction that is perpendicular to $\overrightarrow{OM}$. In this case the integral curves are circles with their center at O which is again a singular point of our differential equation, but a singular point of an altogether different type. It is not a "node", but a so-called "center." There are also other types of singular points (see Chapter V, §6), some of which are illustrated in figures 21 and 22. The differential equation $dy/dx = y/x$ has no closed integral curves. In contrast, the differential equation $dy/dx = -x/y$ has only closed integral curves. There are also possible integral curves that wind in the form of a spiral around a singular point, which in this case is called a focus.

Of extreme importance in various applications is the case of a so-called limit cycle, namely a closed integral curve, around which other integral curves wind spirally. Many other cases of the mutual arrangement of the integral curves are possible, and also their position with respect to the singular points. All problems concerning the forms and positions of the integral curves of a differential equation, and also the number, character, and mutual arrangement of its singular points, belong to the qualitative theory of differential equations. As the name implies, the qualitative theory of differential equations leaves aside the direct integration of the differential equation "in finite form," as well as methods of approximation or of numerical integration. The basic object of the qualitative theory is in essence the topology of the field of directions and the system of integral curves of the given differential equation.

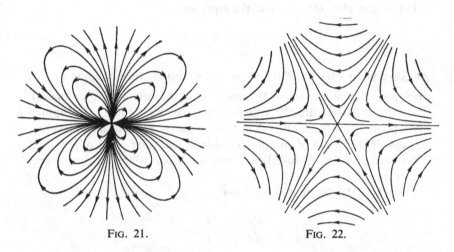

Fig. 21.   Fig. 22.

The qualitative approach to differential equations, including such problems as the existence of closed integral curves, in particular all problems connected with the existence, number, and occurrence of limit cycles, was dictated in the first instance by problems of mechanics, physics, and technology. These problems arose first in connection with the investigations of Poincaré in celestial mechanics and cosmogony which, as we have mentioned above, were also the cause for the topological researches of the French geometer. The circle of topological problems in the theory of differential equations has occupied one of the central places in the outstanding investigations by Soviet scholars in the theory of oscillations and radiotechnology; we think here of the school of L. I. Mandel'stam, or of the school of A. A. Andronov that has developed from it, which constitutes one of the most important centers of research in the qualitative and essentially topological theory of differential equations. Another center of research in the qualitative theory of differential equations by essentially topological methods is the school of V. V. Stepanov and V. V. Nemyckiĭ at the University of Moscow. In papers of mathematicians in Leningrad, Sverdlovsk and Kazan on problems of

§5. VECTOR FIELDS

the qualitative theory of differential equations topological method also play a prominent role.

The theory of differential equations leads to the study of vector fields not only in the plane but also in many-dimensional manifolds; even the simplest systems of several differential equations are interpreted geometrically as fields of directions in many-dimensional Euclidean spaces. The introduction of so-called first integrals of an equation means a selection among the set of all integral curves of those that lie in a certain manifold defined by the given first integral. Every dynamical system (in the classical meaning of this word) gives rise, generally speaking, to the many-dimensional manifold of its possible states (see §3) and to a system of differential equations whose integral curves, filling the given phase space, represent possible motions of the given system. Each individual one of these motions is determined by some set of initial conditions. Therefore, a fundamental object of study in this case is the field of directions and the system of trajectories on the given manifold. Numerous applications, especially in recent years, have made it understandable that the qualitative theory of differential equations should be developed in its widest aspects, with a consequent development of topology also as the basis of this theory. Precisely these mechanical, physical, and even astronomical topics have caused the rapid growth of contemporary topology which forms such a significant part of the general development of mathematics in the present half century. The reader who wishes to acquaint himself with the topological problems in the theory of differential equations and its concrete physical and technological aspects can consult the well-known book by A. A. Andronov and S. E. Haĭkin "Theory of oscillations," an English-language edition of which was published by the Princeton University Press, Princeton, N.J., 1949.

As an example of a concrete problem in the theory of vector fields on many-dimensional manifolds let us consider the problem of the algebraic number of singular points of such a vector field.

Suppose that a smooth manifold is given. For simplicity we shall imagine a smooth closed surface. Let us assume that at every point of this manifold a tangent vector to it is given that depends continuously on the point both in length and in direction. The singular points of such a vector field are those points of the manifold at which the vector associated with them is the null vector, i.e., where there is no definite direction. We shall assume that each of these points is isolated. In the case of a closed manifold, this means that there is only a finite number of singular points. (Otherwise part of these points would be condensed near a certain limit point, which by the continuity of the field would also be a singular point but not isolated.)

We can define the index of an isolated singular point, a concept that in a certain sense is similar to the concept of multiplicity of a root of an algebraic equation. In order to define the index we surround the given singular point by an arbitrary closed curve C that "isolates it" (in the plane simply a circle), i.e., a curve that does not pass through any singular point and contains in its interior only the one singular point in question. At all points of C the direction of the field is uniquely determined. For simplicity we shall assume that a neighborhood of our singular point including C is flat (in the general case we can transform the neighborhood that interests us together with the field given on it into a plane). When we go around the curve in the positive direction, the angle that the direction of the field forms with an arbitrary fixed direction returns to its initial value increased as a result of our going around the closed curve by a certain summand of the form $2k\pi$, where k is a certain well-defined integer. This number is called the *index of the singular point* or the *winding number* of the field along C. Note that it does not depend on the special choice of the closed curve that isolates the given singular point. In figures 21 and 22, we have illustrated singular points with the indices —2 and +3, respectively. In a similar but more complicated manner, we can also define the index for a singular point of a field of vectors (directions) defined on an n-dimensional manifold for $n > 2$. Now the following remarkable theorem was proved in 1926 by the German mathematician Hopf: If on a given manifold a continuous vector field is defined having only a finite number of singular points, then the sum of the indices or, as one says, the algebraic number of these singular points, does not depend on the field and is always equal to the Euler characteristic of the manifold.

From Hopf's theorem it follows that vector fields without singularities are possible only on manifolds whose Euler characteristic is equal to zero; it turns out that on such manifolds a vector field without singularities can always be constructed. Thus, among all the closed surfaces only on the torus and the so-called one-sided torus (the Klein bottle) can vector fields without singularities be constructed.

Closely connected with the theory of vector fields is the theory of continuous mappings of manifolds into themselves and particularly the results concerning the existence of fixed points for such mappings. A point x is called a fixed point of a given mapping f if its image under this mapping coincides with the point itself, i.e., if

$$f(x) = x.$$

In order to explain the character of this connection, let us consider the simplest case, namely the case of a continuous mapping f of a circle K into itself. By joining each point x of K to its image $f(x)$, we obtain a

§5. VECTOR FIELDS

vector $\vec{u_x} = \overrightarrow{x, f(x)}$. This vector is the null vector if and only if $f(x) = x$, i.e, if x is a fixed point of the given mapping. Let us show that such a point actually exists. For this purpose we assume the contrary and determine the winding number of our vector field along the circumference C of the circle K.

Under a continuous transformation of our field, its winding number along C can obviously change only continuously. But since it is an integer, it must remain constant. Hence it follows that the winding number of the field along C is equal to 1. Indeed, since every point of K is mapped into the same circle, the vector u_x of a point x on the circumference C (which according to our assumption is not the null vector) points into the interior of the circle and therefore forms an acute angle with the radius Ox, which we regard as a vector pointing to the center O.

Now let us subject the directions of all the vectors u_x for the points x lying on C to a continuous transformation. This transformation consists of turning each of these vectors through such an acute angle that it comes to point in the direction of the center O. As we have just said, the winding number of the field along C does not change in this process. But as a result of this transformation, our original field goes over on C into the field of radial vectors, which obviously has the winding number one. Thus, our initial field also had the winding number one along the circumference C.

In virtue of the continuity of the original vector field its winding numbers along two circumferences with one and the same centre O and radii of slightly different length have one and the same value.* Therefore the winding number of the field along all circles with center O lying within K has one and the same value, namely one. But since by assumption the vector u_x is defined and different from zero for all points of the circle and among them for its center O, the winding number of the field along a circumference of sufficiently small radius with center at O is certainly equal to zero. So we have arrived at a contradiction and have proved that under a continuous mapping of a circle into itself there is always at least one fixed point. This theorem is a special case of a very important theorem of Brouwer which states that under every continuous mapping of an n-dimensional sphere into itself there is always at least one fixed point.

In recent years the problem of the existence of fixed points under mappings of one type or another has been studied in detail and forms an essential part of the topology of manifolds.

---

* In virtue of the assumption that the field in question, which is everywhere defined and different from zero, has no fixed points under f, we can also talk of its winding number along an arbitrary curve within K.

§6. The Development of Topology

The topology of closed surfaces is the only domain of topology that was more or less worked out at the end of the last century. The construction of this theory was connected with the development of the theory of functions of a complex variable in the course of the 19th century. The latter, which forms one of the most significant phenomena in the history of mathematics during the past century, has been built up by several distinct methods. One of the most fruitful in the sense of understanding the essence of the phenomena to be studied was the geometric method of Riemann. Riemann's method, which showed very convincingly that it is impossible in the general theory of functions of a complex variable to restrict ourselves to single-valued functions only, led to the construction of the so-called Riemann surfaces. In the simplest case of algebraic functions of a complex variable, these surfaces always turn out to be closed orientable surfaces. The investigation of their topological properties is in a certain sense equivalent to the investigation of the given algebraic function. Further development of Riemann's idea is due to Poincaré, Klein, and their followers and led to the discovery of unexpected and deep connections between the theory of functions, the topology of closed surfaces, and non-Euclidean geometry, namely the theory of the group of motions in a Lobačevskiĭ plane.* Thus, topology from the first showed itself to be organically related to a whole group of problems of fundamental importance connected with very diverse domains of mathematics.

In the further development of this circle of problems, it turned out that the topology of surfaces alone was insufficient, that the solution of problems in n-dimensional topology was necessary. The first of these was the problem of the topological invariance of the number of dimensions of a space. This problem consists in proving the impossibility of a topological mapping of an n-dimensional Euclidean space into an m-dimensional for $n \neq m$. This difficult problem was solved in 1911 by Brouwer.† In connection with its solution, new topological methods were discovered that led to a rapid construction of the beginnings of the theory of continuous mappings of many-dimensional manifolds and the theory of vector fields on them. All these investigations were found to involve the first fundamental concepts of the so-called set-theoretical topology

---

* In connection with this see the book by A. I. Markuševič "Theory of analytic functions," Gostehizdat, 1950.

† Actually, for the development of the theory of functions of a complex variable it was necessary to solve an even more difficult problem, namely to prove that the topological image of an n-dimensional domain lying in an n-dimensional space is always again a domain. This problem was also solved by Brouwer.

§6. THE DEVELOPMENT OF TOPOLOGY

that arise on the basis of the general theory of sets constructed by Cantor in the last quarter of the 19th century.

In set-theoretical topology the very object of investigation, i.e., the class of geometric figures under consideration, is extremely wide and comprises if not altogether all sets in Euclidean spaces, then at least all the closed sets. Scholars of many countries collaborated in the rapid development of the new set-theoretical direction of topology, but above all we must mention the Polish topological school.

An essentially new direction of development of set-theoretical topology was taken in papers of Soviet topologists; in particular the outstanding Soviet mathematician P. S. Uryson (1898–1924), who met an untimely death, developed the general theory of dimension which laid the foundation of a classification of very general point sets by a fundamental criterion, namely the number of dimensions. This classification turned out to be extremely fruitful and involved entirely new points of view in the study of the most general geometric forms.* Uryson's ideas, as developed in his theory of dimension, were a stepping stone for the remarkable work of L. A. Ljusternik (jointly with L. G. Šnirel'man) on the variational calculus.

These papers contain, apart from other results, an exhaustive positive solution of a famous problem of Poincaré on the existence of three closed geodesic lines without multiple points on every surface homeomorphic to a sphere.

On the other hand, P. S. Aleksandrov, on the basis of the theory of dimension, transferred the algebraic methods of combinatorial topology to the realm of set theory and this led in turn to new directions of topological investigations in which mathematicians of the Soviet Union, including the younger generation, hold a leading place.†

As regards the proper combinatorial topology, after the papers of Poincaré and Brouwer, approximately in 1915, there begins a group of

---

* The inductive definition of dimension of sets proposed by Uryson can be regarded as the fullest development of Lobačevskiĭ's idea of dissection as the fundamental geometric operation. In a rough approximation it amounts to this. A set is zero-dimensional if it can be represented in the form of a sum of arbitrarily small parts no two of which are in contact with one another. A set is n-dimensional if it can be "dissected" by $(n - 1)$-dimensional subsets into arbitrarily small parts no two of which are in contact and if this cannot be achieved with sets of dimensions smaller than $n - 1$. (A precise definition of contact as it is understood in contemporary topology will be given in §7.)

† Here we must refer to the so-called homological theory of dimension of P. S. Aleksandrov, to the related remarkable constructions of L. N. Pontrjagin and to further developments of the homological theory of dimension in papers by M. S. Bokšteĭn, V. G. Boltjanskiĭ and particularly K. A. Sitnikov. The duality law of L. S. Pontrjagin will be discussed later.

investigations by American topologists, Veblen, Birkhoff, Alexander, and Lefschetz. They achieved very remarkable results. For example, Alexander proved the topological invariance of the Betti numbers and also a very important duality theorem that served as a starting point for the subsequent investigations of L. S. Pontrjagin; Lefschetz gave a certain formula for the algebraic number of fixed points under arbitrary continuous mappings of manifolds and so laid the foundation of the general algebraic theory of continuous mappings that was further developed by Hopf; to Birkhoff our science owes an essential advance in the theory of dynamical systems in its purely topological aspect, its metrical aspect, etc.. Further very deep developments of the topology of manifolds and their continuous mappings were obtained in papers by Hopf, who proved along with many other results the existence of an infinite number of continuous mappings of a three-dimensional sphere onto a two-dimensional one, which are essentially different from one another in the sense that no two of these mappings can be carried into one another by a continuous change. So Hopf became the founder of a new direction, the so-called homotopic topology. Recently a powerful new impetus has come to homotopic topology, as to the whole of combinatorial topology, from the work of the new French topological school (Leray, Serre, and others).

As we have mentioned earlier, the fundamental investigations of Uryson were the beginning of the activities of Soviet mathematicians in the domain of topology. These investigations were concerned with set-theoretical topology, but already at the end of the 1920's Soviet topology comprised also combinatorial topology in the range of its interests. This came about in a rather original manner, namely by the application of combinatorial methods to the study of closed sets, i.e., to objects of a very general nature. On this foundation there arose one of the most remarkable geometric discoveries of the present century, the statement and proof by L. S. Pontrjagin of his general law of duality, which establishes deep and in a certain sense exhaustive connections between the topological structure of a given closed set in an n-dimensional Euclidean space and the parts of the space complementary to it. In connection with his duality law, Pontrjagin constructed a general theory of characters of commutative groups and this led him to further investigations in the domain of the general topological and the classical continuous Lie groups, a domain that has been completely transformed by the work of Pontrjagin. Subsequently Pontrjagin and his pupils made a number of notable investigations on the topology of manifolds and their continuous mappings (Z. G. Boltjanskiĭ, M. M. Postnikov, and others). In these investigations a new method was applied, the so-called ∇-homology (cohomology) introduced into combinatorial topology by A. N. Kolmogorov and, independently, by Alexander. This method, which now occupies the first place in the whole of homotopic topology, has made it possible to continue Pontrjagin's duality theory in the most diverse directions, and this has led to the duality theorems of A. N. Kolmogorov (and Alexander), P. S. Aleksandrov, and K. A. Sitnikov, which belong to the most significant results of contemporary topology. The same method has also found important applications in very recent papers by L. A. Ljusternik on the calculus of variations.

§7. Metric and Topological Spaces

At the beginning of our account, we have talked of adjacency (of different parts of a given figure) as of a fundamental topological concept, and we have defined continuous transformations as those that preserve this relation. However, we have not given a rigorous definition of this fundamental concept; to do this in sufficient generality we have to use concepts of set theory. This will be our task in the present section; we shall finally solve it by introducing the concept of a topological space.

The theory of sets made it possible to give the concept of a geometrical figure a breadth and generality that were inaccessible in the so-called "classical" mathematics. The object of a geometrical, in particular a topological, investigation now becomes an arbitrary point set, i.e., an arbitrary set whose elements are points of an n-dimensional Euclidean space. Between points of an n-dimensional space a *distance* is defined: namely, the distance between the points $A = (x_1, x_2, \cdots, x_n)$ and $B = (y_1, y_2, \cdots, y_n)$ is by definition equal to the nonnegative number

$$\rho(A, B) = \sqrt{(x_1 - y_1)^2 + (x_2 - y_2)^2 + \cdots + (x_n - y_n)^2}.$$

The concept of distance permits us to define adjacency first between a set and a point, and then between two sets. We say that a point A is an adherent point of the set M if M contains points whose distance from A is less than any preassigned positive number. Obviously every point of the given set is an adherent point of it, but there may be points that do not belong to the given set and are adherent to it. Let us take, for example, the open interval (0, 1) on the numerical line, i.e., the set of all points lying between 0 and 1; the points 0 and 1 themselves do not belong to this interval, but are adherent to it, since in the interval (0, 1) there are points arbitrarily near to zero and points arbitrarily near to one. A set is called *closed* if it contains all its adherent points. For example the closed interval [0, 1] of the numerical line, i.e., the set of all points x satisfying the inequality $0 \leqslant x \leqslant 1$, is closed. Closed sets in a plane and all the more in a space of three or more dimensions can have an extremely complicated structure; indeed, they form the main study object of the set theoretical topology of an n-dimensional space.

Next we say that two sets P and Q adjoin one another if at least one of them contains adherent points of the other. From the preceding it follows that two closed sets can adjoin only when they have at least one point in common; but, for example, the intervals [0, 1] and (1, 2), which do not have common points, adjoin because the point 1 which belongs to [0, 1] is at the same time an adherent point of (1, 2). Now we can say that a set R is divided ("dissected") by a set S lying in it, or that S is a "section" of $R - S$ consisting of all the points of R that do not belong to S can be represented as the sum of two nonadjoining sets.

Thus, Lobačevskiĭ's ideas on adjacency and dissection of sets receive in contemporary topology a rigorous and highly general expression. We have already seen how Uryson's definition of dimension of an arbitrary set (see the remark in §6) is founded on these ideas; the statement of this definition now becomes completely rigorous. The same applies to the definition of a continuous mapping or transformation; a mapping f of a set X onto a set Y is called continuous if adjacency is preserved under this mapping, i.e., if the fact that a certain point A of X is an adherent point of an arbitrary subset P of Y implies that that image $f(A)$ of A is an adherent point of the image $f(P)$ of P. Finally a one-to-one mapping of a set X onto a set Y is called *topological* if it is continuous and if its inverse mapping of Y onto X is also continuous. These definitions give an accurate basis to all that has been said in the first sections of the present account.

However, set theoretical topology is not restricted to the degree of generality that is achieved by considering as geometrical figures all the point sets. It is natural to introduce the concept of distance not only between points of an arbitrary Euclidean space but also between other objects that do not appear to refer at all to geometry.

Let us consider, for example, the set of all continuous functions defined, say, on the interval [0, 1]. We can define the distance $\rho(f, g)$ between the two functions f and g as the maximum of the expression $|f(x) - g(x)|$, when x ranges over the whole segment [0, 1]. This "distance" has all the basic properties of distance between two points in space: $\rho(f, g)$ between the two functions f and g is equal to zero if and only if the functions coincide, i.e., if $f(x) = g(x)$ for every point x; further, the distance is obviously symmetrical, i.e., $\rho(f, g) = \rho(g, f)$; finally, it satisfies the so-called triangle axiom: For any three functions f_1, f_2, f_3 we have $\rho(f_1, f_2) + \rho(f_2, f_3) \geqslant \rho(f_1, f_3)$. It is customary to say that the so-defined distance turns our set of functions into a metric space (usually denoted by C). By a metric space we understand more generally, a set of arbitrary objects

§7. METRIC AND TOPOLOGICAL SPACES

that are to be called *points* of the metric space if between any two points there is defined a *distance*, a nonnegative number satisfying the "axioms of distance" just stated.

Now when an arbitrary metric space is given, we can talk of adherent points of its subsets and consequently of adjacency of its subsets to one another and of topological concepts in general (closed sets, continuous mappings, and further concepts to be introduced on the basis of these simplest ones). This course opens up an extensive and extremely fruitful field of application for topological and general geometrical ideas to ranges of mathematical objects where it would appear completely impossible to talk of any kind of geometry. Let us give an illustrative example.

We take again the differential equation (2)

$$\frac{dy}{dx} = F(x, y).$$

If $y = \phi(x)$ for $0 \leqslant x \leqslant 1$ is a solution of this equation that assumes at $x = 0$ the value $y = 0$, say, then the function $\phi(x)$ obviously satisfies the integral equation

$$\phi(x) = \int_0^x F[x, \phi(x)]\, dx. \tag{3}$$

Now we consider the integral $G(f) = \int_0^x F[x, f(x)]\, dx$, where $0 \leqslant x \leqslant 1$ and $f(x)$ is an arbitrary continuous function defined on the interval [0, 1]. This integral is a certain continuous function $g(x)$ also defined on [0, 1]. So the expression $G(f) = \int_0^x F[x, f(x)]\, dx$ associates with every function f a function $g = G(f)$; in other words, we have a mapping G, easily seen to be continuous, of the metric space C into itself. How can we characterize here a function $\phi(x)$ (there may be several of them) that is a solution of the equation (2) or the equation (3) equivalent to it? Obviously under our mapping it goes over into itself; i.e., it is a fixed point of our mapping G. Now it turns out that such a fixed point of the mapping G actually exists, as follows from a very general theorem on fixed points of continuous mappings of metric spaces that was proved in 1926 by P. S. Aleksandrov and V. V. Nemyckiĭ. Nowadays the study of various metric spaces whose points are functions of one kind or another (such spaces are called functional spaces) is a constantly used tool of analysis, and the study of functional spaces by methods which are partly topological, but mainly algebraic, in a wide sense of the word, forms the content of functional analysis (see Chapter XIX).

Functional analysis, as was mentioned in the introductory chapter, occupies an extremely prominent place in the contemporary mathematical scene in view of the variety of its connections with all sorts of other parts of mathematics and its value in natural science, above all in theoretical physics. The investigation of topological properties of functional spaces is closely connected with the calculus of variations and the theory of partial differential equations (investigations by Ljusternik, Morse, Leray, Schauder, Krasnosel'skiĭ, and others). Problems on the existence of fixed points under continuous mappings of functional spaces play an important role in these investigations.

The topology of functional and general metric spaces is not the last word in generality in contemporary topological theories. The fact of the matter is that in metric spaces the fundamental topological concept of adjacency is introduced on the basis of distance between points, which in turn is not a topological concept. The problem therefore arises of a direct, axiomatic definition of adjacency. Thus we are led to the concept of a topological space, the most general concept of present-day topology.

A *topological space* is a set of objects of an arbitrary nature (which are called points of the space) in which for every subset its *adherent* points are given in one way or another. Furthermore, a few natural conditions are supposed to be satisfied, the so-called axioms of a topological space (for example, every point of the given set is an adherent point of it, an adherent point of the sum of two sets is an adherent point of at least one of the summands, etc.). Profound work is going on at present in the theory of topological spaces; in its development the Soviet mathematicians P. S. Uryson, P. S. Aleksandrov, A. N. Tihonov, and others have taken a leading part. Of the latest results in the theory of topological spaces we must mention one of fundamental value: The young mathematician Ju. M. Smirnov has found necessary and sufficient conditions for the metrizability of a topological space, i.e., conditions under which a distance between the points of the space can be defined such that the "topology" which the space carries can be regarded as generating this concept of distance; in other words, such that the adherent points of all possible sets in the metric space obtained are the same as those defined initially in the given topological space.

Suggested Reading

P. S. Aleksandrov, *Combinatorial topology*, 3 vols., Graylock Press, Albany, N. Y., vol. 1, 1956; vol. 2, 1957; vol. 3, 1960.
S. S. Cairns, *Introductory topology*, Ronald Press, New York, 1961.
D. W. Hall and G. L. Spencer, *Elementary topology*, Wiley, New York, 1955.
P. J. Hilton and S. Wiley, *Homology theory: an introduction to algebraic topology*, Cambridge University Press, New York, 1960.

SUGGESTED READING

J. G. Hocking and G. S. Young, *Topology*, Addison-Wesley, Reading, Mass., 1961.

J. L. Kelley, *General topology*, Van Nostrand, New York, 1955.

S. Lefschetz, *Introduction to topology*, Princeton University Press, Princeton, N. J., 1949.

M. H. A. Newman, *Elements of the topology of plane sets of points*, 2nd ed., Cambridge University Press, New York, 1951.

L. S. Pontrjagin, *Foundations of combinatorial topology*, Graylock Press, Albany, N. Y., 1952.

CHAPTER **XIX**

FUNCTIONAL ANALYSIS

The rise and spread of functional analysis in the 20th century had two main causes. On the one hand it became desirable to interpret from a uniform point of view the copious factual material accumulated in the course of the 19th century in various, often hardly connected, branches of mathematics. The fundamental concepts of functional analysis were formed and crystalized under various aspects and for various reasons. Many of the fundamental concepts of functional analysis emerged in a natural fashion in the process of development of the calculus of variations, in problems on oscillations (in the transition from the oscillations of systems with a finite number of degrees of freedom to oscillations of continuous media), in the theory of integral equations, in the theory of differential equations both ordinary and partial (in boundary problems, problems on eigenvalues, etc.) in the development of the theory of functions of a real variable, in operator calculus, in the discussion of problems in the theory of approximation of functions, and others. Functional analysis permitted an understanding of many results in these domains from a single point of view and often promoted the derivation of new ones. In recent decades the preparatory concepts and apparatus were then used in a new branch of theoretical physics—in quantum mechanics.

On the other hand, the investigation of mathematical problems connected with quantum mechanics became a crucial feature in the further development of functional analysis itself: It created, and still creates at the present time, fundamental branches of this development.

Functional analysis has not yet reached its completion by far. On the contrary, undoubtedly in its further development the questions and requirements of contemporary physics will have the same significance for it as classical mechanics had for the rise and development of the differential and integral calculus in the 18th century.

It is impossible here to include in this chapter all, or even only all the fundamental, problems of functional analysis. Many important branches exceed the limitations of this book. Nevertheless, by confining ourselves to certain selected problems, we wish to acquaint the reader with some fundamental concepts of functional analysis and to illustrate as far as possible the connections of which we have spoken here. These problems were analyzed mainly at the beginning of the 20th century on the basis of the classical papers of Hilbert, who was one of the founders of functional analysis. Since then functional analysis has developed very vigorously and has been widely applied in almost all branches of mathematics; in partial differential equations, in the theory of probability, in quantum mechanics, in the quantum theory of fields, etc. Unfortunately these further developments of functional analysis cannot be included in our account. In order to describe them we would have to write a separate large book, and therefore, we restrict ourselves to one of the oldest problems, namely the theory of eigenfunctions.

§1. n-Dimensional Space

In what follows we shall make use of the fundamental concepts of n-dimensional space. Although these concepts have been introduced in the chapters on linear algebra and on abstract spaces, we do not think it superfluous to repeat them in the form in which they will occur here. For scanning through this section it is sufficient that the reader should have a knowledge of the foundations of analytic geometry.

We know that in analytic geometry of three-dimensional space a point is given by a triplet of numbers (f_1, f_2, f_3), which are its coordinates. The distance of this point from the origin of coordinates is equal to $\sqrt{f_1^2+f_2^2+f_3^2}$. If we regard the point as the end of a vector leading to it from the origin of coordinates, then the length of the vector is also equal to $\sqrt{f_1^2+f_2^2+f_3^2}$. The cosine of the angle between nonzero vectors leading from the origin of coordinates to two distinct points $A(f_1, f_2, f_3)$ and $B(g_1, g_2, g_3)$ is defined by the formula

$$\cos \phi = \frac{f_1 g_1 + f_2 g_2 + f_3 g_3}{\sqrt{f_1^2+f_2^2+f_3^2}\sqrt{g_1^2+g_2^2+g_3^2}}.$$

From trigonometry we know that $|\cos \phi| \leq 1$. Therefore we have the inequality

$$\frac{|f_1 g_1 + f_2 g_2 + f_3 g_3|}{\sqrt{f_1^2+f_2^2+f_3^2}\sqrt{g_1^2+g_2^2+g_3^2}} \leq 1,$$

§1. n-DIMENSIONAL SPACE

and hence always

$$(f_1g_1 + f_2g_2 + f_3g_3)^2 \leq (f_1^2 + f_2^2 + f_3^2)(g_1^2 + g_2^2 + g_3^2). \tag{1}$$

This last inequality has an algebraic character and is true for any arbitrary six numbers (f_1, f_2, f_3) and (g_1, g_2, g_3), since any six numbers can be the coordinates of two points of space. All the same, the inequality (1) was obtained from purely geometric considerations and is closely connected with geometry, and this enables us to give it an easily visualized meaning.

In the analytic formulation of a number of geometric relations, it often turns out that the corresponding facts remain true when the triplet of numbers is replaced by n numbers. For example, our inequality (1) can be generalized to $2n$ numbers $(f_1, f_2, \cdots, f_n)$ and $(g_1, g_2, \cdots, g_n)$. This means that for any arbitrary $2n$ numbers $(f_1, f_2, \cdots, f_n)$ and $(g_1, g_2, \cdots, g_n)$ an inequality analogous to (1) is true, namely:

$$(f_1g_1 + f_2g_2 + \cdots + f_ng_n)^2 \leq (f_1^2 + f_2^2 + \cdots + f_n^2)(g_1^2 + g_2^2 + \cdots + g_n^2).$$
$$\tag{1'}$$

This inequality, of which (1) is a special case, can be proved purely analytically.* In a similar way many other relations between triplets of numbers derived in analytic geometry can be generalized to n numbers. This connection of geometry with relations between numbers (numerical relations) for which the cited inequality is an example becomes particularly lucid when the concept of an n-dimensional space is introduced. This concept was introduced in Chapter XVI. We repeat it here briefly.

A collection of n numbers $(f_1, f_2, \cdots, f_n)$ is called a point or *vector* of n-dimensional space (we shall more often use the latter name). The vector $(f_1, f_2, \cdots, f_n)$ will from now on be abbreviated by the single letter f.

Just as in three-dimensional space on addition of vectors their components are added, so we define the sum of the vectors

$$f = \{f_1, f_2, \cdots, f_n\} \quad \text{and} \quad g = \{g_1, g_2, \cdots, g_n\}$$

as the vector $\{f_1 + g_1, f_2 + g_2, \cdots, f_n + g_n\}$ and we denote it by $f + g$.

The product of the vector $f = \{f_1, f_2, \cdots, f_n\}$ by the number λ is the vector $\lambda f = \{\lambda f_1, \lambda f_2, \cdots, \lambda f_n\}$.

The length of the vector $f = \{f_1, f_2, \cdots, f_n\}$, like the length of a vector in three-dimensional space, is defined as $\sqrt{f_1^2 + f_2^2 + \cdots + f_n^2}$.

---
* See Chapter XVI.

The angle ϕ between the two vectors $f = \{f_1, f_2, \cdots, f_n\}$ and $g = \{g_1, g_2, \cdots, g_n\}$ in n-dimensional space is given by its cosine in exactly the same way as the angle between vectors in three-dimensional space. For it is defined by the formula*

$$\cos \phi = \frac{f_1 g_1 + f_2 g_2 + \cdots + f_n g_n}{\sqrt{f_1^2 + f_2^2 + \cdots + f_n^2} \sqrt{g_1^2 + g_2^2 + \cdots + g_n^2}}. \qquad (2)$$

The scalar product of two vectors is the name for the product of their lengths by the cosine of the angle between them. Thus, if $f = \{f_1, f_2, \cdots, f_n\}$ and $g = \{g_1, g_2, \cdots, g_n\}$, then since the lengths of the vectors are $\sqrt{f_1^2 + f_2^2 + \cdots + f_n^2}$ and $\sqrt{g_1^2 + g_2^2 + \cdots + g_n^2}$, respectively, their scalar product, which is denoted by (f, g), is given by the formula

$$(f, g) = f_1 g_1 + f_2 g_2 + \cdots + f_n g_n. \qquad (3)$$

In particular, the condition of orthogonality (perpendicularity) of two vectors is the equation $\cos \phi = 0$; i.e., $(f, g) = 0$.

By means of the formula (3) the reader can verify that the scalar product in n-dimensional space has the following properties:

1. $(f, g) = (g, f)$.
2. $(\lambda f, g) = \lambda (f, g)$.
3. $(f, g_1 + g_2) = (f, g_1) + (f, g_2)$.
4. $(f, f) \geqslant 0$, and the equality sign holds for $f = 0$ only, i.e., when $f_1 = f_2 = \cdots = f_n = 0$.

The scalar product of a vector f with itself (f, f) is equal to the square of the length of f.

The scalar product is a very convenient tool in studying n-dimensional spaces. We shall not study here the geometry of an n-dimensional space but shall restrict ourselves to a single example.

As our example we choose the theorem of Pythagoras in n-dimensional space: The square of the hypotenuse is equal to the sum of the squares of the sides. For this purpose we give a proof of this theorem in the plane which is easily transferred to the case of an n-dimensional space.

Let f and g be two perpendicular vectors in a plane. We consider the right-angled triangle constructed on f and g (figure 1). The hypotenuse of this triangle is equal in length to the vector $f + g$. Let us write down in vector form the theorem of Pythagoras in our notation. Since the square of the length of a vector is equal to the scalar product of the vector with

---

* The fact that $|\cos \phi| \leqslant 1$ follows from the inequality (1').

§1. n-DIMENSIONAL SPACE

itself, Pythagoras' theorem can be written in the language of scalar products as follows:

$$(f+g, f+g) = (f,f) + (g,g).$$

The proof immediately follows from the properties of the scalar product. In fact,

$$(f+g, f+g) = (f,f) + (f,g) + (g,f) + (g,g),$$

and the two middle summands are equal to zero owing to the orthogonality of f and g.

In this proof we have only used the definition of the length of a vector, the perpendicularity of vectors, and the properties of the scalar product. Therefore nothing changes in the proof when we assume that f and g are two orthogonal vectors of an n-dimensional space. And so Pythagoras' theorem is proved for a right-angled triangle in n-dimensional space.

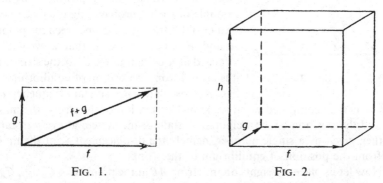

FIG. 1.   FIG. 2.

If three pairwise orthogonal vectors f, g and h are given in n-dimensional space, then their sum $f + g + h$ is the diagonal of the right-angled parallelepiped constructed from these vectors (figure 2) and we have the equation

$$(f+g+h, f+g+h) = (f,f) + (g,g) + (h,h),$$

which signifies that the square of the length of the diagonal of a parallelepiped is equal to the sum of the squares of the lengths of its edges. The proof of this statement, which is entirely analogous to the one given earlier for Pythagoras' theorem, is left to the reader. Similarly, if in an n-dimensional space there are k pairwise orthogonal vectors $f^1, f^2, \cdots, f^k$ then the equation

$$(f^1 + f^2 + \cdots + f^k, f^1 + f^2 + \cdots + f^k)$$
$$= (f^1, f^1) + (f^2, f^2) + \cdots + (f^k, f^k), \qquad (4)$$

which is just as easy to prove, signifies that the square of the length of the diagonal of a "k-dimensional parallelepiped" in n-dimensional space is also equal to the sum of the squares of the lengths of its edges.

§2. Hilbert Space (Infinite-Dimensional Space)

Connection with n-dimensional space. The introduction of the concept of n-dimensional space turned out to be useful in the study of a number of problems of mathematics and physics. In its turn this concept gave the impetus to a further development of the concept of space and to its application in various domains of mathematics. An important role in the development of linear algebra and of the geometry of n-dimensional spaces was played by problems of small oscillations of elastic systems.

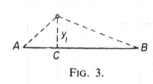

Fig. 3.

Let us consider the following classical example of such a problem (figure 3). Let AB be a flexible string spanned between the points A and B. Let us assume that a weight is attached at a certain point C to the string. If it is moved from its position of equilibrium, it begins to oscillate with a certain frequency ω, which can be computed when we know the tension of the string, the mass m and the position of the weight. The state of the system at every instant is then given by a single number, namely the displacement y_1 of the mass m from the position of equilibrium of the string.

Now let us place n weights on the string AB at the points C_1, C_2, $\cdots$, C_n. The string itself is taken to be weightless. This means that its mass is so small that compared with the masses of the weights it can be neglected. The state of such a system is given by n numbers y_1, y_2, $\cdots$, y_n equal to the displacements of the weights from the position of equilibrium. The collection of numbers y_1, y_2, $\cdots$, y_n can be regarded (and this turns out to be useful in many respects) as a vector $(y_1, y_2, \cdots, y_n)$ of an n-dimensional space.

The investigation of the small oscillations that take place under these circumstances turns out to be closely connected with fundamental facts of the geometry of n-dimensional spaces. We can show, for example, that the determination of the frequency of the oscillations of such a system can be reduced to the task of finding the axes of a certain ellipsoid in n-dimensional space.

Now let us consider the problem of the small oscillations of a string spanned between the points A and B. Here we have in mind an idealized string, i.e., an elastic thread having a finite mass distributed continuously

§2. HILBERT SPACE (INFINITE-DIMENSIONAL SPACE)

along the thread. In particular, by a homogeneous string we understand one whose density is constant.

Since the mass is distributed continuously along the string, the position of the string can no longer be given by a finite set of numbers $y_1, y_2, \cdots, y_n$, and instead the displacement $y(x)$ of every point x of the string has to be given. Thus, the state of the string at each instant is given by a certain function $y(x)$.

The state of a thread with n weights attached at the points with the abscissas $x_1, x_2, \cdots, x_n$, is represented graphically by a broken line with n members (figure 4), so that when the number of weights is increased, then the number of segments of the broken line

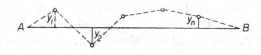

Fig. 4.

increases correspondingly. When the number of weights grows without bound and the distance between adjacent weights tends to zero, we obtain in the limit a continuous distribution of mass along the thread, i.e., an idealized string. The broken line that describes the position of the thread with weights then goes over into a curve describing the position of the string (figure 5).

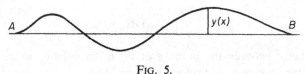

Fig. 5.

So we see that there exists a close connection between the oscillations of a thread with weights and the oscillations of a string. In the first problem the position of the system was given by a point or vector of an n-dimensional space. Therefore it is natural to regard the function $f(x)$ that describes the position of the oscillating string in the second case as a vector or a point of a certain infinite-dimensional space. A whole series of similar problems leads to the same idea of considering a space whose points (vectors) are functions $f(x)$ given on a certain interval.*

---

* As another such problem let us consider the electrical oscillations set up in a series of connected electrical circuits (figure 6).

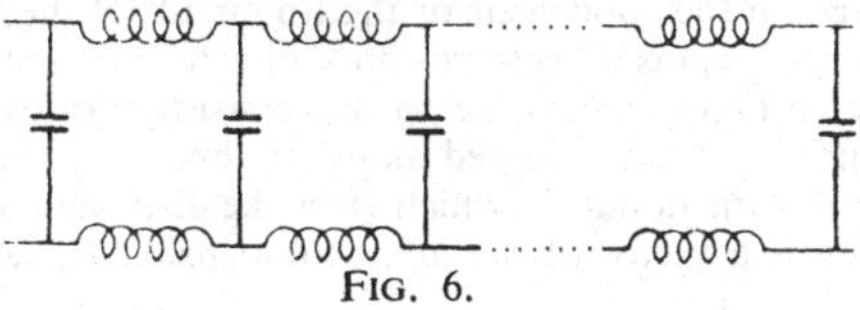

Fig. 6.

This example of oscillation of a string, to which we shall return again in §4, suggests to us how we shall have to introduce the fundamental concepts in an infinite-dimensional space.

Hilbert space. Here we shall discuss one of the most widespread concepts of an infinite-dimensional space of the greatest importance for the applications, namely the concept of the Hilbert space.

A vector of an n-dimensional space is defined as a collection of n numbers f_i, where i ranges from 1 to n. Similarly a vector of an infinite-dimensional space is defined as a function $f(x)$, where x ranges from a to b.

Addition of vectors and multiplication of a vector by a number is defined as addition of the functions and multiplication of the function by a number.

The length of a vector f in an n-dimensional space is defined by the formula

$$\sqrt{\sum_{i=1}^{n} f_i^2}$$

Since for functions the role of the sum is taken by the integral, the length of the vector $f(x)$ of a Hilbert space is given by the formula

$$\sqrt{\int_a^b f^2(x)\,dx}. \tag{5}$$

The distance between the points f and g in an n-dimensional space is defined as the length of the vector $f - g$, i.e., as

$$\sqrt{\sum_{i=1}^{n} (f_i - g_i)^2}.$$

Similarly the "distance" between the elements $f(t)$ and $g(t)$ in a functional space is equal to

$$\sqrt{\int_a^b [f(t) - g(t)]^2\,dt}.$$

The state of such a series can be expressed by the set of n numbers $u_1, u_2, \cdots, u_n$, where u_i is the voltage on the condensor of the ith circuit of the chain. The collection of the n numbers $(u_1, \cdots, u_n)$ is a vector of an n-dimensional space.

Now let us imagine a two-wire line, i.e., a line consisting of two conductors having finite capacity and inductance, distributed along the line. The electric state of the line is expressed by a certain function $u(x)$, which gives the distribution of the voltage along the line. This function is a vector of the infinite-dimensional space of functions given on the interval (a, b).

§2. HILBERT SPACE (INFINITE-DIMENSIONAL SPACE)

The expression $\int_a^b [f(t) - g(t)]^2 \, dt$ is called the mean-square deviation of the functions $f(t)$ and $g(t)$. Thus, the mean-square deviation of two elements of Hilbert space is taken to be a measure of their distance.

Let us now proceed to the definition of the angle between vectors. In an n-dimensional space the angle ϕ between the vectors $f = \{f_i\}$ and $g = \{g_i\}$ is defined by the formula

$$\cos \phi = \frac{\sum_{i=1}^n f_i g_i}{\sqrt{\sum_{i=1}^n f_i^2} \sqrt{\sum_{i=1}^n g_i^2}}.$$

In an infinite-dimensional space the sums are replaced by the corresponding integrals and the angle ϕ between the two vectors f and g of Hilbert space is defined by the analogous formula

$$\cos \phi = \frac{\int_a^b f(t) g(t) \, dt}{\sqrt{\int_a^b f^2(t) \, dt} \sqrt{\int_a^b g^2(t) \, dt}}. \tag{6}$$

This expression can be regarded as the cosine of a certain angle ϕ, provided the fraction on the right-hand side is an absolute value less than one, i.e., if

$$\left| \int_a^b f(t) g(t) \, dt \right| < \sqrt{\int_a^b f^2(t) \, dt} \sqrt{\int_a^b g^2(t) \, dt}. \tag{7}$$

This inequality in fact holds for two arbitrary functions $f(t)$ and $g(t)$. It plays an important role in analysis and is known as the Cauchy-Bunjakovskiĭ inequality. Let us prove it.

Let $f(x)$ and $g(x)$ be two functions, not identically equal to zero, given on the interval (a, b). We choose arbitrary numbers λ and μ and form the expresson

$$\int_a^b [\lambda f(x) - \mu g(x)]^2 \, dx.$$

Since the function $[\lambda f(x) - \mu g(x)]^2$ under the integral sign is nonnegative, we have the following inequality

$$\int_a^b [\lambda f(x) - \mu g(x)]^2 \, dx \geq 0;$$

i.e.,

$$2\lambda\mu \int_a^b f(x) g(x) \, dx \leq \lambda^2 \int_a^b f^2(x) \, dx + \mu^2 \int_a^b g^2(x) \, dx.$$

For brevity we introduce the notation

$$\left| \int_a^b f(x) g(x) \, dx \right| = C, \quad \int_a^b f^2(x) \, dx = A, \quad \int_a^b g^2(x) \, dx = B. \tag{8}$$

In this notation the inequality can be rewritten as follows:*

$$2\lambda\mu C \leqslant \lambda^2 A + \mu^2 B. \tag{9}$$

This inequality is valid for arbitrary values of λ and μ; in particular we may set

$$\lambda = \sqrt{\frac{C}{A}}, \mu = \sqrt{\frac{C}{B}}. \tag{10}$$

Substituting these values of λ and μ in (9), we obtain

$$\frac{C}{\sqrt{AB}} \leqslant 1.$$

When we replace A, B and C by their expressions in (8), we finally obtain the Cauchy-Bunjakovskiĭ inequality.

In geometry the scalar product of vectors is defined as the product of their lengths by the cosine of the angle between them. The lengths of the vectors f and g in our case are equal to

$$\sqrt{\int_a^b f^2(x)\, dx} \quad \text{and} \quad \sqrt{\int_a^b g^2(x)\, dx},$$

and the cosine of the angle between them is defined by the formula

$$\cos \phi = \frac{\int_a^b f(x)\, g(x)\, dx}{\sqrt{\int_a^b f^2(x)\, dx} \sqrt{\int_a^b g^2(x)\, dx}}.$$

When we multiply out these expressions, we arrive at the following formula for the scalar product of two vectors of Hilbert space:

$$(f, g) = \int_a^b f(x)\, g(x)\, dx. \tag{11}$$

From this formula it is clear that the scalar product of the vector f with itself its the square of its length.

If the scalar product of the nonzero vectors f and g is equal to zero, it means that $\cos \phi = 0$, i.e., that the angle ϕ ascribed to them by our definition is 90°. Therefore functions f and g for which

$$(f, g) = \int_a^b f(x)\, g(x)\, dx = 0,$$

are called orthogonal.

Pythagoras' theorem (see §1) holds in Hilbert space as in an n-dimen-

---

* For C we have to take the modulus of the integral because of the arbitrary sign of λ or μ.

§3. EXPANSION BY ORTHOGONAL SYSTEMS

sional space. Let $f_1(x), f_2(x), \cdots, f_N(x)$ be N pairwise orthogonal functions and
$$f(x) = f_1(x) + f_2(x) + \cdots + f_N(x).$$

Then the square of the length of f is equal to the sum of the squares of the lengths of $f_1, f_2, \cdots, f_N$.

Since the lengths of vectors in Hilbert space are given by means of integrals, Pythagoras' theorem in this case is expressed by the formula

$$\int_a^b f^2(x)\,dx = \int_a^b f_1^2(x)\,dx + \int_a^b f_2^2(x)\,dx + \cdots + \int_a^b f_N^2(x)\,dx. \tag{12}$$

The proof of this theorem does not differ in any respect from the one given previously (§1) for the same theorem in n-dimensional space.

So far we have not made precise what functions are to be regarded as vectors in Hilbert space. For such functions we have to take all those for which $\int_a^b f^2(x)\,dx$ has a meaning. It might appear natural to confine ourselves to continuous functions for which $\int_a^b f^2(x)\,dx$ always exists. However, the theory of Hilbert space becomes more complete and natural if $\int_a^b f^2(x)\,dx$ is interpreted in a generalized sense, namely as a Lebesgue integral (see Chapter XV).

This extension of the concept of integrals (and correspondingly of the class of functions to be discussed) is necessary for functional analysis in the same way as a strict theory of the real numbers is necessary for the foundation of the differential and integral calculus. Thus, the generalization of the ordinary concept of an integral that was created at the beginning of the 20th century in connection with the development of the theory of functions of a real variable turned out to be quite essential for functional analysis and the branches of mathematics connected with it.

§3. Expansion by Orthogonal Systems of Functions

Definition and examples of orthogonal systems of functions. If in a plane two arbitrary mutually perpendicular vectors e_1 and e_2 of unit length are chosen (figure 7), then every vector of the same plane can be decomposed in the directions of these two vectors, i.e., can be represented in the form

$$f = a_1 e_1 + a_2 e_2,$$

where a_1 and a_2 are the numbers equal to the projections of the vector f in the direction of the axis of e_1 and e_2. Since

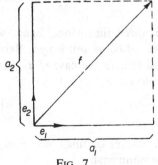

FIG. 7.

the projection of f on an axis is equal to the product of the length of f by the cosine of the angle between f and the axis, we can write, remembering the definition of the scalar product,

$$a_1 = (f, e_1),$$
$$a_2 = (f, e_2).$$

Similarly if in a three-dimensional space any three mutually perpendicular vectors e_1, e_2, e_3 of unit length are chosen, then every vector f in this space can be written in the form

$$f = a_1 e_1 + a_2 e_2 + a_3 e_3,$$

where

$$a_k = (f, e_k) \; (k = 1, 2, 3).$$

In Hilbert space we can also consider systems of pairwise orthogonal vectors of the space, i.e., functions $\phi_1(x), \phi_2(x), \cdots, \phi_n(x), \cdots$.

Such systems of functions are called orthogonal and play an important role in analysis. They occur in very diverse problems of mathematical physics, integral equations, approximate computations, the theory of functions of a real variable, etc. The ordering and unification of the concepts relating to such systems formed one of the motivations that led at the beginning of the 20th century to the creation of the general concept of a Hilbert space.

Let us give a precise definition. A system of functions

$$\phi_1(x), \phi_2(x), \cdots, \phi_n(x), \cdots$$

is called *orthogonal* if any two functions of the system are orthogonal, i.e., if

$$\int_a^b \phi_i(x) \phi_k(x) \, dx = 0 \quad \text{for} \quad i \neq k. \tag{13}$$

In three-dimensional space we required that the vectors of the system should be of unit length. Recalling the definition of length of a vector we see that in the case of Hilbert space this requirement can be written as follows:

$$\int_a^b \phi_k^2(x) \, dx = 1. \tag{14}$$

A system of functions satisfying the conditions (13) and (14) is called orthonormal.

§3. EXPANSION BY ORTHOGONAL SYSTEMS

Let us give examples of such systems of functions.

1. On the interval $(-\pi, \pi)$ we consider the sequence of functions

$$1, \cos x, \sin x, \cos 2x, \sin 2x, \cdots, \cos nx, \sin nx, \cdots.$$

Any two functions of this sequence are orthogonal to each other. This can be verified by the simple computation of the corresponding integrals. The square of the length of a vector in Hilbert space is the integral of the square of the function. Thus, the squares of the lengths of the vectors of the sequence

$$1, \cos x, \sin x, \cos 2x, \sin 2x, \cdots, \cos nx, \sin nx, \cdots$$

are the integrals

$$\int_{-\pi}^{\pi} dx = 2\pi, \quad \int_{-\pi}^{\pi} \cos^2 nx\, dx = \pi, \quad \int_{-\pi}^{\pi} \sin^2 nx\, dx = \pi,$$

i.e., the vectors of our sequence are orthogonal, but not normalized. The length of the first vector of the sequence is equal to $\sqrt{2\pi}$, and all the others are of length $\sqrt{\pi}$. When we divide every vector by its length, we obtain the orthonormal system of trigonometric functions

$$\frac{1}{\sqrt{2\pi}}, \frac{\cos x}{\sqrt{\pi}}, \frac{\sin x}{\sqrt{\pi}}, \frac{\cos 2x}{\sqrt{\pi}}, \frac{\sin 2x}{\sqrt{\pi}}, \cdots, \frac{\cos nx}{\sqrt{\pi}}, \frac{\sin nx}{\sqrt{\pi}}, \cdots.$$

This system is historically one of the first and most important examples of orthogonal systems. It appeared in the works of Euler, D. Bernoulli, and d'Alembert in connection with problems on the oscillations of strings. The study of it plays an essential role in the development of the whole of analysis.*

The appearance of the orthogonal system of trigonometrical functions in connection with problems on oscillations of strings is not accidental. Every problem on small oscillations of a medium leads to a certain system of orthogonal functions that describe the so-called characteristic oscillations of the given system (see §4). For example, in connection with problems on the oscillations of a sphere there appear the so-called spherical functions, in connection with problems on the oscillations of a circular membrane or a cylinder there appear the so-called cylinder functions, etc.

2. We can give an example of an orthogonal system of functions in

---

* See Chapter XII, §1.

which every function is a polynomial. Such an example is the sequence of Legendre polynomials

$$P_n(x) = \frac{1}{2^n n!} \frac{d^n(x^2-1)^n}{dx^n},$$

i.e., $P_n(x)$ is (apart from a constant factor) the nth derivative of $(x^2-1)^n$. Let us write down the first few polynomials of this sequence:

$$P_0(x) = 1;$$
$$P_1(x) = x;$$
$$P_2(x) = \tfrac{1}{2}(3x^2 - 1);$$
$$P_3(x) = \tfrac{1}{2}(5x^3 - 3x).$$

Obviously $P_n(x)$ is a polynomial of degree n. We leave it to the reader to convince himself that these polynomials are an orthogonal sequence on the interval $(-1, 1)$.

The general theory of orthogonal polynomials (the so-called orthogonal polynomials with weights) was developed in the second half of the 19th century by the famous Russian mathematician P. L. Čebyšev.

Expansion by orthogonal systems of functions. Just as in three-dimensional space every vector can be represented in the form of a linear combination of three pairwise orthogonal vectors e_1, e_2, e_3 of unit length

$$f = a_1 e_1 + a_2 e_2 + a_3 e_3,$$

so in a functional space there arises the problem of the decomposition of an arbitrary function f in a series with respect to an orthonormal system of functions, i.e., of the representation of f in the form

$$f(x) = a_1\phi_1(x) + a_2\phi_2(x) + \cdots + a_n\phi_n(x) + \cdots. \tag{15}$$

Here the convergence of the series (15) to the function f has to be understood in the sense of the distance between elements in Hilbert space. This means that the mean-square deviation of the partial sum of the series

$$S_n(t) = \sum_{k=1}^{n} a_k \phi_k(t)$$

from the function $f(t)$ tends to zero for $n \to \infty$; i.e.,

$$\lim_{n\to\infty} \int_a^b [f(t) - S_n(t)]^2\, dt = 0. \tag{16}$$

This convergence is usually called "convergence in the mean."

§3. EXPANSION BY ORTHOGONAL SYSTEMS

Expansions in various systems of orthogonal functions often occur in analysis and are an important method for the solution of problems of mathematical physics. For example, if the orthogonal system is the system of trigonometric functions on the interval $(-\pi, \pi)$

$$1, \cos x, \sin x, \cos 2x, \sin 2x, \cdots, \cos nx, \sin nx, \cdots,$$

then this expansion is the classical expansion of a function in a trigonometric series*

$$f(x) = a_0 + a_1 \cos x + b_1 \sin x + a_2 \cos 2x + b_2 \sin 2x + \cdots.$$

Let us assume that an expansion (15) is possible for every function f of a Hilbert space and let us find its coefficients a_n. For this purpose we multiply both sides of the equation scalarly by one and the same function ϕ_m of our system. We obtain the equation

$$(f, \phi_m) = a_1(\phi_1, \phi_m) + a_2(\phi_2, \phi_m) + \cdots + a_m(\phi_m, \phi_m)$$
$$+ a_{m+1}(\phi_{m+1}, \phi_m) + \cdots,$$

in virtue of the fact that $(\phi_m, \phi_n) = 0$ for $m \neq n$ and $(\phi_m, \phi_m) = 1$, this determines the value of the coefficient a_m

$$a_m = (f, \phi_m) \quad (m = 1, 2, \cdots).$$

We see that, as in ordinary three-dimensional space (see the beginning of this section), the coefficients a_m are equal to the projections of the vector f in the direction of the vectors ϕ_k.

Recalling the definition of the scalar product we see that the coefficients of the expansion of $f(x)$ by the normal orthogonal system of functions $\phi_1(x), \phi_2(x), \cdots, \phi_n(x), \cdots$

$$f(x) = a_1\phi_1(x) + a_2\phi_2(x) + \cdots + a_n\phi_n(x) + \cdots \quad (17)$$

are determined by the formulas

$$a_m = \int_a^b f(t)\phi_m(t)\, dt. \quad (18)$$

As an example let us consider the normal orthogonal trigonometric system of functions mentioned previously:

$$\frac{1}{\sqrt{2\pi}}, \frac{\cos x}{\sqrt{\pi}}, \frac{\sin x}{\sqrt{\pi}}, \frac{\cos 2x}{\sqrt{\pi}}, \frac{\sin 2x}{\sqrt{\pi}}, \cdots.$$

---

* Such a decomposition often occurs in various problems of physics in the decomposition of an oscillation into its harmonic constituents. See Chapter VI, §5.

Then

$$f(x) = \frac{a_0}{2} + \sum_{n=1}^{\infty} (a_n \cos nx + b_n \sin nx),$$

where

$$a_0 = \frac{1}{\pi} \int_{-\pi}^{\pi} f(x)\, dx, \quad a_n = \frac{1}{\pi} \int_{-\pi}^{\pi} f(x) \cos nx\, dx,$$

$$b_n = \frac{1}{\pi} \int_{-\pi}^{\pi} f(x) \sin nx\, dx.$$

So we have obtained the formula for the computation of the coefficients of the expansion of a function in trigonometric series, assuming of course that this expansion is possible.*

We have established the form of the coefficients of the expansion (18) of the function $f(x)$ by an orthogonal system of functions under the assumptions that this expansion holds. However, an infinite orthogonal system of functions $\phi_1, \phi_2, \cdots, \phi_n, \cdots$ may turn out to be insufficient for every function of a Hilbert space to have such an expansion. For such an expansion to be possible, the system of orthogonal functions must satisfy an additional condition, namely the so-called condition of completeness.

An orthogonal system of functions is called *complete* if it is impossible to add to it even one function, not identically equal to zero, that is orthogonal to all the functions of the system.

It is easy to give an example of an incomplete orthogonal system. For this purpose we choose an arbitrary orthogonal system, for example that of the trigonometric functions, and remove one of the functions of the system, for example cos x. The remaining infinite system of functions

$$1, \sin x, \cos 2x, \sin 2x, \cdots, \cos nx, \sin nx, \cdots$$

is orthogonal as before, but of course it is not complete, since the function cos x which we have excluded is orthogonal to all the functions of the system.

If a system of functions is incomplete, then not every function of a Hilbert space can be expanded by it. For if we attempt to expand by such a system a nonzero function $f_0(x)$ that is orthogonal to all the functions of the system, then by (18) all the coefficients turn out to be zero, whereas the function $f_0(x)$ is not equal to zero.

The following theorem holds: If a complete orthonormal system of

---

* On trigonometric series see also Chapter XII, §7.

§3. EXPANSION BY ORTHOGONAL SYSTEMS

functions in a Hilbert space $\phi_1(x)$, $\phi_2(x)$, $\cdots$, $\phi_n(x)$, $\cdots$, is given, then every function $f(x)$ can be expanded in a series by functions of this system*

$$f(x) = a_1\phi_1(x) + a_2\phi_2(x) + \cdots + a_n\phi_n(x) + \cdots.$$

Here the coefficients a_n of the expansion are equal to the projections of the vectors f on the elements of the normal orthogonal system

$$a_n = (f, \phi_n) = \int_a^b f(x)\,\phi_n(x)\,dx.$$

Pythagoras' theorem in Hilbert space, which was established in §2, enables us to find an interesting relation between the coefficients a_k and the function $f(x)$. We denote by $r_n(x)$ the difference between $f(x)$ and the sum of the first n terms of its series; i.e.,

$$r_n(x) = f(x) - [a_1\phi_1(x) + \cdots + a_n\phi_n(x)].$$

The function $r_n(x)$ is orthogonal to $\phi_1(x)$, $\phi_2(x)$, $\cdots$, $\phi_n(x)$. Let us verify for example that it is orthogonal to $\phi_1(x)$, i.e., that $\int_a^b r_n(x)\,\phi_1(x)\,dx = 0$. We have

$$\int_a^b r_n(x)\,\phi_1(x)\,dx = \int_a^b [f(x) - a_1\phi_1(x) - a_2\phi_2(x) - \cdots - a_n\phi_n(x)]\,\phi_1(x)\,dx$$

$$= \int_a^b f(x)\,\phi_1(x)\,dx - a_1 \int_a^b \phi_1^2(x)\,dx.\dagger$$

Since $a_1 = \int_a^b f(x)\,\phi_1(x)\,dx$, and $\int_a^b \phi_1^2(x)\,dx = 1$, it follows from this that $\int_a^b r_n(x)\phi_1(x)\,dx = 0$.

Thus, in the equation

$$f(x) = a_1\phi_1(x) + a_2\phi_2(x) + \cdots + a_n\phi_n(x) + r_n(x) \qquad (19)$$

the individual terms on the right-hand side are orthogonal to each other. Hence, by Pythagoras' theorem as formulated in §1, the square of the length of $f(x)$ is equal to the sum of the squares of the lengths of the summands of the right-hand side in (19); i.e.,

$$\int_a^b f^2(x)\,dx = \int_a^b [a_1\phi_1(x)]^2\,dx + \cdots + \int_a^b [a_n\phi_n(x)]^2\,dx + \int_a^b r_n^2(x)\,dx.$$

---

* This series is related to its sum in the sense defined in formula (16).
† The remaining integrals are equal to zero, because the functions $\phi_k(x)$ are orthogonal to each other.

Since the system of functions $\phi_1, \phi_2, \cdots, \phi_n$ is normalized [equation (14)], we have

$$\int_a^b f^2(x)\,dx = a_1^2 + a_2^2 + \cdots + a_n^2 + \int_a^b r_n^2(x)\,dx. \tag{20}$$

The series $\sum_{k=1}^{\infty} a_k \phi_k(x)$ converges in the mean. This means that

$$\int_a^b [f(x) - a_1\phi_1(x) - \cdots - a_n\phi_n(x)]^2\,dx \to 0,$$

i.e., that

$$\int_a^b r_n^2(x)\,dx \to 0.$$

But then we obtain from the formula (20) the equation

$$\sum_{k=1}^{\infty} a_k^2 = \int_a^b f^2(x)\,dx,* \tag{21}$$

which states that the integral of the square of a function is equal to the sum of the squares of the coefficients of its expansion by a closed orthogonal system of functions. If the condition (21) holds for an arbitrary function of the Hilbert space, it is called the condition of completeness.

We wish to draw attention to the following important question. Which numbers a_k can be the coefficients of the expansion of a function in Hilbert space? The equation (21) asserts that for this purpose the series $\sum_{k=1}^{\infty} a_k^2$ must converge. Now it turns out that this condition is also sufficient; i.e., a sequence of numbers a_k is the sequence of coefficients of the expansion by an orthogonal system of functions in Hilbert space if and only if the series $\sum_{k=1}^{\infty} a_k^2$ converges.

We remark that this fundamental theorem holds if Hilbert space is interpreted as the collection of all functions with integrable square in the sense of Lebesgue (see §2). If we were to confine ourselves in Hilbert space, for example, to the continuous functions, then the solution of the problem as to which numbers a_k can be the coefficients of an expansion would become unnecessarily complicated.

The arguments given here are only one of the reasons that have led to the use of an integral in a generalized (Lebesgue) sense in the definition of Hilbert space.

---

* Geometrically, this means that the square of the length of a vector in Hilbert space is equal to the sum of the squares of its projections onto a complete system of mutually orthogonal directions.

§4. Integral Equations

§4. Integral Equations

In this section the reader will become acquainted with one of the most important and, historically, one of the first branches of functional analysis, namely the theory of integral equations, which has also played an essential role in the subsequent development of functional analysis. Quite apart from internal requirements of mathematics [for example, boundary problems for partial differential equations (Chapter VI)], various problems of physics were of great importance in the development of the theory of integral equations. Side by side with differential equations, the integral equations are, in the 20th century, one of the most important means of the mathematical investigation of various problems of physics. In this section we shall give a certain amount of information concerning the theory of integral equations. The facts we shall explain here are closely connected and have essentially sprung up (directly or indirectly) in connection with the study of small oscillations of elastic systems.

The problem of small oscillations of elastic systems. We return to the problem of small oscillations discussed in §2. Let us find equations that describe such oscillations. For the sake of simplicity we assume that we are dealing with the oscillation of a linear elastic system. As examples of such systems we can take, say, a string of length l (figure 8) or an elastic rod (figure 9). We shall assume that in the position of equilibrium our elastic system is situated along the segment Ol of the x-axis. We apply a unit force at the point x. Under the action of this force all the points of

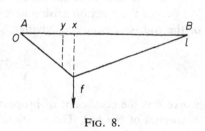

FIG. 8.

the system receive a certain displacement. The displacement arising at the point y (figure 8) is denoted by $k(x, y)$.

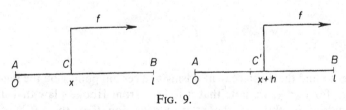

FIG. 9.

The function $k(x, y)$ is a function of two points: the point x at which the force is applied, and the point y at which we measure the displacement. It is called the influence function (Green's function).

From the law of conservation of energy, we can deduce an important property of the Green's function $k(x, y)$, namely the so-called reciprocity law: The displacement arising at the point y under the action of a force applied at the point x is equal to the displacement arising at the point x under the action of the same force applied at the point y. In other words, this means that

$$k(x, y) = k(y, x). \tag{22}$$

Let us find, for example, the Green's function for the longitudinal oscillations of an elastic rod (in figure 8 we have illustrated other transverse displacements). We consider a rod AB of length fixed at the ends (figure 9). At the point C we apply a force f acting in the direction of B. Under the action of this force the rod is deformed and the point C is shifted into the position C'. We denote the magnitude of the shift of C by h. Let us find the value of h. By means of h we can then find the shift at an arbitrary point y. For this purpose we shall make use of Hooke's law, which states that the force is proportional to the relative extension (i.e., to the ratio of the amount of displacement to the length). A similar relation holds for compressions.

Under the action of the force f the part AC of the rod is stretched. We denote the reaction arising here by T_1. At the same time the part CB of the rod is compressed, giving rise to a reaction T_2. By Hooke's law

$$T_1 = \kappa \frac{h}{x}, \quad T_2 = \kappa \frac{h}{l - x},$$

where κ is the coefficient of proportionality that characterizes the elastic properties of the rod. The position of equilibrium of the forces acting at the point C gives us

$$f = \kappa \frac{h}{x} + \kappa \frac{h}{l - x}, \quad \text{i.e.,} \quad f = \frac{\kappa l h}{x(l - x)}.$$

Hence

$$h = \frac{f}{\kappa l} x(l - x).$$

In order to find the displacement arising at a certain point y on the segment AC, i.e., for $y < x$, we note that it follows from Hooke's law that under an extension of the rod the relative extension (i.e., the ratio of the displacement of the point to its distance from the fixed end) does not depend on the position of the point. We denote the displacement of the point y by k.

§4. INTEGRAL EQUATIONS

Then by comparing the relative displacements at the points x and y we obtain
$$\frac{k}{y} = \frac{h}{x};$$
hence
$$k = h\frac{y}{x} = \frac{f}{\kappa_l} y(l-x) \quad \text{for} \quad y < x.$$

Similarly, if the point lies on the segment CB ($y > x$), we obtain
$$k = h\frac{l-y}{l-x} = \frac{f}{\kappa_l} x(l-y).$$

Bearing in mind that the Green's function $k(x, y)$ is the displacement at the point y under the action of a unit force applied at the point x, we see that on the longitudinal oscillations of an elastic rod the Green's function has the form
$$k(x, y) = \begin{cases} \dfrac{1}{\kappa_l} y(l-x) & \text{for} \quad y < x, \\ \dfrac{1}{\kappa_l} x(l-y) & \text{for} \quad y > x. \end{cases}$$

In a more or less similar way we could have found the Green's function for a string. If the tension of the string is T and the length l, then under the action of a unit force applied at the point x the string assumes the form illustrated in figure 8, and the displacement $k(x, y)$ at the point y is given by the formula
$$k(x, y) = \begin{cases} \dfrac{1}{T_l} x(l-y), & \text{for} \quad x < y, \\ \dfrac{1}{T_l} y(l-x), & \text{for} \quad x > y, \end{cases}$$
which coincides with the Green's function for the rod which we have derived.

In terms of the Green's function we can express the displacement of the system from its position of equilibrium provided that it is acted upon by a continuously distributed force of density $f(y)$. Since on an interval of length Δy there acts a force $f(y) \Delta y$, which we can regard approximately as concentrated at the point y, under the action of this force at the point x there arises a displacement $k(x, y) f(y) \Delta y$. The displacement under the action of the whole load is approximately equal to the sum
$$\sum k(x, y) f(y) \Delta y.$$

Passing to the limit for $\Delta y \to 0$ we see that the displacement $u(x)$ at the point x under the action of the force $f(y)$ distributed along the system is given by the formula

$$u(x) = \int_a^b k(x, y) f(y)\, dy. \tag{23}$$

Let us assume that our elastic system is not subject to the action of external forces. If it is displaced from its position of equilibrium, it then begins to move. These motions are called the free oscillations of the system.

Now let us write down in terms of the Green's function $k(x, y)$ the equation that the free oscillations of the elastic system in question have to obey. For this purpose we denote by $u(x, t)$ the displacement from the position of equilibrium at the point x and the instant of time t. Then the acceleration of x at the time t is equal to $\partial^2 u(x, t)/\partial t^2$.

If ρ is the linear density of the field, i.e., $\rho\, dy$ the mass of the element of length dy, then we obtain by a fundamental law of mechanics the equation of motion by replacing in (23) the force $f(y)\, dy$ by the product of the mass and the acceleration $[\partial^2 u(y, t)/\partial t^2]\, \rho\, dy$ taken with the opposite sign.

Thus, the equation of the free oscillations has the form

$$u(x, t) = -\int_a^b k(x, y) \frac{\partial^2 u(y, t)}{\partial t^2}\, \rho\, dy.$$

An important role in the theory of oscillations is played by the so-called harmonic oscillations of the elastic system, i.e., the motions for which

$$u(x, t) = u(x) \sin \omega t.$$

They are characterized by the fact that every fixed point performs harmonic oscillations (moves according to a sinusoidal law) with a certain frequency ω, and that this frequency is one and the same for all the points x.

Later on we shall see that every free oscillation is composed of harmonic oscillations.

We set

$$u(x, t) = u(x) \sin \omega t$$

in the equation of the free oscillations and cancel $\sin \omega t$. Then we obtain the following equation to determine the function $u(x)$

$$u(x) = \rho \omega^2 \int_a^b k(x, y)\, u(y)\, dy. \tag{24}$$

§4. INTEGRAL EQUATIONS

Such an equation is called a homogeneous integral equation for the function $u(x)$.

Obviously the equation (24) has for every ω the uninteresting solution $u(x) \equiv 0$, which corresponds to the state of rest. Those values of ω for which there exist other solutions of the equation (24), different from zero, are called the eigenfrequencies of the system.

Since nonzero solutions do not exist for every value of ω, the system can perform free oscillations only with definite frequencies. The smallest of these is called the fundamental tone of the system, and the remaining ones are overtones.

Now it turns out that for every system there exists an infinite sequence of eigenfrequencies, the so-called frequency spectrum

$$\omega_1, \omega_2, \cdots, \omega_n, \cdots$$

The nonzero solution $u_n(x)$ of the equation (24) corresponding to the the eigenfrequency ω_n gives us the form of the corresponding characteristic oscillation.

For example, if the elastic system is a string stretched between the points O and l and fastened at these points, then the possible frequencies of the characteristic oscillations of the system are equal to

$$a\frac{\pi}{l}, 2a\frac{\pi}{l}, 3a\frac{\pi}{l}, \cdots, na\frac{\pi}{l}, \cdots,$$

where a is a coefficient depending on the density and the tension of the

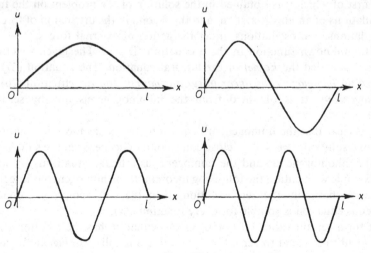

Fig. 10.

string, namely, $a = \sqrt{T/\rho}$. The fundamental tone is here $\omega_1 = a(\pi/l)$, and the overtones are $\omega_2 = 2\omega_1$, $\omega_3 = 3\omega_1$, $\cdots$, $\omega_n = n\omega_1$. The form of the corresponding harmonic oscillations is given by the equation

$$u_n(x) = \sin \frac{n\pi}{l} x$$

and are illustrated for $n = 1, 2, 3, 4$ in figure 10.

So far we have discussed free oscillations of elastic systems. Now if an exterior harmonic force acts on the elastic system during the motion, then, in determining the harmonic oscillations under the action of this force, we arrive at the function $u(x)$ at the so-called inhomogeneous integral equation

$$u(x) = \rho\omega^2 \int_a^b k(x, y) u(y) \, dy + h(x). \tag{25}$$

Properties of integral equations. Previously we have become acquainted with examples of integral equations

$$f(x) = \lambda \int_a^b k(x, y) f(y) \, dy \tag{26}$$

and

$$f(x) = \lambda \int_a^b k(x, y) f(y) \, dy + h(x), \tag{27}$$

the first of which was obtained in the solution of the problem on the free oscillations of an elastic system, and the second in the discussion of forced oscillations, i.e., oscillations under the action of external forces.

The unknown function in these equations is $f(x)$. The given function $k(x, y)$ is called the *kernel* of the integral equation. The equation (27) is called an *inhomogeneous linear integral equation*, and the equation (26) is *homogeneous*. It is obtained from the inhomogeneous one by setting $h(x) = 0$.

It is clear that the homogeneous equation always has the zero solution, i.e., the solution $f(x) = 0$. A close connection exists between the solutions of the inhomogeneous and the homogeneous integral equations. By way of example we mention the following theorem: If the homogeneous integral equation has only the zero solution, then the corresponding inhomogeneous equation is soluble for every function $h(x)$.

If for a certain value λ a homogeneous equation has the solution $f(x)$, not identically equal to zero, then this value λ is called an *eigenvalue* and the corresponding solution $f(x)$ an *eigenfunction*. We have seen earlier

§4. INTEGRAL EQUATIONS

that when an integral equation describes the free oscillations of an elastic system, then the eigenvalues are closely connected with the frequencies of the oscillations of the system (namely $\lambda = \rho\omega^2$). The eigenfunctions then give the form of the corresponding harmonic oscillations.

In the problems on oscillations it followed from the law of conservation of energy that

$$k(x, y) = k(y, x). \qquad (28)$$

A kernel satisfying the condition (28) is called *symmetric*.

The eigenfunctions and eigenvalues of an equation with a symmetric kernel have a number of important properties. One can prove that such an equation always has a sequence of real eigenvalues

$$\lambda_1, \lambda_2, \cdots, \lambda_n, \cdots.$$

To every eigenvalue there correspond one or several eigenfunctions. Here eigenfunctions corresponding to distinct eigenvalues are always orthogonal to each other.*

Thus, for every integral equation with a symmetric kernel the system of eigenfunctions is an orthogonal system of functions. There arises the question of when this system is complete, i.e., when can every function of the Hilbert space be expanded in a series by a system of eigenfunctions of the integral equation. In particular, if the equation

$$\int_a^b k(x, y) f(y) \, dy = 0 \qquad (29)$$

is satisfied for $f(y) \equiv 0$ only, then the system of eigenfunctions of the integral equation

$$\lambda \int_a^b k(x, y) f(y) \, dy = f(x)$$

is a complete orthogonal system.†

Thus, every function $f(x)$ with integrable square can in this case be expanded in a series by eigenfunctions. By discussing various types of integral equations, we obtain a general and powerful method of proving

---

* The latter statement will be proved in the next section.

† In the case when $k(x, y)$ is the Green's function of an elastic system, the equation (29) assumes a simple physical meaning. In fact [see formula (23)] we have seen that under the action of a force $f(y)$ distributed along the system the displacement of the system from the position of equilibrium is expressed by the formula $u(x) = \int_b^a k(x, y) f(y) \, dy$. Thus, the condition (29) signifies that every nonzero force takes the system out of its position of equilibrium.

that various important orthogonal systems are closed, i.e., that the functions are expandable in series by orthogonal functions. By this method we can prove the completeness of the system of trigonometric functions, of cylinder functions, spherical functions, and many other important systems of functions.

The fact that an arbitrary function can be expanded in a series by eigenfunctions means in the case of oscillations that every oscillation can be decomposed into a sum of harmonic oscillations. Such a decomposition yields a method that is widely applicable in solving problems on oscillations in various domains of mechanics and physics (oscillations of elastic bodies, acoustic oscillations, electromagnetic waves, etc.).

The development of the theory of linear integral equations gave the impetus to the creation of the general theory of linear operators of which the theory of linear integral equations forms an organic part. In the last few decades the general methods of the theory of linear operators have vigorously contributed to the further development of the theory of integral equations.

§5. Linear Operators and Further Developments of Functional Analysis

In the preceding section we have seen that problems on the oscillations of an elastic system lead to the search for the eigenvalues and eigenfunctions of integral equations. Let us note that these problems can also be reduced to the investigation of the eigenvalues and eigenfunctions of linear differential equations.* Many other physical problems also lead to the task of computing the eigenvalues and eigenfunctions of linear differential or integral equations.

Let us give one more example. In modern radio technology the so-called wave guides are widely used for the transmission of electromagnetic oscillations of high frequencies, i.e., hollow metallic tubes in which electromagnetic waves are propagated. It is known that in a wave guide only electromagnetic oscillations of not too large a wave length can be propagated. The search for the critical wave length amounts to a problem on the eigenvalues of a certain differential equation.

Problems on eigenvalues occur, moreover, in linear algebra, in the theory of ordinary differential equations, in questions of stability, etc.

So it became necessary to discuss all these related problems from one single point of view. This common point of view is the general theory of linear operators. Many problems on eigenfunctions and eigenvalues in various concrete cases came to be fully understood only in the light of

---

* See Chapter VI, §5.

§5. LINEAR OPERATORS AND FURTHER DEVELOPMENTS 253

the general theory of operators. Thus, in this and a number of other directions the general theory of operators turned out to be a very fruitful research tool in those domains of mathematics in which it is applicable.

In the subsequent development of the theory of operators, quantum mechanics played a very important role, since it makes extensive use of the methods of the theory of operators. The fundamental mathematical apparatus of quantum mechanics is the theory of the so-called self-adjoint operators. The formulation of mathematical problems arising in quantum mechanics was and still is a powerful stimulus for the further development of functional analysis.

The operator point of view on differential and integral equations turned out to be extremely useful also for the development of practical methods for approximate solutions of such equations.

Fundamental concepts of the theory of operators. Let us now proceed to an explanation of the fundamental definitions and facts in the theory of operators.

In analysis we have come across the concept of a function. In its simplest form this was a relation that associates with every number x (the value of the independent variable) a number y (the value of the function). In the further development of analysis it became necessary to consider relations of a more general type.

Such more general relations are discussed, for example, in the calculus of variations (Chapter VIII), where we associated with every function a number. If with every function a certain number is associated, then we say that we are given a functional. As an example of a functional we can take the association between an arbitrary function $y = f(x)$ ($a \leqslant x \leqslant b$) and the arc length of the curve represented by it. We obtain another example of a functional if we associate with every function $y = f(x)$ ($a \leqslant x \leqslant b$) its definite integral $\int_a^b f(x)\, dx$.

If we regard $f(x)$ as a point of an infinite-dimensional space, then a functional is simply a function of the points of the infinite-dimensional space. From this point of view the problems of the calculus of variations concern the search for maxima and minima of functions of the points of an infinite-dimensional space.

In order to define what we mean by a continuous functional it is necessary to define first what we mean by proximity of two points of an infinite-dimensional space. In §2 we gave the distance between two functions $f(x)$ and $g(x)$ (points of an infinite-dimensional space) as

$$\sqrt{\int_a^b [f(x) - g(x)]^2\, dx}.$$

This method of assigning a distance in infinite-dimensional space is often used, but of course it is not the only possible one. In other problems other methods of giving the distance between functions may turn out to be better. We may point, for example, to the problem of the theory of approximation of functions (see Chapter XII, §3), where the distance between functions, which characterizes the measure of proximity of the two functions $f(x)$ and $g(x)$, is given, for example, by the formula

$$\max |f(x) - g(x)|.$$

Other methods of giving a distance between functions are used in the investigation of functionals in the calculus of variations. Distinct methods of giving the distance between functions lead us to distinct infinite-dimensional spaces.

Thus, various infinite-dimensional (functional) spaces differ from each other by their set of functions and by the definition of distance between them. For example, if we take the set of all functions with integrable square and define distance as

$$\sqrt{\int_a^b [f(x) - g(x)]^2 \, dx},$$

then we arrive at the Hilbert space that was introduced in §2; but if we take the set of all continuous functions and define distance as $\max |f(x) - g(x)|$, then we obtain the so-called space (C).

In the discussion of integral equations we come across expressions of the form

$$g(x) = \int_a^b k(x, y) f(y) \, dy.$$

For a given kernel $k(x, y)$ this equation indicates a rule by which every function $f(x)$ is set in correspondence with another function $g(x)$.

This kind of a correspondence that relates with one function f another function g is called an *operator*.

We shall say that we are given a linear operator A in a Hilbert space if we have a rule by which we associate with every function f another function g. The correspondence need not be given for all the functions of the Hilbert space. In that case the set of those functions f for which there exists the function $g = Af$ is called the *domain of definition* of the operator A (similar to the domain of definition of a function in ordinary analysis). The correspondence itself is usually denoted as follows:

$$g = Af. \tag{30}$$

§5. LINEAR OPERATORS AND FURTHER DEVELOPMENTS 255

The linearity of the operator means that the sum of the functions f_1 and f_2 is associated with the sum of Af_1 and Af_2, and the product of f and a number λ with the function λAf; i.e.,

$$A(f_1 + f_2) = Af_1 + Af_2 \qquad (31)$$

and

$$A(\lambda f) = \lambda Af. \qquad (32)$$

Occasionally continuity is also postulated for linear operators; i.e., it is required that the convergence of a sequence of functions f_n to a function f should imply that the sequence Af_n should converge to Af.

Let us give examples of linear operators.

1. Let us associate with every function $f(x)$ the function $g(x) = \int_a^x f(t)\,dt$, i.e., the indefinite integral of f. The linearity of this operator follows from the ordinary properties of the integral, i.e., from the fact that the integral of the sum is equal to the sum of the integrals and that a constant factor can be taken out of the integral sign.

2. Let us associate with every differentiable function $f(x)$ its derivative $f'(x)$. This operator is usually denoted by the letter D; i.e.,

$$f'(x) = D f(x).$$

Observe that this operator is not defined for all the functions of the Hilbert space but only for those that have a derivative belonging to the Hilbert space. These functions form, as we have said previously, the domain of definition of this operator.

3. The examples 1 and 2 were examples of linear operators in an infinite-dimensional space. But examples of linear operators in finite-dimensional spaces have occurred in other chapters of this book. Thus, in Chapter III affine transformations were investigated. If an affine transformation of a plane of space leaves the origin of coordinates fixed, then it is an example of a linear operator in a two-dimensional, or three-dimensional, space. The linear transformations of an n-dimensional space introduced in Chapter XVI now appear as linear operators in n-dimensional space.

4. In the integral equations, we have already met a very important and widely applicable class of linear operators in a functional space, namely the so-called integral operators. Let us choose a certain definite function $k(x, y)$. Then the formula

$$g(x) = \int_a^b k(x, y) f(y)\,dy$$

associates with every function f a certain function g. Symbolically we can write this transformation as follows:

$$g = Af.$$

The operator A in this case is called an integral operator. We could mention many other important examples of integral operators.

In §4 we spoke of the inhomogeneous integral equation

$$f(x) = \lambda \int_a^b k(x, y) f(y)\, dy + h(x).$$

In the notation of the theory of operators this equation can be rewritten as follows

$$f = \lambda A f + h, \qquad (33)$$

where λ is a given number, h a given function (a vector of an infinite-dimensional space), and f the required function. In the same notation the homogeneous equation can be written as follows:

$$f = \lambda A f. \qquad (34)$$

The classical theorems on integral equations, such as, for example, the theorem formulated in §4 on the connection between the solvability of the inhomogeneous and the corresponding homogeneous integral equation, are not true for every operator equation. However, one can indicate certain general conditions to be imposed on the operator A under which these theorems are true.

These conditions are stated in topological terms and express that the operator A should carry the unit sphere (i.e., the set of vectors whose length does not exceed 1) into a compact set.

Eigenvalues and eigenvectors of operators. The problem of eigenvalues and eigenfunctions of an integral equation to which we were led by problems on oscillations can be formulated as follows: to find the values λ for which there exists a nonzero function f satisfying the equation

$$f(x) = \lambda \int_a^b k(x, y) f(y)\, dy.$$

As before, this equation can be written as follows:

$$f = \lambda A f$$

or

$$A f = \frac{1}{\lambda} f. \qquad (35)$$

§5. LINEAR OPERATORS AND FURTHER DEVELOPMENTS 257

Now we shall understand by A an arbitrary linear operator. Then a vector f satisfying the equation (35) is called an eigenvector of the operator A, and the number $1/\lambda$ the corresponding eigenvalue.

Since the vector $(1/\lambda)f$ coincides in direction with the vector f (differs from f only by a numerical factor), the problem of finding eigenvectors can also be stated as the problem of finding nonzero vectors f that do not change direction under the transformation A.

This way of looking at the eigenvalues enables us to unify the problem of eigenvalues of integral equations (if A is an integral operator), differential equations (if A is a differential operator), and the problem of eigenvalues in linear algebra (if A is a linear transformation in finite-dimensional space; see Chapter VI and Chapter XVI). In the case of three-dimensional space this problem arises in the search for the so-called principal axes of an ellipsoid.

In the case of integral equations a number of important properties of the eigenfunctions and eigenvalues (for example the reality of the eigenvalues, the orthogonality of the eigenfunctions, etc.) are consequences of the symmetry of the kernel, i.e., of the equation $k(x, y) = k(y, x)$.

For an arbitrary linear operator A in a Hilbert space the analogue of of this property is the so-called self-adjointness of the operator.

The condition for an operator A to be self-adjoint in the general case is that for any two elements f_1 and f_2 the equation

$$(Af_1, f_2) = (f_1, Af_2)$$

holds, where (Af_1, f_2) denotes the scalar product of the vector Af_1 and the vector f_2.

In problems of mechanics the condition of self-adjointness of an operator is usually a consequence of the law of conservation of energy. Therefore it is satisfied for operators connected with, say, oscillations for which there is no loss (dissipation) of energy.

The majority of operators that occur in quantum mechanics are also self-adjoint.

Let us verify that an integral operator with a symmetric kernel $k(x, y)$ is self-adjoint. In fact, in this case Af_1 is the function $\int_a^b k(x, y) f_1(y)\, dy$. Therefore the scalar product (Af_1, f_2), which is equal to the integral of the product of this function with f_2, is given by the formula

$$(Af_1, f_2) = \int_a^b \int_a^b k(x, y) f_1(y) f_2(x)\, dy\, dx.$$

Similarly

$$(f_1, Af_2) = \int_a^b \int_a^b k(x, y) f_2(y) f_1(x)\, dy\, dx.$$

The equation $(Af_1, f_2) = (f_1, Af_2)$ is an immediate consequence of the symmetry of the kernel $k(x, y)$.

Arbitrary self-adjoint operators have a number of important properties that are useful in the applications of these operators to the solution of a variety of problems. Indeed, the eigenvalues of a self-adjoint linear operator are always real and the eigenfunctions corresponding to distinct eigenvalues are orthogonal to each other.

Let us prove, for example, the last statement. Let λ_1 and λ_2 be two distinct eigenvalues of the operator A, and f_1 and f_2 eigenvectors corresponding to them. This means that

$$Af_1 = \lambda_1 f_1,$$
$$Af_2 = \lambda_2 f_2. \tag{36}$$

We form the scalar product of the first equation (36) by f_2, and of the second by f_1. Then we have

$$(Af_1, f_2) = \lambda_1(f_1, f_2),$$
$$(Af_2, f_1) = \lambda_2(f_2, f_1). \tag{37}$$

Since the operator A is self-adjoint, we have $(Af_1, f_2) = (Af_2, f_1)$. When we subtract the second equation (37) from the first, we obtain

$$0 = (\lambda_1 - \lambda_2)(f_1, f_2).$$

Since $\lambda_1 \neq \lambda_2$, we have $(f_1, f_2) = 0$, i.e., the eigenvectors f_1 and f_2 are orthogonal.

The investigation of self-adjoint operators has brought clarity into many concrete problems and questions connected with the theory of eigenvalues. Let us dwell in more detail on one of them, namely on the problem of the expansion by eigenfunctions in the case of a continuous spectrum.

In order to explain what a continuous spectrum means, let us turn again to the classical example of the oscillation of a string. Earlier we have shown that for a string of length l the characteristic frequencies of oscillations can assume the sequence of values.

$$a\frac{\pi}{l}, 2a\frac{\pi}{l}, \cdots, na\frac{\pi}{l}, \cdots.$$

Let us plot the points of this sequence on the numerical axis $O\lambda$. When we increase the length of the string l, the distance between any two adjacent points of the sequence will decrease, and they will fill the numerical axis

§5. LINEAR OPERATORS AND FURTHER DEVELOPMENTS

more densely. In the limit, when $l \to \infty$, i.e., for an infinite string, the the eigenfrequencies fill the whole numerical semiaxis $\lambda \geqslant 0$. In this case we say that the system has a continuous spectrum.

We have already said that for a string of length l the expansion in a series by eigenfunctions is an expansion in a series by sines and cosines of $n(\pi/l)x$; i.e., in a trigonometric series

$$f(x) = \frac{a_0}{2} + \sum a_n \cos n \frac{\pi}{l} x + b_n \sin n \frac{\pi}{l} x.$$

For the case of an infinite string we can again show that a more or less arbitrary function can be expanded by sines and cosines. However, since the eigenfrequencies are now distributed continuously along the numerical line, this is not an expansion in a series, but in a so-called Fourier integral

$$f(x) = \int_{-\infty}^{+\infty} [A(\lambda) \cos \lambda x + B(\lambda) \sin \lambda x] \, d\lambda.$$

The expansion in a Fourier integral was already well known and widely used in the 19th century in the solutions of various problems of mathematical physics.

However, in more general cases with a continuous spectrum* many problems referring to an expansion of functions by eigenfunctions were not properly clarified. Only the creation of the general theory of self-adjoint operators brought the necessary clarity to these problems.

Let us mention still another set of classical problems that have been solved on the basis of the general theory of operators. The discussion of oscillations involving dissipation (scattering) of energy belongs to such problems.

In this case we can again look for free oscillations of the system in the form $u(x) \phi(t)$. However, in contrast to the case of oscillations without dissipation of energy, the function $\phi(t)$ is not simply $\cos \omega t$, but has the form $e^{-kt} \cos \omega t$, where $k > 0$. Thus, the corresponding solution has the form $u(x)e^{-kt} \cos \omega t$. In this case every point x again performs oscillations (with frequency ω), however the oscillations are damped because for $t \to \infty$ the amplitude of these oscillations containing the factor e^{-kt} tends to zero.

It is convenient to write the characteristic oscillations of the system in the complex form $u(x)e^{-i\lambda t}$, where in the absence of friction the number λ is real and in the presence of friction λ is complex.

---

* As examples we can take the oscillations of an inhomogeneous elastic medium and also many problems of quantum mechanics.

The problem of the oscillations of a system with dissipation of energy again leads to a problem on eigenvalues, but this time not for self-adjoint operators. A characteristic feature here is the presence of complex eigenvalues indicative of the damping of the free oscillations.

Using a method of the theory of operators in conjuntion with methods of the theory of analytic functions M. V. Keldyš investigated this class of problems in 1950–1951 and proved for it the completeness of the system of eigenfunctions.

Connection of functional analysis with other branches of mathematics and quantum mechanics. We have already mentioned that the creation of quantum mechanics gave a decisive impetus to the development of functional analysis. Just as the rise of the differential and integral calculus in the 18th century was dictated by the requirements of mechanics and classical physics, so the development of functional analysis was, and still is, the result of the vigorous influence of contemporary physics, principally of quantum mechanics. The fundamental mathematical apparatus of quantum mechanics consists of the branches of mathematics relating essentially to functional analysis. We can only briefly indicate the connections existing here, because an explanation of the foundations of quantum mechanics exceeds the framework of this book.

In quantum mechanics the state of the system is given in its mathematical description by a vector of Hilbert space. Such quantities as energy, impulse, and moment of momentum are investigated by means of self-adjoint operators. For example, the possible energy levels of an electron in an atom are computed as eigenvalues of the energy operator. The differences of these eigenvalues give the frequencies of the emitted quantum of light and thus define the structure of the radiation spectrum of the given substance. The corresponding states of the electron are here described as eigenfunctions of the energy operator.

The solution of problems of quantum mechanics often requires the computation of eigenvalues of various (usually differential) operators. In some complicated cases the precise solution of these problems turns out to be practically impossible. For an approximate solution of these problems the so-called perturbation theory is widely used, which enables us to find from the known eigenvalues and functions of a certain self-adjoint operator A the eigenvalues of an operator A_1 slightly different from it. We mention that the perturbation theory has not yet received a full mathematical foundation, which is an interesting and important mathematical problem.

Independently of the approximate determination of eigenvalues, we can often say a good deal about a given problem by means of qualitative investigation. This investigation proceeds in problems of quantum mechanics on the basis of the symmetries existing in the given case. As examples of such symmetries we can take the properties of symmetry of crystals, spherical symmetry in an atom, symmetry with respect to rotation, and others. Since the symmetries form a group (see Chapter XX), the group methods (the so-called representation theory of groups) enables us to answer a number of problems without computation. As examples we may mention the classification of atomic spectra, nuclear transformations, and other problems. Thus, quantum mechanics makes extensive use of the mathematical apparatus of the theory of self-adjoint operators. At the same time the continued contemporary development of quantum mechanics leads to a further development of the theory of operators by placing new problems before this theory.

The influence of quantum mechanics and also the internal mathematical developments of functional analysis have had the effect that in recent years algebraic problems and methods have played a significant role in functional analysis. This intensification of algebraic tendencies in contemporary analysis can well be compared with the growth of the value of algebraic methods in contemporary theoretical physics in comparison with the methods of physics of the 19th century.

In conclusion, we wish to emphasize once more that functional analysis is one of the rapidly developing branches of contemporary mathematics. Its connections and applications in contemporary physics, differential equations, approximate computations, and its use of general methods developed in algebra, topology, the theory of functions of a real variable, etc., make functional analysis one of the focal points of contemporary mathematics.

Suggested Reading

N. Dunford and J. T. Schwartz, *Linear operators*. I. *General theory*, Interscience, New York, 1958.

I. M. Gel'fand and Z. Ja. Šapiro, *Representations of the rotation group of three-dimensional space and their applications*, Amer. Math. Soc. Translations Series 2, vol. 2, 1956, 207-316.

A. N. Kolmogorov and S. V. Fomin, *Elements of the theory of functions and functional analysis*. Vol. 1, *Metric and normed spaces*. Vol. 2, *Measure, Lebesgue integrals and Hilbert space*. Graylock, New York, 1957/1961.

L. D. Landau and E. M. Lifšic, *Course of theoretical physics*. Vol. 3, *Quantum mechanics*, Pergamon, New York, 1958/1960.

F. Riesz and B. Sz.-Nagy, *Functional analysis*, Frederick Ungar, New York, 1955.

A. E. Taylor, *Introduction to functional analysis*, Wiley, New York, 1958.

CHAPTER **XX**

GROUPS AND OTHER ALGEBRAIC SYSTEMS

§1. Introduction

In Chapter IV, which deals with the algebra of polynomials, we have already talked of the main lines of development of algebra, its place among other mathematical disciplines, and of the changes in the views on the very subject-matter of algebra. The aim of the present chapter is to give the reader an idea of those new algebraic theories that have sprung up in the last century, but have only been fully developed in the present one and have made a deep impact on the contemporary mathematical research.

Contemporary, as well as classical, algebra is the study of operations, of rules of computation. But it is not restricted to the study of properties of operations on numbers, since it strives to study the properties of operations on elements of a far more general nature. This tendency is dictated by practical requirements. For example, in mechanics we add up forces, velocities, or rotations. In linear algebra (see Chapter XVI), whose ideas and methods have wide application in practical calculations, the domains of operations are matrices, linear transformations, or vectors of an n-dimensional space.

The theory of groups plays a particularly prominent role in contemporary algebra, and a large part of this chapter is devoted to it. Among other algebraic theories, we shall dwell on the theory of hypercomplex systems, which is a necessary and important stage in the historical process of the development of the concept of number. Of course, these two theories do not exhaust by any means the content of contemporary algebra, but they illustrate rather well its ideas and methods.

The theory of groups has arisen from the necessity of finding an apparatus for investigating such important regularities of the real world as, for example, symmetry.

A knowledge of the symmetry properties of geometric bodies or other mathematical or physical objects sometimes gives us a key to the clarification of their structure. However, although the concept of symmetry is altogether intuitive, an accurate and general description of what symmetry is, and in particular a quantitative account of the properties of symmetry, requires use of the apparatus of the theory of groups.

The theory of groups arose rather long ago, at the end of the 18th and the beginning of the 19th century. Originally it was developed only as an auxiliary apparatus in problems on the solution by radicals of equations of higher degree. This was due to the fact that precisely in this problem it was first observed that properties of equivalence, of symmetry of the roots of the equation, are fundamental for the solution of the whole problem. In the course of the 19th and 20th centuries the important role of the laws of symmetry appeared in many other branches of science; geometry, crystallography, physics, and chemistry. This led to a wide propagation of the methods and results of the theory of groups. Since every domain of application presented its own peculiar problems to the theory of groups, the growing number of these domains also exerted the opposite effect, in giving rise to new branches of the theory of groups, and the result of all this is that the contemporary theory of groups, which is a single entity in its essential concepts, actually splits into a number of more or less independent disciplines: the general theory of groups, the theory of finite groups, the theory of continuous groups, of discrete groups of transformations, the theory of representations and characters of groups, and so forth. In their gradual evolution, the methods and concepts of the theory of groups turned out to be important not only for the investigation of the laws of symmetry but also for the solution of many other problems.

In our time the concept of a group has become one of the most important general concepts of modern mathematics, and the theory of groups has assumed a conspicuous place among the mathematical disciplines. Outstanding contributors to the development of the theory of groups and its applications were E. S. Fedorov, O. Ju. Šmidt, and L. S. Pontrjagin. The researches of Soviet mathematicians in the realm of group theory occupy a leading place in the present-day development of this theory.

§2. Symmetry and Transformations

The simplest forms of symmetry. We begin with an account of the simplest forms of symmetry with which the reader is familiar from everyday

§2. SYMMETRY AND TRANSFORMATIONS

life. One of these is the mirror symmetry of geometric bodies or the symmetry with respect to a plane.

A point A in space is called symmetrical to a point B with respect to a plane α (figure 1) if the plane intersects the segment AB perpendicularly at its midpoint. We also say that B is the mirror image of A in the plane α. A geometric body is called symmetric with respect to a plane if the plane divides the body into two parts each of which is the mirror image of the other in the plane. The plane itself is then called a plane of symmetry of the body. Mirror symmetry is often encountered in nature. For example, the form of the human body, or of the body of birds or animals, usually has a plane of symmetry.

Symmetry with respect to a line is defined in a similar way. We say that the points A, B lie symmetrically with respect to a line if the line intersects the segment AB at its midpoint and is perpendicular to AB (figure 2).

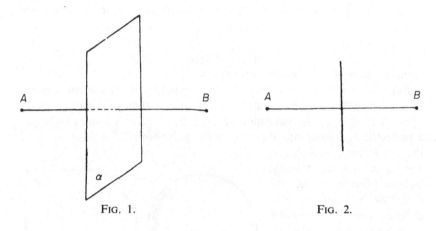

FIG. 1.   FIG. 2.

A geometric body is said to be symmetrical with respect to a line or to have this line as an axis of symmetry of order 2 if for every point of the body the symmetrical point also belongs to the body.

A body having an axis of symmetry of order 2 comes into coincidence with itself when the body is rotated around this axis by a half rotation, i.e., by an angle of 180°.

The concept of an axis of symmetry can be generalized in a natural way. A line is called an axis of symmetry of order n for a given body if the body comes into coincidence with itself on rotation around the axis by an angle $1/n\, 360°$. For example, a regular pyramid whose base is a regular n-gon has the line joining the vertex of the pyramid to the center of the base (figure 3) as an axis of symmetry of order n.

A line is called an axis of rotation of a body if the body comes into coincidence with itself on rotation around the axis by an arbitrary angle. For example, the axis of a cylinder or a cone, or any diameter of a sphere, is an axis of rotation. An axis of rotation is also an axis of symmetry of every order.

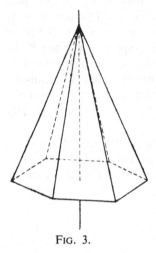

FIG. 3.

Finally, an important type of symmetry is symmetry with respect to a point or central symmetry. Points A and B are called symmetrical with respect to a center O if the segment joining A and B is bisected at O. A body is called symmetrical with respect to a center O if all its points fall into pairs of points symmetrical with respect to O. Examples of centrally symmetric bodies are the sphere and the cube, whose centers are their center of symmetry (figure 4).

A knowledge of all the planes, axes, and centers of symmetry of a body gives a fairly complete idea of its symmetry properties.

But the concept of symmetry has a meaning not only when applied to geometric figures. For example, the statement that in the polynomial $x_1^3 + x_2^3 + x_3^3 + x_4^3$ the variables x_1, x_2, x_3, x_4 occur symmetrically has a perfectly clear meaning; also that in the polynomial $x_1^3 + x_2^2 + x_3^2 + x_4^3$ the variables x_1 and x_4, x_2 and x_3 occur symmetrically, whereas for example, the variables x_1 and x_2 play different roles. The number of such examples could easily be increased. This prompts us to raise the important question: What is symmetry in general and how can we take account mathematically of the relation of symmetry?

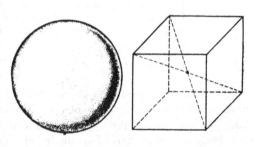

FIG. 4.

Now it turns out that a precise answer to this question is connected with the concept of transformation, which has already occurred many times in this book, right from the very first chapters. In order to be in a position to give a general definition of symmetry comprising such heterogeneous cases as the symmetry of spatial bodies and the symmetry of polynomials, it is necessary to formulate the concept of transformation in a very general way.

§2. SYMMETRY AND TRANSFORMATIONS

Transformations. Let M denote a finite or infinite collection of completely arbitrary objects. For example, M may be the set of numbers 1, 2, ⋯, n, the set of independent variables x_1, x_2, x_3, x_4, or the set of all points of a plane. If with every element of M a well-defined element of the same set is associated, then we say that a transformation of M is given. Every transformation of a finite set M can be given by means of a table consisting of two rows: In the upper row we write the names of the elements of M in an arbitrary order and below each of them we write the name of the element corresponding to it. For example, the table

$$\begin{pmatrix} 1 & 2 & 3 & 4 \\ 2 & 3 & 2 & 1 \end{pmatrix}$$

denotes the transformation of the set of numbers 1, 2, 3, 4 in which the numbers 1, 2, 3, 4 go over, respectively, into the numbers 2, 3, 2, 1. When we set out in the upper row the numbers 1, 2, 3, 4 in the order 3, 4, 1, 2, then we can write the same transformation also in form of the table

$$\begin{pmatrix} 3 & 4 & 1 & 2 \\ 2 & 1 & 2 & 3 \end{pmatrix}.$$

If the set M is infinite, but its elements can be counted (enumerated), then the transformation can be given in a similar way be setting out the elements in a single row (for example, if M is the set of all natural numbers 1, 2, 3, ...).

In studying transformations it is necessary to introduce a comprehensive notation for them. We shall denote transformations simply by letters A, B, etc., and if some transformation of the set M is denoted by the letter A, then we denote by mA, where m is an arbitrary element of M, the image of the element m, i.e., that element into which M goes over, under the transformation A. Suppose, for example, that

$$A = \begin{pmatrix} 1 & 2 & 3 & 4 \\ 2 & 3 & 2 & 1 \end{pmatrix}; \text{ then } 1A = 2,\ 2A = 3,\ 3A = 2,\ 4A = 1.$$

Let us indicate some transformations that play an important role in geometry.

We draw an arbitrary line a in space and associate with every point P of space the point Q obtained by rotating the point P around the axis a by a fixed angle ϕ (figure 5). In this way we have defined a transformation of the set of all points of space, the so-called rotation of space by the angle ϕ around the axis a.

Observe that the word "rotation" in mechanics denotes a certain process as a result of which the points of the body assume a new position. Here we have used the term "rotation" in the sense of a transformation of space. We abstract from the actual process of motion and consider only its final result, namely the correspondence between the initial and the final position of the points.

Another important transformation of space is the parallel shift of all the points in a given direction by a given distance. From figure 6, in which

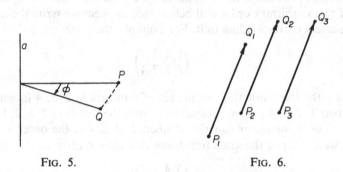

Fig. 5.   Fig. 6.

we have indicated for arbitrary points P_1, P_2, P_3 the corresponding points Q_1, Q_2, Q_3, it is clear that when we know the corresponding point of only one point of space in a parallel shift, then we can find the corresponding points for all other points of space.

Earlier we have defined the concepts of a plane of symmetry and of an axis and center of symmetry of a figure in space. To each of these concepts there corresponds a definite transformation of space: a reflection in the plane, a rotation around the line, and a reflection with respect to the center. For example, a reflection in a plane is the transformation in which every point of space is associated with its symmetrical point with respect to the plane. A rotation around the line and a reflection with respect to the center are similarly defined.

So far we have talked of transformations of space. The corresponding transformations of a plane: rotation of the plane around a point by a given angle, a parallel shift of the plane in itself in a given direction, and a reflection with respect to a line lying in the plane, all these are similarly defined and are even more intuitive than the corresponding transformations of space.

One-to-one transformations. In discussing all possible transformations of one and the same set, we must first of all observe the fundamental difference between one-to-one transformations of the set onto itself and transformations that are not one-to-one. A transformation A of a set M

§2. SYMMETRY AND TRANSFORMATIONS

is called a one-to-one transformation of the set onto itself if not only to every element of M there corresponds a definite unique element of M (this is part of the definition of transformation) but if also for every element y of M there exists one and only one element x that goes over into y. In other words, a transformation A is one-to-one if the "equation" $xA = y$ has one and only one "solution" x in M for every y in M.

All the transformations of space considered here, reflections, rotations and translations, are one-to-one, since in these cases not only is there for every point X a point into which X goes over but there is also a unique point that goes over into X.

It is easy to give examples to the contrary; thus, the transformation of the set of numbers 1, 2, 3, 4, given by the table

$$\begin{pmatrix} 1 & 2 & 3 & 4 \\ 2 & 1 & 2 & 3 \end{pmatrix}$$

is not one-to-one, since in it no number goes over into 4. The transformation of the set of all natural numbers 1, 2, 3, $\cdots$, given by the table

$$\begin{pmatrix} 1 & 2 & 3 & 4 & 5 & 6 & \cdots \\ 1 & 1 & 2 & 2 & 3 & 3 & \cdots \end{pmatrix}$$

is also not one-to-one. Although here for every number n there is the number $2n$ that goes over into it, the number $2n$ is not the only one having this property, since $2n - 1$ also goes over into n. For transformations given by tables it is very easy to establish a criterion under with the transformation is one-to-one. For this it is obviously necessary and sufficient that the lower line of the table should contain every element of the set once and once only. Occasionally in mathematics one discusses transformations that are not one-to-one. For example, the great importance of the operation of projecting a space onto a plane is well known. This transformation is not one-to-one, because in it every point is the projection not of one but of a whole series of points of space. But in the majority of cases it is convenient to deal only with one-to-one transformations; these transformations, in particular, play a fundamental role when physical processes are considered under which the elements of the system in question are not merged with one another, not annihilated and not created.

Henceforth in talking of transformations we shall tacitly assume that they are one-to-one; they are also often called permutations, especially when we are dealing with transformations of a finite set.

For every (one-to-one) transformation A of a set M onto itself, we can easily define an inverse transformation A^{-1}. If A carries an arbitrary element x of M into y, then the transformation carrying y into x is called the inverse transformation to A and is denoted by A^{-1}. For example, if

$$A = \begin{pmatrix} 1 & 2 & 3 & 4 \\ 2 & 3 & 4 & 1 \end{pmatrix},$$

then

$$A^{-1} = \begin{pmatrix} 2 & 3 & 4 & 1 \\ 1 & 2 & 3 & 4 \end{pmatrix} = \begin{pmatrix} 1 & 2 & 3 & 4 \\ 4 & 1 & 2 & 3 \end{pmatrix};$$

if A is a rotation of space around an axis by an angle ϕ, then A^{-1} is the rotation around the same axis by the angle ϕ in the opposite direction, etc.

Occasionally it happens that the inverse transformation coincides with the given one. In particular, reflections with respect to a plane or a point in space have this property. So has the permutation

$$A = \begin{pmatrix} 2 & 1 & 4 & 3 \\ 1 & 2 & 3 & 4 \end{pmatrix}, \quad \text{since} \quad A^{-1} = \begin{pmatrix} 1 & 2 & 3 & 4 \\ 2 & 1 & 4 & 3 \end{pmatrix} = \begin{pmatrix} 2 & 1 & 4 & 3 \\ 1 & 2 & 3 & 4 \end{pmatrix}.$$

Note that we cannot speak of an inverse transformation for those that are not one-to-one, because an individual element may be such that no elements or several elements go over into it.

The general definition of symmetry. In mathematics and its applications it is very rarely necessary to consider all transformations of a given set. The fact is that the sets themselves are rarely thought of as merely the collections of their elements completely disconnected from one another. This is natural, because the sets that are discussed in mathematics are abstract images of real collections, whose elements always stand in an infinite variety of interrelations with each other, and of connections with what is going on beyond the limits of the set in question. But in mathematics it is convenient to abstract from the major part of these connections and to preserve and take into account the most essential one. This compels us in the first instance to consider only such transformations of sets as do not destroy the relevant connections of one kind or another between their elements. These are often called admissible transformations or *automorphisms* with respect to the relevant connections between the elements of the set. For example, for points of space the concept of distance between two points is important. The presence of this concept forges a link between points which consists in the fact that any two points stand at a definite distance from one another. Transformations that do not destroy these connections are the same as those under which the

§2. SYMMETRY AND TRANSFORMATIONS

distance between points remains unchanged. These transformations are called "motions" of space.

With the help of the concept of automorphism it is not difficult to give a general definition of symmetry. Suppose that a certain set M is given, in which definite connections between the elements are to be taken into account, and that P is a certain part of M. We say that P is symmetrical or invariant with respect to the admissible transformation A of M if A carries every element of P again into an element of P. Therefore, a symmetry of P is characterized by the collection of admissible transformations of the containing set M that transform P into itself. The concept of symmetry of a body in space falls entirely under this definition. The role of the set M is played by the whole space, the role of admissible transformations by the "motions," the role of P by the given body. The symmetry of P is therefore characterized by the collection of motions under which P coincides with itself.

The reflections, parallel shifts, and rotations of space around a given line that we have discussed are special cases of motions, because distances between points obviously remain unchanged under these transformations. A more detailed investigation shows that every motion of a plane is either a parallel shift or a rotation around a center or a reflection in a line or a combination of a reflection in a line with a parallel shift along that line. Similarly, every motion of space is either a parallel shift or a rotation around an axis or a spiral motion, i.e., a rotation around an axis combined with the shift along this axis, or a reflection in a plane combined with, possibly, a shift along the plane of reflection or a rotation around an axis perpendicular to this plane.

Parallel shifts, rotations, and spiral motions of space are called proper motions or motions of the first kind. The remaining "motions" (including reflections) are known as improper motions or motions of the second kind. In a plane, motions of the first kind are parallel shifts and rotations, whereas reflections in a line and reflections combined with a rotation or a translation are motions of the second kind.

It is easy to imagine how transformations that are motions of the first kind can be obtained as a result of a continuous motion of space or of a plane in itself. Motions of the second kind cannot be obtained in this way, because this is prevented by the mirror reflection that occurs in their formation.

One often says that the plane is symmetrical in all its parts or that all points of the plane are equivalent. In the strict language of transformations this statement means that every point of the plane can be superimposed on any other point by means of a suitable "motion."

The cases of symmetry of bodies or figures discussed previously are also comprised under the general definition of symmetry. For example, a body that is symmetrical with respect to a plane α comes into coincidence with itself on reflection in the plane α; a body that is symmetrical with respect to a center O comes into coincidence with itself under reflection in O. Therefore, the degree of symmetry of a body or of a spatial figure can be completely characterized by the collection of all motions of space of the first and second kind that bring the body or the figure into coincidence with itself. The greater and more diverse this collection of motions, the higher is the degree of symmetry of the body or figure. If, in particular, this collection contains no motions except the identity transformation, then the body can be called unsymmetrical.

The degree of symmetry of a square in a plane is characterized by the collection of motions of the plane that bring the square into coincidence with itself. But if the square coincides with itself, then the point of intersection of its diagonals must also coincide with itself. Therefore the required motions leave the center of the square invariant, and so they are either rotations around the center or reflections in lines passing through the center. From figure 7 we can easily read that the square $ABCD$ is symmetrical with respect to the rotations around its center O by angles that are multiples of $90°$ and also with respect to reflections in the diagonals AC, BD and the lines KL, MN. These eight motions characterize the symmetry of the square.

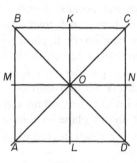

Fig. 7.

The collection of symmetries of a rectangle reduces to a rotation around the center by $180°$ and a reflection in the lines that join the midpoints of opposite sides; and the set of symmetries of a parallelogram (figure 8) consists only of the rotations around the center by angles that are multiples of $180°$, i.e., of reflections in the center and the identity transformation.

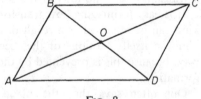

Fig. 8.

Previously we have given an algebraic example of symmetry; we mentioned that the concept of symmetry of a polynomial in several variables also has a meaning.

Let us discuss how the symmetry of a polynomial can be characterized.

§3. GROUPS OF TRANSFORMATIONS

We shall say that the permutation of the variables

$$A = \begin{pmatrix} x_1, x_2, \cdots, x_n \\ x_{i_1}, x_{i_2}, \cdots, x_{i_n} \end{pmatrix}$$

or briefly

$$A = \begin{pmatrix} 1, 2, \cdots, n \\ i_1, i_2, \cdots, i_n \end{pmatrix}$$

has been made in the polynomial $F(x_1, x_2, \cdots, x_n)$ if everywhere in the polynomial the letter x_1 has been replaced by x_{i_1}, x_2 by x_{i_2}, etc. The polynomial so obtained will be denoted by FA. Thus, if $F = x_1^2 - 2x_2 + x_3 - x_4$,

$$A = \begin{pmatrix} 1 & 2 & 3 & 4 \\ 3 & 1 & 4 & 2 \end{pmatrix}, \quad \text{then} \quad FA = x_3^2 - 2x_1 + x_4 - x_2.$$

The symmetry of the given polynomial is characterized by the collection of those permutations of the variables that, when carried out on the polynomial, leave it unchanged. For example, the symmetry of the polynomial $x_1^3 + 2x_2 + x_3^3 + 2x_4$ is characterized by the four permutations:

$$\begin{pmatrix} 1 & 2 & 3 & 4 \\ 1 & 2 & 3 & 4 \end{pmatrix}, \begin{pmatrix} 1 & 2 & 3 & 4 \\ 3 & 2 & 1 & 4 \end{pmatrix}, \begin{pmatrix} 1 & 2 & 3 & 4 \\ 1 & 4 & 3 & 2 \end{pmatrix}, \begin{pmatrix} 1 & 2 & 3 & 4 \\ 3 & 4 & 1 & 2 \end{pmatrix},$$

and the symmetry of the polynomial $x_1^3 + 2x_2 + x_3^3 + x_4$ is characterized by the two permutations:

$$\begin{pmatrix} 1 & 2 & 3 & 4 \\ 1 & 2 & 3 & 4 \end{pmatrix} \quad \text{and} \quad \begin{pmatrix} 1 & 2 & 3 & 4 \\ 3 & 2 & 1 & 4 \end{pmatrix}.$$

§3. Groups of Transformations

Multiplication of transformations. In studying properties of transformations it is easy to observe that certain transformations can be constructed from others. For example, a spiral motion is composed of a rotation around the axis and a shift along the axis. This process of forming new transformations from given ones is called multiplication of transformations. When we apply to an arbitrary element x of a set M some transformation A and then apply the transformation B to the new element xA, we obtain the element $(xA)B$. The transformation that carries x immediately into $(xA)B$ is called the product of A and B and is denoted by AB. Therefore, by definition, we have

$$x(AB) = (xA)B.$$

XX. GROUPS AND OTHER ALGEBRAIC SYSTEMS

Example:

$$\begin{pmatrix} 1 & 2 & 3 & 4 \\ 2 & 3 & 4 & 1 \end{pmatrix} \begin{pmatrix} 1 & 2 & 3 & 4 \\ 3 & 4 & 1 & 2 \end{pmatrix} = \begin{pmatrix} 1 & 2 & 3 & 4 \\ 4 & 1 & 2 & 3 \end{pmatrix}.$$

Since the first permutation carries 1 into 2 and the second, 2 into 4, therefore the resulting permutation must carry 1 into 4, and so forth. Here are a few more examples:

$$\begin{pmatrix} 1 & 2 & 3 & 4 \\ 3 & 1 & 4 & 2 \end{pmatrix} \begin{pmatrix} 1 & 2 & 3 & 4 \\ 1 & 3 & 2 & 4 \end{pmatrix}^{-1} = \begin{pmatrix} 1 & 2 & 3 & 4 \\ 3 & 1 & 4 & 2 \end{pmatrix} \begin{pmatrix} 1 & 2 & 3 & 4 \\ 1 & 3 & 2 & 4 \end{pmatrix} = \begin{pmatrix} 1 & 2 & 3 & 4 \\ 2 & 1 & 4 & 3 \end{pmatrix};$$

$$\begin{pmatrix} 1 & 2 & 3 & 4 \\ 2 & 1 & 4 & 3 \end{pmatrix} \begin{pmatrix} 1 & 2 & 3 & 4 \\ 3 & 1 & 2 & 4 \end{pmatrix} = \begin{pmatrix} 1 & 2 & 3 & 4 \\ 1 & 3 & 4 & 2 \end{pmatrix}; \begin{pmatrix} 1 & 2 & 3 & 4 \\ 3 & 1 & 2 & 4 \end{pmatrix} \begin{pmatrix} 1 & 2 & 3 & 4 \\ 2 & 1 & 4 & 3 \end{pmatrix} = \begin{pmatrix} 1 & 2 & 3 & 4 \\ 4 & 2 & 1 & 3 \end{pmatrix}.$$

The last two examples show that the multiplication of transformations is, as we say, a noncommutative operation: Its result depends on the order of the factors. This is also easily verified for the multiplication of motions of a plane. Suppose, for example, that A is a rotation of the plane by 90° around the origin O, and B a parallel shift by a unit length along the x-axis. Let us find the image of O under the transformations AB and BA. By definition we have (figure 9)

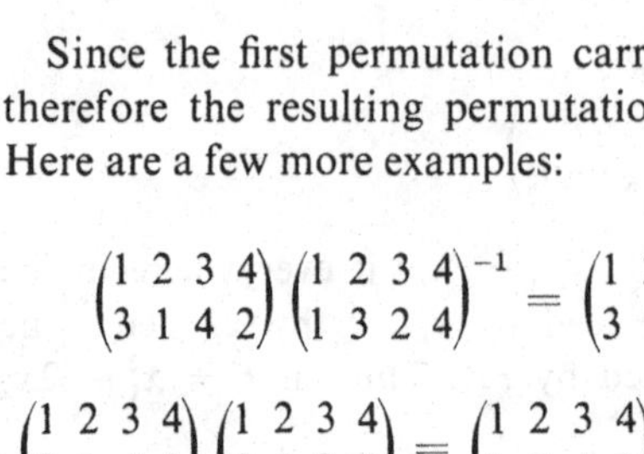

$$O(AB) = (OA) B = OB = M,$$
$$O(BA) = (OB) A = MA = N,$$

i.e., $AB \neq BA$.

FIG. 9.

For a closer understanding of the geometric nature of the transformation BA, let us consider the point P. We have

$$P(BA) = (PB) A = QA = P,$$

i.e., the point P remains unchanged under the transformation BA. Starting out from this it is easy to show that BA is simply a rotation of the plane by 90° around P. Similarly

$$Q(AB) = (QA) B = PB = Q,$$

and AB is the rotation of the plane by 90° around Q.

The multiplication of motions of the plane or of space generally follows rather complicated rules. However, in two important cases the rules of multiplication are very simple. First, when we multiply rotations of a plane around one and the same point or rotations of space around one and the

§3. GROUPS OF TRANSFORMATIONS

same line by the angles ϕ and ψ, then the resulting transformation is the corresponding rotation by the angle $\phi + \psi$. Second, when we multiply parallel shifts characterized by the vectors $\overrightarrow{MN}$ and $\overrightarrow{NP}$, then the product is also a parallel shift characterized by the vector $\overrightarrow{MP}$, i.e., the sum of the original vectors.

The very term "multiplication" of transformations points to a certain analogy between the multiplication of numbers and the multiplication of transformations. However, this analogy is incomplete. For example, for the multiplication of numbers we have the commutative law. But we have already seen that in the multiplication of transformations this law may be violated. The second fundamental law of arithmetic, namely the associative law, is completely preserved for transformations. In fact, for arbitrary transformations A, B, C of a set M we have the equation $A(BC) = (AB) C$.

For if m is an arbitrary element of M, then

$$m[A(BC)] = (mA)(BC) = [(mA) B] C = [m(AB)] C = m[(AB) C].$$

The associative law enables us, instead of speaking of the two products $A(BC)$ and (AB) of the transformations A, B, C, to speak only of the single product $A(BC) = (AB) C = ABC$. The same law shows that the product of four or more transformations does not depend on the distribution of parentheses.

Furthermore, among transformations there is the one that plays the role of the number 1, this is the identity or unit transformation E, which leaves every element of M unchanged. Clearly, $AE = EA = A$, whatever the transformation A.

We mention the following important fact: The product of one-to-one transformations is also one-to-one. For in order to find the element x of M that is carried by AB into a given element a, it is sufficient to find the element x_1 that is carried by B into a and then to find the element x_2 that is carried by A into x_1. Since $x_2(AB) = (x_2A) B = x_1A = a$, then x_2 is the required element x.

The product of a transformation A and the inverse transformation A^{-1} is the unit transformation; i.e.,

$$AA^{-1} = A^{-1}A = E.$$

This follows immediately from the definition of the inverse transformation.

The example discussed previously of the multiplication of a parallel shift of a plane and a rotation shows that properties of a product of transformations are not always easily discerned starting from properties of the factors. However, the product of the transformations of the form

$C = B^{-1}AB$ is an important exception: The properties of C are here very simply connected with the properties of A and B. For if an element m of M is carried by A into n, then the element mB, which is "shifted" by means of B, is carried by C into the "shifted" element nB.

Proof: $(mB)B^{-1}AB = mAB = nB$.

The transformation $B^{-1}AB$ is said to be obtained from A by transforming it by B or to be conjugate to A by means of B.

Let us transform, for example, a rotation P_0 of a plane around the point O by means of a translation V. By the preceding rule, in order to find the pairs of initial and final positions of points for the transformed motions $C = V^{-1}P_0V$, we have to shift by means of V the corresponding pairs of points for the transformation P_0. Since the point O in the rotation P_0 remains unchanged (figure 10), the point OV will remain unchanged under the transformation C. Furthermore, if a point M is carried by P into N, then the shifted point MV will be carried by C into the point NV. From figure 10 it is then clear that the transformation C is a rotation around the point OV by the same angle ϕ as the rotation P.

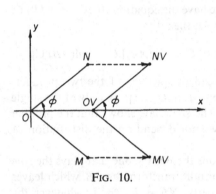

Fig. 10.

Similarly, it can be shown that if a translation of the plane characterized by the vector $\overrightarrow{MN}$ is transformed by means of a rotation P_0 by the angle ϕ, then we obtain again a translation of the plane, characterized by a different vector.

The previous rule for finding the transformation $B^{-1}AB$ can be formulated in a very elegant way, when the transformations are given by tables. Suppose that

$$A = \begin{pmatrix} 1 & 2 & \cdots & n \\ a_1 & a_2 & \cdots & a_n \end{pmatrix}, \quad B = \begin{pmatrix} 1 & 2 & \cdots & n \\ b_1 & b_2 & \cdots & b_n \end{pmatrix},$$

then

$$B^{-1}AB = \begin{pmatrix} b_1 & b_2 & \cdots & b_n \\ 1 & 2 & \cdots & n \end{pmatrix} \begin{pmatrix} 1 & 2 & \cdots & n \\ a_1 & a_2 & \cdots & a_n \end{pmatrix} \begin{pmatrix} 1 & 2 & \cdots & n \\ b_1 & b_2 & \cdots & b_n \end{pmatrix} = \begin{pmatrix} b_1 & b_2 & \cdots & b_n \\ b_{a_1} & b_{a_2} & \cdots & b_{a_n} \end{pmatrix};$$

i.e., in order to transform a permutation A by means of permutation B,

§3. GROUPS OF TRANSFORMATIONS

we have to subject all the elements of the upper and of the lower row of A to the transformation specified by B. For example, if

$$A = \begin{pmatrix} 1 & 2 & 3 & 4 & 5 \\ 3 & 5 & 4 & 1 & 2 \end{pmatrix}, \quad B = \begin{pmatrix} 1 & 2 & 3 & 4 & 5 \\ 2 & 5 & 1 & 3 & 4 \end{pmatrix},$$

then

$$B^{-1}AB = \begin{pmatrix} 1B & 2B & 3B & 4B & 5B \\ 3B & 5B & 4B & 1B & 2B \end{pmatrix} = \begin{pmatrix} 2 & 5 & 1 & 3 & 4 \\ 1 & 4 & 3 & 2 & 5 \end{pmatrix} = \begin{pmatrix} 1 & 2 & 3 & 4 & 5 \\ 3 & 1 & 2 & 5 & 4 \end{pmatrix}.$$

Note that although in general the product of two transformations depends on the order of the factors, in individual cases the products AB and BA may be one and the same. Then the transformations A and B are called permutable or commuting. If $AB = BA$, then

$$B^{-1}AB = B^{-1}BA = A.$$

Thus, the transformation of a permutation by means of another one commuting with it does not change the given permutation.

Groups of transformations. The set of transformations that characterizes the symmetry of a certain figure cannot be arbitrary, it must necessarily have the following properties:

1. The product of two transformations belonging to the set also belongs to the set.
2. The identity transformation belongs to the set.
3. If a transformation belongs to the set, then the inverse transformation also belongs to the set.

These properties turn out to be very important for the study of transformations; in view of this, every set of one-to-one transformations of a set that has these three properties is called a *group of transformations* of M, independently of the fact whether this set characterizes the symmetry of a certain figure or not.

From the point of view of algebra, the properties 1–3 are very important, since they enable us, starting from certain transformations A, B, C, $\cdots$, belonging to a given set, to form various new transformations of the form $ABAC$, $A^{-1}BCB^{-1}$ and so forth, and the properties 1–3 guarantee that all the transformations so obtained do not carry us beyond the limits of the given set of transformations.

The number of transformations that form a group is called the *order of the group*; it may be finite or infinite. Accordingly, groups are divided into finite and infinite. Earlier we discussed the group of symmetry of a square in a plane. This group turned out to consist altogether of eight transformations. On the other hand, the infinite set of points A_i of the plane, illustrated in figure 11, is transformed into itself by the following motions of the plane: translations along the axis OA in either of the two directions by distances that are multiples of OA; reflections in the dotted lines; reflection in the axis OA. Hence it is clear that the group of symmetries of this figure is infinite.

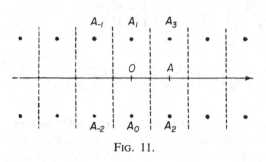

FIG. 11.

The collection of transformations that preserve a certain object, i.e., characterize its symmetry, is always a group. This method of giving groups in the form of symmetry is one of most significance. Very important groups can be obtained by this principle. Of first importance among these are the groups of motions of a plane and of space. The symmetry groups of the regular polyhedra are also of great interest. It is known that in space there exist altogether five types of regular polyhedra (with 4, 6, 8, 12 and 20 faces). When we take an arbitrary regular polyhedron and consider all the motions of space that bring the given polyhedron into coincidence with itself, we obtain a group, namely the symmetry group of the polyhedron. If instead of all the motions we consider only the motions of the first kind that carry the polyhedron into coincidence with itself, then we obtain again a group that is part of the full group of symmetries of the polyhedron. This group is called the group of rotations of the polyhedron. Since in a superposition of the polyhedron with itself, its center is also superimposed on itself, all motions that occur in the group of symmetries of the polyhedron leave the center of the polyhedron unchanged and can therefore only be either rotations around axes passing through the center or reflections in planes passing through the center or, finally, reflections in such planes combined with rotations around axes passing through the center and perpendicular to these planes.

With the help of these remarks it is easy to find all the groups of symmetry and the groups of rotations of the regular polyhedra. In Table 1 we have given the order of the symmetry groups and the rotation groups of the regular polyhedron. All these groups are finite.

§3. GROUPS OF TRANSFORMATIONS

Table 1.

Number of faces	4	6	8	12	20
Order of the symmetry group	24	48	48	120	120
Order of the rotation group	12	24	24	60	60

Permutation groups. Of all the transformation groups, historically, the first to be studied in mathematics were the permutation groups of polynomials in the variables $x_1, x_2, \cdots, x_n$. The investigation of these groups is closely connected with the problem of solving equations of higher degrees by radicals. Obviously, the collection of all permutations of the variables that do not change the values of one or several polynomials in these variables is a group. Polynomials that are unchanged under all permutations of the variables are called symmetric polynomials. For example, $x_1 + x_2 + \ldots + x_n$ is a symmetric polynomial. Accordingly the set of all permutations of a given set of variables is called the *symmetric* group of the permutations of this set.

The number of the variables to be permuted is called the degree of the symmetric group. Instead of the permutations of the variables $x_1, \ldots, x_n$, we can simply consider the permutations of the numbers $1, 2, \ldots, n$. Since every permutation of these numbers can be written in the form

$$\begin{pmatrix} 1 & 2 & \ldots & n \\ a_1 & a_2 & \ldots & a_n \end{pmatrix},$$

where $a_1, a_2, \cdots, a_n$ are the numbers $1, 2, \cdots, n$ written in some order, the number of all permutations of n elements; i.e., the order of the symmetric group, is equal to $n! = 1 \cdot 2 \cdot 3 \cdots n$. This order increases very rapidly with n and the group of permutations of 10 variables is already of the order 3,628,800.

Let us consider the polynomial

$$F(x_1, \cdots, x_n)$$
$$= (x_2 - x_1)(x_3 - x_1) \cdots (x_n - x_1)(x_3 - x_2) \cdots (x_n - x_2) \cdots (x_n - x_{n-1}). \quad (1)$$

It is clear that every permutation of the variables either leaves the value of the polynomial F unchanged or changes its sign only. Permutations of the first kind are called even. Permutations that change the sign of F are called odd. The set of even permutations forms the symmetry group of the polynomial (1). It is called the alternating group.

The product of two even permutations is even, because even permutations form a group. The product of two odd permutations is an even permutation.

For if A and B are odd permutations, then

$$FAB = (FA)B = (-F)B = -(-F) = F.$$

In the same way it can be shown that the product of an even and an odd permutation is an odd permutation and that the permutation inverse to an even or an odd permutation is a permutation of the same parity.

An example of an odd permutation is

$$S = \begin{pmatrix} 1, 2, 3, \cdots, n \\ 2, 1, 3, \cdots, n \end{pmatrix},$$

which interchanges the elements 1 and 2.

Decomposition of permutations into cycles. In studying permutation groups it is very helpful to represent permutations in the form of products of so-called cycles. By definition, the symbol $(m_1, m_2, \cdots, m_k)$ denotes the permutation that carries m_1 into m_2, m_2 into m_3, $\cdots$, m_{k-1} into m_k, and m_k again into m_1 and leaves all the remaining elements of the set in question unchanged. For example, if we consider permutations of the numbers 1, 2, 3, 4, 5, then

$$(1, 2, 3, 4, 5) = \begin{pmatrix} 1 & 2 & 3 & 4 & 5 \\ 2 & 3 & 4 & 5 & 1 \end{pmatrix}, \quad (3, 5) = \begin{pmatrix} 1 & 2 & 3 & 4 & 5 \\ 1 & 2 & 5 & 4 & 3 \end{pmatrix}.$$

A permutation of the form $(m_1, m_2, \cdots, m_k)$ is called *cyclic* or a *cycle* of length k, and $m_1, m_2, \cdots, m_k$ are called the elements of the cycle. The unit permutation can be written in the form of cycles $(1) = (2) = \cdots$ of length 1. Cycles of length 2 are called *transpositions*. When we permute the elements of a cycle in cyclic order, we obtain the same permutation, for example $(1, 2, 3) = (2, 3, 1) = (3, 1, 2)$, $(5, 6) = (6, 5)$.

It is easy to verify that cycles without common elements, for example $(2, 3)$ and $(1, 4, 5)$, are permutable, so that in multiplying such cycles we need not take the order of the factors in the product into account.

The significance of cycles in the general theory is based on the following theorem: Every permutation can be represented in the form of a product of cycles without common elements, and this representation is unique to within the order of the factors.

§3. GROUPS OF TRANSFORMATIONS

The proof of the theorem is immediately clear from the method of such a representation. Suppose that we wish to decompose the permutation.

$$A = \begin{pmatrix} 1 & 2 & 3 & 4 & 5 & 6 \\ 4 & 5 & 6 & 3 & 2 & 1 \end{pmatrix}.$$

We see that A carries 1 into 4, 4 into 3, 3 into 6, and 6 into 1. As a result we have a first factor (1, 4, 3, 6). Of the remaining numbers we consider the 2 and note that A carries 2 into 5, 5 into 2. Therefore the second factor is (2, 5). Since all numbers are now accounted for, we have

$$\begin{pmatrix} 1 & 2 & 3 & 4 & 5 & 6 \\ 4 & 5 & 6 & 3 & 2 & 1 \end{pmatrix} = (1, 4, 3, 6)(2, 5). \tag{2}$$

It is also possible to decompose permutations into cycles with common elements, but this is not unique. For example,

$$(a_1, a_2, \cdots, a_n) = (a_1, a_2)(a_1, a_3) \cdots (a_1, a_n)$$
$$= (a_2, a_3)(a_2, a_4) \cdots (a_2, a_n)(a_2, a_1). \tag{3}$$

Let us show that every cycle of length 2 is an odd permutation. We have already seen this for the cycle (1, 2). But every cycle (i, j) is $S^{-1}(1, 2) S$, where S is an arbitrary permutation

$$\begin{pmatrix} 1, 2, \cdots \\ i, j, \cdots \end{pmatrix}$$

carrying 1 into i and 2 into j. The permutation $S^{-1}(1, 2) S$ is an odd permutation, because (1, 2) is odd, and S and S^{-1} are simultaneously even or odd.

According to (3) a cycle of length $m + 1$ can be represented as a product of m odd permutations. Therefore a cycle of length $m + 1$ is an odd permutation, when $m + 1$ is even, and even, when $m + 1$ is odd. This enables us to compute rapidly the parity of permutations whose decomposition into cycles is known. Specifically, the permutation

$$\begin{pmatrix} 1 & 2 & 3 & 4 & 5 & 6 \\ 4 & 5 & 6 & 3 & 2 & 1 \end{pmatrix}$$

is even since by (2) it is the product of two odd permutations.

Subgroups. A part of a group that is itself a group is called a *subgroup* of the given group. Thus, the alternating group of permutations of the variables $x_1, x_2, \cdots, x_n$ is a subgroup of the symmetric group. The set of proper motions of a plane is a group, which is a subgroup of the group of all proper and improper motions of the plane.

From the formal point of view the unit (identity) transformation forms a subgroup by itself. Equally, every group can be regarded as a subgroup of itself. But almost always groups contain many other subgroups apart from these trivial ones. A knowledge of all the subgroups of a given group gives a fairly complete idea of the internal structure of the group.

One of the most extensively used methods of forming subgroups is that of giving so-called generators of the subgroup.

Let $A_1, A_2, \cdots, A_m$ be arbitrary transformations belonging to a group G. The set H of all transformations that can be obtained by multiplying the given permutations and their inverses among each other arbitrarily often is a group. For the unit transformation belongs to this set, since it can be represented in the form $A_1 A_1^{-1}$. Next, if the transformations B and C can be represented as such products, then by multiplying these products we obtain the required representation for BC. Finally, if B is expressed as such a product, e.g., $B = A_1^{-1} A_2 A_1 A_1 A_2^{-1}$, then B^{-1} can also be represented in the required product form since $B^{-1} = A_2 A_1^{-1} A_1^{-1} A_2^{-1} A_1$.

The group H obviously is a subgroup of G and is called the subgroup generated by the transformations $A_1, \cdots, A_m$, and these transformations $A_1, \cdots, A_m$ are called the generators of H. It can happen that H coincides with G, and in this case $A_1, \cdots, A_m$ are called the generators of the whole group G. It is easy to verify in examples that one and the same subgroup may be generated by several distinct systems of generators.

A subgroup generated by a single transformation A is called cyclic. Its elements are transformations

$$E, A, AA, AAA, \cdots, A^{-1}, A^{-1}A^{-1}, A^{-1}A^{-1}A^{-1}, \cdots,$$

which are naturally called the powers of A. In fact:

$$E = A^0, A = A^1, AA = A^2, \cdots, A^{-1}A^{-1} = A^{-2},$$
$$A^{-1}A^{-1}A^{-1} = A^{-3}, \cdots.$$

It is easy to show, as in the ordinary arithmetic, that

$$A^m A^n = A^{m+n} \quad \text{and} \quad (A^m)^n = A^{mn}. \tag{4}$$

A transformation is called *periodic* if some positive power of it is the identity transformation. The smallest positive exponent of the power to which a periodic transformation must be raised in order to obtain the

§3. GROUPS OF TRANSFORMATIONS

identity is called the *order of the transformation*. We also say that a transformation that is not periodic is of infinite order.

Let us consider some examples. Let A be a rotation of the plane around the point O by $360°/n$, where n is a given positive integer greater than 1. Then A^2 is the rotation by the angle $2(360°/n)$, A^3 the rotation by $3(360°/n)$, A^{n-1} the rotation by $(n-1)(360°/n)$, and A^n the rotation by $360°$, i.e., the identity transformation. This shows that the rotation by $360°/n$ is a periodic transformation of order n.

Let A be a shift of the plane along a certain line. Then $A^2, A^3, \cdots$ are also shifts along this line by twice the distance, three times, and so forth. Therefore, no positive power of A is the identity transformation, and the order of A is infinite.

The elements of the cyclic group generated by A are

$$\cdots, A^{-2}, A^{-1}, E, A, A^2, \cdots. \tag{5}$$

If A is a transformation of infinite order, then all the transformations in the sequence (5) are distinct, and the group is infinite. For otherwise we would have an equation of the form $A^k = A^l$ ($k < l$), hence $A^{l-k} = E$ ($l - k > 0$), and this contradicts the fact that A is not periodic.

Now let us assume that A is a periodic transformation of order m. Then

$$A^m = E, A^{m+1} = A, A^{m+2} = A^2, \cdots, A^{m-1} = A^{-1}, A^{m-2} = A^{-2}, \cdots;$$

i.e., the sequence (5) consists precisely of the transformations $E, A, A^2, \cdots, A^{m-1}$ repeated periodically. They are distinct from one another, since if we had $A^k = A^l$ ($0 \leqslant k < l < m$), then we would have $A^{l-k} = E$ ($0 < l - k < m$), in contradiction to the choice of m. Consequently, the cyclic group generated by a transformation of order m contains precisely m distinct transformations.

A group in which all elements commute with one another is called commutative or *Abelian*, in honor of the Norwegian mathematician Abel who discovered the great importance of these groups for the theory of equations

The formulas (4) show that the powers of one and the same transformation always commute with one another: $A^m A^n = A^n A^m = A^{m+n}$. Therefore, cyclic groups are always Abelian.

In the arithmetic of numbers, apart from the multiplication, great importance also attaches to the operation of division. In the theory of groups, as a consequence of the fact that multiplication need not be commutative, we have to speak of two divisions: on the right and on the left. The solution of the equation $Ax = B$, where A, B are given trans-

XX. GROUPS AND OTHER ALGEBRAIC SYSTEMS

formations, is naturally called the right quotient, and the solution of the equation $yA = B$ is the left quotient on division of B by A. When we multiply both sides of the first equation by A^{-1} on the left and both sides of the second by A^{-1} on the right, we obtain: $x = A^{-1}B$, $y = BA^{-1}$. Thus, we can regard $A^{-1}B$ or BA^{-1} as the "quotient" of the transformations B and A.

In numerous examples we have seen that in groups, in general, $AB = BA$. The "quotient" $(AB)(BA)^{-1}$ or $(BA)^{-1}(AB)$ can be taken to be a "measure" of the noncommutativity of the permutations A and B. The second of these expressions, namely $(BA)^{-1}(AB) = A^{-1}B^{-1}AB$, is called the *commutator* of A and B and denoted by (A, B). From the formula

$$(A, B) = A^{-1}B^{-1}AB$$

it follows that the commutator can be represented as the quotient on "division" of the conjugate transformation $B^{-1}AB$ by A.

For example, if A is a translation of the plane, then a conjugate transformation is also a translation, and the quotient of two translations obviously is a translation. Therefore the commutator of a translation and an arbitrary motion of the plane is a translation. Now let A be a rotation around a certain point O by an angle ϕ, and B a rotation or translation. Then the conjugate transformation is again a rotation by the angle ϕ, but around another point O'. Therefore the commutator (A, B) in this case is the product of a rotation around O by the negative angle ϕ and the rotation around O' by the positive angle ϕ. From figure 12 it is clear that the resulting transformation is a translation by the distance $2 \cdot \overline{OO'} \sin \phi/2$ in the direction at an angle $\pi/2 - \phi/2$ with the segment $\overline{OO'}$.

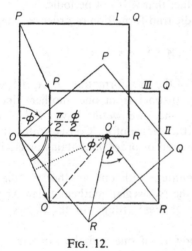

Fig. 12.

Thus, we have arrived at the interesting fact that for a plane the commutator of any two motions of the first kind is a parallel shift or the identity transformation. Since $(A, B) = E$ signifies that $AB = BA$, every noncommutative group of motions of the first kind in the plane contains parallel shifts.

The subgroup generated by the commutators of all elements of a group G is called the *commutator subgroup* or *derived group* of G. Recalling the relevant definitions, we can say that the derived group of G consists

§4. FEDOROV GROUPS

precisely of those elements that can be represented in the form of products of commutators. Since in a plane the commutator of any two motions of the first kind is a parallel shift, and products of parallel shifts are again parallel shifts, we can say that the derived group of the group of motions of the first kind in a plane consists only of parallel shifts.

The derived group of an Abelian group consists only of the identity transformation, because from $AB = BA$ it follows that $(A, B) = E$.

Let G be the symmetric group of all permutations of the numbers $1, 2, \cdots, n$. Let us show that the commutator of any two permutations A, B is always an even permutation. Indeed, the permutation AB, BA and consequently also $(BA)^{-1}$, always have the same parity; but then the commutator $(A, B) = (BA)^{-1}(AB)$, as the product of permutations of equal parity, is an even permutation.

We have seen that the derived group of the symmetric group consists of even permutations only. It is easy to show that in fact it coincides with the whole alternating group.

The derived group of a group G is often denoted by G'; the derived group of the derived group of G is called the second derived group of G and is denoted by G''. Repeating this process we can define the derived group of arbitrary order of a group G.

If among the derived groups of a group G at least one (hence all subsequent ones) consists of the identity transformation only, then the group G is called *solvable*. This name has arisen in the theory of equations, where *solvability* of a group corresponds to solvability of an equation by radicals. The group of motions of the first kind in a plane is solvable, because its second derived group is the identity. The symmetric groups of degree 2, 3, and 4 are solvable, because their first, second, and third derived groups, respectively, are the identity. In contrast, the symmetric groups of degree 5 and higher are not solvable, since it can be shown that their second derived group coincides with the first and is different from the identity.

§4. Fedorov Groups (Crystallographic Groups)

The symmetry groups of finite plane figures. As we have already seen, the symmetry of a figure or a body is characterized by the group of motions of the plane or space that bring the figure into coincidence with itself.

The symmetry groups of finite plane figures are the easiest of these groups to find.* For suppose that a finite plane figure is given and that

---

* Finiteness is to be understood in the sense that the whole figure lies in a bounded part of the plane, for example, within a certain circle.

this figure is brought into coincidence with itself by a certain motion A. Then the center of gravity O of the figure must also be brought into coincidence with itself by A; i.e., A is either a rotation around O or a reflection in a line passing through O. Thus, the symmetry group of an arbitrary finite plane figure can consist only of rotations around its center of gravity and of reflections in lines passing through this center.

Let us discuss a number of different cases that can arise in studying symmetry groups of a finite plane figure.

1. The symmetry group K_1 consists only of the unit (identity) transformation. This is the symmetry group of an arbitrary unsymmetric figure (figure 13).

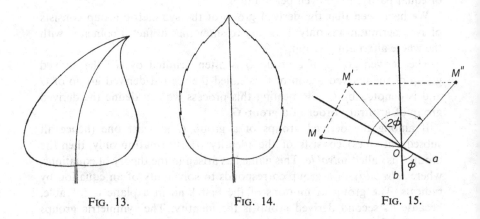

Fig. 13.   Fig. 14.   Fig. 15.

2. The symmetry group K_2 consists of the unit transformation and a reflection in a single line (figure 14).

Observe that if a group K contains reflections in two lines a, b passing through O and forming an angle ϕ between them, then the product of these reflections is a rotation around O by the angle 2ϕ (figure 15). Hence it is clear that the group K_2 is the only symmetry group not containing rotations.

3. The symmetry group K_3 consists of rotations only, and among them there are no rotations by arbitrarily small angles. In that case there is among the rotations of K_3 a rotation by a smallest positive angle. Let this angle be $\alpha°$. We shall show that every other rotation contained in the group is a multiple of $\alpha°$. We denote the number of degrees in such a rotation by β and find the integer h, for which $h\alpha° \leqslant \beta° < (h+1)\alpha°$, so that $0 \leqslant \beta° - h\alpha° < \alpha°$. The group K_3, which contains rotations by $\alpha°$

§4. FEDOROV GROUPS

and $\beta°$, also contains a rotation by $\beta° - h\alpha°$. But $0 \leqslant \beta° - h\alpha° < \alpha°$, and the group does not contain positive rotations by less than $\alpha°$. Therefore $\beta° - h\alpha° = 0$; i.e., $\beta° = h\alpha°$. In particular, since the group K_3 contains the rotation by 360°, we have $n\alpha° = 360°$, for a certain *integer* n, so that $\alpha° = 360°/n$.

Thus, the group K_3 consists of the rotations by $0°$, $360°/n$, $2(360°/n)$, $\cdots$, $(n-1)(360°/n)$. By giving to n the values 2, 3, 4, $\cdots$, we obtain all types of groups K_3.

In figure 16 we have illustrated figures whose symmetry groups consist only of rotations around O by angles that are multiples of $360°/n$, for $n = 19$, $n = 3$.

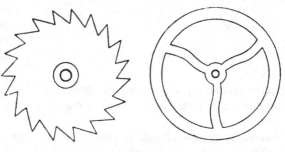

FIG. 16.

4. The symmetry group K_4 consists only of rotations, but contains arbitrarily small rotations. Then a rotation of arbitrary angle α can be made up with any degree of accuracy from rotations belonging to the group K_4. Of course, we are interested here only in closed figures, i.e., such that include their boundary points (see Chapter XVII, §9). It is easy to establish that for closed figures the group K_4 contains rotations by any angle ϕ. This is the case of directed circular symmetry (illustrated by a circumference, a circular annulus, and so forth) provided with a definite sense of direction (figure 17). Here not only the figure must come into coincidence under all admissible transformations but also its directional sense, and this excludes reflections in a line.

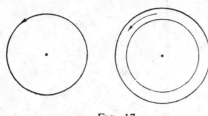

FIG. 17.

It now remains to discuss the mixed cases when the symmetry group K contains both rotations and reflections. Without going into the proof, which would be quite simple, we only state the result: Apart from the groups K_1 through K_4 there only exist groups of the following two types.

5. The symmetry group K_5 consists of n reflections in lines passing

through O and dividing the plane into $2n$ equal angles, and of rotations by angles that are multiples of $360°/n$. For example, regular n-gons (figure 18) have such a symmetry group.

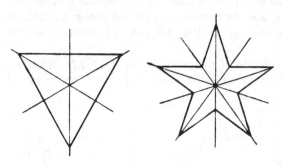

Fig. 18.

6. The symmetry group K_6 consists of all rotations around O and of reflections in all lines passing through the center O. This is the case of complete circular symmetry, which can be illustrated by the symmetry of an unorientated circumference or unorientated annulus.

Symmetry groups of infinite plane figures. The task of finding all possible symmetry groups of infinite plane figures is more complicated. Of course, in practice we are never given a whole infinite plane. However, often a piece of a plane happens to be covered with figures so fine that compared with them the piece appears to be infinitely large. For example, the smoothly polished plane surface of a piece of steel is covered with figures of microscopic dimensions. The regularity of these figures is an indication of the internal homogeneity of the structure of the metal.

Other examples are patterns on wallpaper or tapestries with repeating figures. The art of making such patterns, the art of ornamentation, has been widespread among most nations from antiquity to the present day. In figure 19 we have a specimen of an Egyptian ceiling pattern which dates from the middle of the second millenium B.C.

Fig. 19.

§4. FEDOROV GROUPS

In discussing the symmetry groups of finite figures we were compelled to distinguish between the cases 1, 2, 3, and 5, when the symmetry group does not contain rotations by arbitrarily small angles, and the cases 4 and 6 when the group contains such rotations. In studying the symmetry groups of infinite figures, especially in the three-dimensional case, this division into discrete groups and groups with arbitrarily small transformations becomes even more important. Therefore we shall begin with a more careful discrimination between these cases.

A group of motions of a plane is called *discrete* if every point of the plane can be enclosed in a circle such that every motion of the group either leaves the point unchanged or carries it outside the chosen circle.

In the same way we can find all discrete groups of motions of a plane. All these groups are symmetry groups of plane figures. It is natural here to distinguish three types of discrete symmetry groups:

I. There exists a point in the plane that remains fixed under all symmetry transformations. This type contains the groups K_1, K_2, K_3 and K_5 of our previous list.

II. There are no fixed points in the plane, but there exists a line that is carried into itself under all transformations of the group. This line is called an axis of the group. Symmetry groups of this type occur in ornaments that are set out in the form of an infinite strip (border). Of such groups there exist altogether seven:

1. The symmetry group L_1 consisting only of translations by distances that are multiples of a certain segment a.

2. The group L_2, which is obtained from L_1 by adjoining the rotation by 180° around one of the points on the axis of the group.

3. The group L_3, which is obtained from L_1 by adjoining the reflection in a line perpendicular to the axis of the group.

4. The group L_4, which is obtained from L_1 by adjoining the reflection in the axis.

5. The group L_5, which is obtained from L_1 by adjoining a translation by $a/2$ combined with a reflection in the axis.

6. The group L_6, which is obtained from L_4 by adjoining the reflection in a certain line perpendicular to the axis of the group.

7. The group L_7, which is obtained from L_5 by adjoining the reflection in some line perpendicular to the axis of the group.

Table 2 gives examples of "borders" corresponding to each of the groups L_1 through L_7.

Table 2.

L_1

L_2

L_3

L_4

L_5

L_6

L_7

III. There exists neither a point nor a line in the plane that is carried into itself under all the transformations of the group. Groups of this type are called *plane Fedorov groups*. They are the symmetry groups of infinite plane ornaments. There are altogether 17 of them: five consist of motions of the first kind only, and twelve of motions of the first and second kind.

In Table 3 we have given examples of ornaments corresponding to each of the seventeen plane Fedorov groups; every group consists of precisely those motions that carry an arbitrary flag drawn in the diagram into any other flag of the same diagram.

It is interesting to note that the masters of the art of ornamentation have in practice discovered ornaments with all possible symmetry groups; it fell to the theory of groups to prove that other forms do not exist.

Crystallographic groups. In 1890 the eminent Russian crystallographer and geometer E. S. Fedorov solved by group-theoretical methods one of the fundamental problems of crystallography: to classify the regular systems of points in space. This was the first example of a direct application

§4. FEDOROV GROUPS

of the theory of groups to the solution of an important problem in natural science and made a substantial impact on the development of the theory of groups.

Table 3.

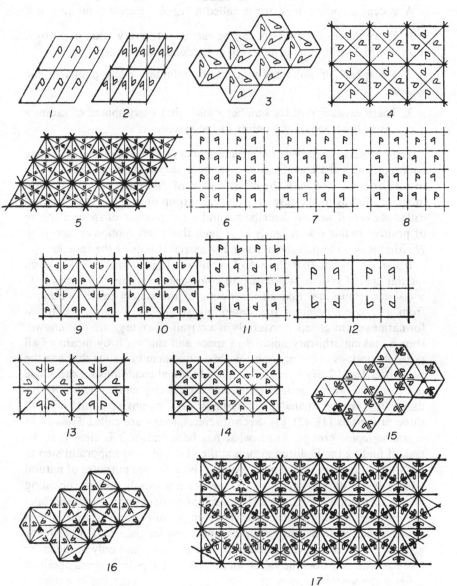

A crystal has the peculiarity that the atoms of which it is composed form in a certain sense a regular system in space. Let us consider the motions of space that carry the points of the system again into points of the system. These motions form a group whose properties enable us to formulate more accurately the concept of a regular point system.

A system of points in space is called a *regular spatial point system* if

1. Every point of the system can be carried into every other point by a motion which brings the system into coincidence with itself;

2. No sphere of finite radius contains infinitely many points of the system;

3. There exists a positive number r such that every sphere of radius r contains at least one of the points of the system.

The problem of studying the structure of crystals turns out to be closely connected with the classification of regular spatial point systems, which in turn is connected with the classification of discrete groups of motion in space. Just as in the case of a plane, a group of motions H of space is called *discrete* if we can describe around every point A of space a sphere of positive radius r with center at A such that every motion occurring in H either leaves the point A fixed or else carries it outside the sphere.

It can be shown that the set of motions of space that bring a given regular spatial system of points into coincidence with itself is necessarily a discrete group and that all the points of the system can be obtained from any given point of the system by subjecting it to all the transformations of the group. Conversely, if a certain discrete group H is known, then by taking arbitrary point A in space and shifting it by means of all possible motions occuring in H we obtain a system of points that has the properties 1 and 2. By means of simple additional conditions we can single out from the discrete groups those that for suitably chosen points A give us, in fact, regular spatial point systems, i.e., systems of points with all three properties (1), (2), (3). Such discrete groups are called *Fedorov* or *crystallographic* groups. From what has been shown it is clear that the task of finding the Fedorov groups is the first and most important step in the study of regular spatial point systems. Now for the purposes of natural science it has proved necessary to consider not merely groups consisting of proper motions only, but also those that contain proper and improper motions, i.e., including reflections. The number of Fedorov groups formed from proper motions only is significantly smaller than that of Fedorov groups composed of proper and improper motions, and only in the latter more general case does the variety of regular spatial point systems obtained exhaust the whole diversity of structure of crystals occurring in nature.

§5. GALOIS GROUPS

It is interesting to note that, in contrast to the plane case we have treated previously, only the theory of groups enable us to analyze this exceptionally large number of possibilities.

The complexity of the space problem compared with the plane one is clear from Table 4.

Table 4. Number of Spatial Fedorov Groups

Groups containing only motions of the first kind 65
Groups containing also motions of the second kind . . . 165
Total 230

Even today a detailed derivation and enumeration of all Fedorov groups in space requires several dozen pages of text. We shall therefore restrict ourselves to reporting these quantitative results and refer the interested reader to the special literature.*

The modern developments of crystallography have made it necessary to introduce a further extension of the concept of symmetry. Such new possibilities in methods are outlined in the book on crystallography by Academician A. V. Šubnikov "Symmetry and anti-symmetry of finite figures," Akad. Nauk SSSR, Moscow, 1951.

§5. Galois Groups

The results explained on the preceding pages give a certain idea of the role played by the theory of groups in the solution of the problem of classification of crystals. However, this problem was not the motivation for the creation of the theory of groups. Approximately a hundred years earlier Lagrange noticed a connection between the symmetry properties of the roots of an algebraic equation and the possibility of solving the equation by radicals. This connection was the object of deep investigations by the famous mathematicians Abel and Galois in the first thirty years of the 19th century; and so they arrived at a solution of the celebrated problem of conditions for the solvability of algebraic equations by radicals. This solution was based entirely on a subtle investigation of properties of permutation groups and was, in fact, the beginning of the theory of groups.

The study of connections between properties of algebraic equations and

---

* A detailed account of the plane discrete groups of motion of the first kind is contained in the book by D. Hilbert and S. E. Cohn-Vossen, *Geometry and the imagination*, Chelsea, New York, 1952. A derivation of the crystallographic groups in space can be found in the fundamental article by E. S. Fedorov, *Symmetry of regular systems of figures* (Collected works, Akad. Nauk SSSR, Moscow, 1949).

properties of groups nowadays forms the object of an extensive theory, which is known as Galois theory.

An idea of the history of the problem and of the significance of Galois theory was given in Chapter IV. However, since Galois theory has played such a decisive role in the development of group theory, we shall repeat here the basic facts of the theory, but in a form more convenient for the purpose of throwing light on the theory of groups itself. The proofs of these facts require many auxiliary concepts and will be omitted.

The group of an algebraic equation. Consider an equation of degree n

$$x^n + a_1 x^{n-1} + \cdots + a_n = 0, \tag{6}$$

whose coefficients are assumed to have given values; for example, certain complex numbers. The set of all quantities that can be obtained from the coefficients of the equation by means of a finite number of the operations of addition, subtraction, multiplication, and division is called the *ground field* or domain of rationality of the equation.

For example, if the equation has rational coefficients, then the domain of rationality consists of all rational numbers; if the equation has the form $x^2 + \sqrt{2} x + 1 = 0$, then the domain of rationality consists of all numbers of the form $a + b \sqrt{2}$, where a, b are rational numbers.

We shall now denote the roots of this equation by $\xi_1, \cdots, \xi_n$. The set of quantities that can be obtained by means of a finite number of the operations of addition, subtraction, multiplication, and division starting out from the roots $\xi_1, \cdots, \xi_n$, is called the *splitting field* of the equation. For example, the splitting field of the equation $x^2 + 1 = 0$ is the set of complex numbers $a + bi$ with rational a, b; and the splitting field of the above equation $x^2 + \sqrt{2} x + 1 = 0$ is the set of numbers of the form $a + bi + c \sqrt{2} + di \sqrt{2}$, where a, b, c, d are rational numbers.

By Viète's formulas the coefficients of the equation are obtained from its roots by means of the operation of addition and multiplication, therefore the splitting field of an equation always contains its ground field. Sometimes these fields coincide.

A one-to-one mapping A of the splitting field onto itself is called an automorphism of the splitting field with respect to the ground field, if for every pair of elements of the splitting field their sum goes over into the sum of their images, and their product into the product, and every element of the ground field goes over into itself. These properties can be described by the formulas

$$(a+b)A = aA + bA, \quad (ab)A = aA \cdot bA, \quad \alpha A = \alpha$$
$$(a, b \in K, \quad \alpha \in P), \tag{7}$$

§5. GALOIS GROUPS

where aA is the image of a; that is, Aa is the element into which a goes over under the mapping A; P is the ground field; and K is the splitting field.

By the general principle, explained in §2, the set of all automorphisms of the splitting field relative to the ground field is a group. This group is called the *Galois group* of the given equation.

To form a more concrete idea of the Galois group, let us note first of all that the automorphisms of the Galois group carry a root of the given equation into another root. For if x is a root of the equation (6), then operating on both sides of the equation with the automorphism A and using the properties (7) we obtain

$$(xA)_n + a_1A(xA)^{n-1} + \cdots + a_nA = 0 \cdot A;$$

since $0 \cdot A = 0$, $a_iA = a_i$, we thus have

$$(xA)^n + a_1(xA)^{n-1} + \cdots + a_n = 0,$$

as required. Consequently, every automorphism A effects a definite permutation of the set of roots of the equation. On the other hand, when we know this permutation, we also know the automorphism, because all the elements of the splitting field are obtained from the roots by means of arithmetical operations only. This shows that instead of the automorphism group we can also consider the group of permutations of the roots of the equation corresponding to it. Hence it follows, in particular, that all Galois groups are finite.

To find the Galois group of a given equation is usually a complicated problem, and only in special cases is the task comparatively easy. Let us consider, for example, the equation (6) with literal coefficients $a_1, \cdots, a_n$. The ground field of this equation is formed by the rational fractions of the coefficients, i.e., the fractions whose numerators and denominators are polynomials in $a_1, \cdots, a_n$. The splitting field is formed by the rational fractions of the roots of the equation $\xi_1, \cdots, \xi_n$, which are connected with the coefficients by the formulas

$$\begin{aligned}
-a_1 &= \xi_1 + \xi_2 + \cdots + \xi_n, \\
a_2 &= \xi_1\xi_2 + \xi_1\xi_3 + \cdots + \xi_{n-1}\xi_n, \\
&\cdots\cdots\cdots\cdots\cdots\cdots\cdots\cdots\cdots\cdots\cdots\cdots \\
(-1)^n a_n &= \xi_1\xi_2 \cdots \xi_n.
\end{aligned} \tag{8}$$

Since the equation (6) is "general," we can regard its roots as independent variables. Then every permutation of these roots gives rise to an automorphism of the splitting field. The formulas (8) show that under every such automorphism the coefficients go over into themselves and, together with them, all rational fractions formed from them also go over into themselves. Thus, the Galois group of the general equation of degree n is essentially the symmetric group of all permutations of n letters.

We can also indicate equations with numerical coefficients that have the symmetric group for their Galois group. For example, it has been shown that the Galois group of the equation

$$1 - \frac{n}{1}x + \frac{n(n-1)}{1 \cdot 2} \cdot \frac{1}{1 \cdot 2} x^2 - \frac{n(n-1)(n-2)}{1 \cdot 2 \cdot 3} \cdot \frac{1}{1 \cdot 2 \cdot 3} x^3$$

$$+ \cdots + (-1)^n \frac{1}{1 \cdot 2 \cdots n} x^n = 0 \qquad (9)$$

for arbitrary n is the symmetric group of permutations of degree n.

General methods are known for constructing equations with any preassigned group as Galois group, but under the condition that the coefficients can be taken to be arbitrary. However, if a construction is required for equations that are required to have rational coefficients, then this is known at present only for individual types of groups. Remarkable progress in this direction has been made by the Soviet mathematician I. R. Šafarevič, who has found methods of constructing equations with rational coefficients having an arbitrary preassigned solvable group as Galois group. In general, however, this problem is still unsolved.

Solvability of equations by radicals. The Galois group of an equation characterizes, as is clear from the definition, the intrinsic symmetry of the roots of the equation. All the most fundamental problems concerning the possibility of reducing the solution of a given equation to that of equations of lower degree and also many other problems can be formulated as problems on the structure of the Galois group; and the Galois group of every equation of degree n is a certain group of permutations of degree n, i.e., an entirely finite object, in which all relationships, at least theoretically, can be found by means of trial and error.

The study of the Galois group is a valuable method of solving problems related to algebraic equations of higher degree. For example, it can be shown that an equation is solvable by radicals if and only if its Galois group is solvable (for the definition of a solvable group see §3). We have already mentioned that the symmetric groups of degree 2, 3, and 4 are solvable. This is in complete accord with the well-known fact that equations of degree 2, 3, and 4 are solvable by radicals. The Galois groups of the "general" equations of degree 5, 6, and so forth, are the symmetric groups of the same degrees. But these groups are not solvable. Hence it

§6. FUNDAMENTAL CONCEPTS OF GENERAL THEORY

follows that the general equations of degree higher than 4 cannot be solved by radicals.

Among the equations that are not solvable by radicals there are also the equations (9) for $n > 4$, because their Galois group is the symmetric group.

§6. Fundamental Concepts of the General Theory of Groups

In the 19th century the theory of groups arose primarily as the theory of transformation groups. However, in the course of time it became more and more clear that the most significant of the results obtained depend only on the fact that transformations can be multiplied and that this operation has a number of characteristic properties. On the other hand, objects were found having nothing to do with transformations, on which a certain operation can be carried out (for the time being we shall call it multiplication) having the same properties as in transformation groups and to which the main theorems of the theory of transformation groups were applicable. As a result, the concept of a group was applied at the end of the last century not only to systems of transformations, but also to systems of arbitrary elements.

General definition of a group. The following definition of a group is generally accepted nowadays: Suppose that with every pair of elements a, b, taken in a definite order, of an arbitrary set G another well-defined element c of the same set is associated. Then we say that an operation is given on the set G. It is customary to introduce special names for operations: addition, multiplication, composition. The element of G that corresponds to the pair a, b is then called the sum, product, and compositum of the elements a, b, and is denoted by $a + b$, ab, $a * b$, respectively. The names "addition" or "multiplication" are used even in cases when the operation in question has nothing to do with the ordinary operations of addition and multiplication of numbers.

A set G together with an operation $*$ defined on it is called a *group* with respect to this operation if the following group axioms are satisfied:

1. For any three elements x, y, z of G

$$x * (y * z) = (x * y) * z \text{ (associative law).}$$

2. Among the elements of G there exists an element e such that for every x of G

$$x * e = e * x = x.$$

3. For every element a of G there exists an element a^{-1} of G such that

$$a * a^{-1} = a^{-1} * a = e.$$

The element e, described in axiom 2, is called the neutral element of the group, and the element a^{-1} whose existence is postulated by axiom 3 is called the inverse of a. If the group operation is called addition or multiplication, then the neutral element is called the zero or the unit element, respectively, and the group axioms assume the form

(1) $x + (y + z) = (x + y) + z$,   (1) $x(yz) = (xy)z$,
(2) $x + 0 = 0 + x = x$,   (2) $xe = ex = x$,
(3) $x + (-x) = (-x) + x = 0$,   (3) $xx^{-1} = x^{-1}x = e$.

In the preceding sections we have discussed many examples of groups. The elements of these groups were transformations, and the group operation was multiplication of transformations. The set of numbers $0, \pm 1, \pm 2, \cdots$ also forms a group under the operation of addition, because the sum of integers is again an integer and addition of integers is associative; the neutral element is the integer 0 and for every number a of our set there is the opposite number $-a$. Another example of a group is the set of all real numbers (except 0) under multiplication. For the product of any two real numbers different from zero is a real number different from zero; the operation of multiplication of real numbers is associative; the neutral element is the number 1; and every nonzero real number a has the inverse $a^{-1} = 1/a$. The number of similar examples could be increased indefinitely.

Although the group operation may be called by different names, let us agree henceforth to call it almost always multiplication. The concepts of a subgroup, of powers of an element of a group, of a cyclic group, of the order of an element of a group are defined exactly as for transformation groups and we shall not repeat this here (see §3). We only mention that an element a of a group is called conjugate to an element b if there is an element x in G such that $b = x^{-1}ax$. Since $a = a^{-1}aa$, every element of a group is conjugate to itself. Furthermore, from $b = x^{-1}ax$ it obviously follows that $xbx^{-1} = a$ or $a = (x^{-1})^{-1}bx^{-1}$, i.e., if a is conjugate to b, then b is conjugate to a. Finally, if $b = x^{-1}ax$ and $c = y^{-1}by$, then

$$c = y^{-1}x^{-1}axy = (xy)^{-1}a(xy).$$

Therefore two elements conjugate to a third are conjugate to each other. These properties show that all elements of a group split into disjoint classes of conjugate elements. Also, if the group is commutative, i.e., $xy = yx$ for every x and y, then conjugate elements coincide and every class of conjugate elements consists of one element only.

§6. FUNDAMENTAL CONCEPTS OF GENERAL THEORY

Isomorphisms. Two aspects can be distinguished in the concept of a group. In order to give a group we have to: (1) indicate what objects are its elements and (2) indicate the law of multiplication of the elements. Accordingly, the study of group properties can be carried out from distinct points of view. We can study connections between individual properties of elements of the group and of sets of them and their properties in relation to the group operation. This point of view is often adopted in studying individual concrete groups; for example, the group of motions of space or a plane. However, we can also study those group properties that are entirely expressed in terms of properties of the group operation. This point of view is characteristic for the *abstract* or *general* theory of groups. It can be expressed more clearly by means of the concept of isomorphism.

Two groups are called *isomorphic* if the elements of one of them can be associated with the elements of the other in such a way that the product of arbitrary elements of the first group is associated with the product of the corresponding elements of the second group. A one-to-one correspondence between elements of two groups that has this property is called an *isomorphism*.

It is easy to see that elements of two groups that correspond to each other under an isomorphism have identical properties with respect to the group operation. Thus, under an isomorphism the neutral element, inverse elements, elements of a given order n, subgroups of one group go over, respectively, into the neutral element, inverse elements, elements of the same order, subgroups of the second group. We can therefore say that the abstract theory of groups studies only those properties of a group that are preserved under isomorphic mappings. For example, from the point of view of the abstract theory of groups the group of all permutations of four elements and the group of proper and improper motions of space that carry a fixed regular tetrahedron into itself have identical properties, because they are isomorphic. In fact, the motions in question carry the vertices of the tetrahedron again into its vertices. The number of these motions is 24. By associating with every motion the permutation of the vertices that it produces, we obtain a one-to-one correspondence between the elements of the two groups which is the required isomorphism.

A remarkable example of an isomorphic mapping is given by the theory of logarithms. By associating with every positive real number its logarithm, we obtain a one-to-one mapping of the set of positive real numbers onto the set of all real numbers. The relation $\log(xy) = \log x + \log y$ shows that this correspondence is an isomorphic mapping of the group of positive real numbers under multiplication with the group of all real numbers under addition. The practical importance of this isomorphism is well known.

XX. GROUPS AND OTHER ALGEBRAIC SYSTEMS

Examples of nonisomorphic groups are finite groups of distinct orders. As we have already mentioned, an abstract group is determined by the law of multiplication of its elements, independent of their nature, so that distinct but isomorphic concretely given groups can be regarded as models of one and the same abstract group.

An abstract group can be given by various methods of which the most natural, at least for finite groups, is by means of the "multiplication table."

For a group of order n whose elements are written down in an arbitrary order, such a multiplication table consists of a square divided into n rows and n columns. In the cell at the intersection of the ith row and the jth column we write down the element that is the product of the element with the number i and that with the number j. This multiplication table for finite groups is sometimes called its Cayley square.

However, in practice it is almost never convenient to give a group by means of the multiplication table, because it is very clumsy.

There are other methods of giving an abstract group. One of them, namely by means of generating elements and defining relations, we have already come across. However, most frequently an abstract group is defined by giving a concrete group isomorphic to it, in particular, a transformation group.

Naturally the problem arises whether every abstract group can be regarded as a transformation group. The following theorem gives us the answer: Every group G is isomorphic to some transformation group of the set of its elements.

For let g be a fixed element of G. We denote by A_g the transformation of the set of elements of G under which to every element x of G there corresponds the element xg. The transformation A_g is one-to-one, because the equation

$$xA_g = xg = a$$

has for every given a the unique solution $x = ag^{-1}$. On the other hand, the product of group elements gh is associated with the product of the corresponding transformations $A_g A_h$, because

$$xA_{gh} = x(gh) = (xg)h = (xA_g)A_h = x(A_g A_h).$$

To the neutral element e of G there corresponds the identity, and to the inverse element g^{-1} the inverse transformation. Therefore the set Γ of all transformations corresponding to the elements of G is a transformation group isomorphic to G. It is easy to verify that if the number of elements of G is greater than 2, then the set Γ does not exhaust all the transformations of G and is only a subgroup of the "symmetric" groups of all transformations of that set.

§6. FUNDAMENTAL CONCEPTS OF GENERAL THEORY

Normal subgroups and factor-groups. Let P and Q be arbitrary collections of elements of some group G. The product of P by Q, symbolically PQ, is the name for the set of those elements of G that can be represented in the form of a product of some element of P by some element of Q. In particular, the product gP, where g is an element of G, is the set of products of g by every element of the set P.

A subgroup H of G is called a *normal* or *invariant subgroup* of G if $gH = Hg$ for every g of G. The sets of the form gH and Hg, where H is an arbitrary subgroup, are called respectively the *right and left cosets* of G with respect to H, containing the element g. Thus we can say that normal subgroups are entirely characterized by the property that for them the left and right cosets corresponding to one and the same element coincide.

If H is a normal subgroup, then the product of two cosets is again a coset, as it is easy to see, in fact: $aH \cdot bH = ab \cdot H$. The subgroup H by itself is a coset corresponding to the unit element or to any of its elements h, since $hH = H$. Multiplication of cosets is associative

$$(aH \cdot bH)cH = (ab \cdot c)H = (a \cdot bc)H = aH(bH \cdot cH).$$

The subgroup H plays the role of the neutral element in this multiplication: $H \cdot aH = eH \cdot aH = (ea)H = aH$, similarly $aH \cdot H = aH$. The coset $a^{-1}H$ is the inverse of aH, since $aH \cdot a^{-1}H = aa^{-1}H = H$. Therefore, by regarding every coset with respect to a normal subgroup as an element of a new set, we see that this set is a group under the operation of multiplication of cosets. This group is called the *factor group* of G with respect to the normal subgroup H and is denoted by G/H.

It is easy to show that for finite groups every coset with respect to an arbitrary subgroup contains as many distinct elements as the subgroup H contains and that distinct cosets have no elements in common. Hence it follows that the number of cosets of a finite group G with respect to its subgroup H is equal to the order of G divided by the order of H; and this implies the important theorem of Lagrange which states that the order of every subgroup of a finite group is a divisor of the order of a group.

From the definition of a normal subgroup it is clear that in Abelian groups every subgroup is normal. The other extreme case consists of the so-called *simple groups* in which no subgroup other than the unit subgroup and the group itself is normal. Apart from Abelian and simple groups, the solvable groups defined in §3 are also very important. It can be shown that solvable groups have a finite chain of normal subgroups $G, G_1, G_2, \cdots, G_k$, the first of which coincides with the given group G, while every subsequent one is contained in its predecessor, the last group being the unit element, and all the factor groups $G/G_1, G_1/G_2, \cdots, G_{k-1}/G_k$ being Abelian.

Homomorphisms. The concept of a factor group is very closely connected with the concept of a homomorphic mapping, which is fundamental for the whole theory of groups.

A single-valued mapping of the set of elements of a group G onto the set of elements of a group H is called a *homomorphism* or *homomorphic mapping* if the product of any two elements of the first group is mapped onto the product of the corresponding elements of the second.

If for every element x of G we denote the corresponding element of H by x', then a homomorphic mapping can be characterized by the property

$$(x_1 x_2)' = x'_1 x'_2 .$$

From the definitions of a homomorphism and an isomorphism it is clear that an isomorphic mapping is necessarily one-to-one, whereas a homomorphic mapping is single-valued only in one direction: To every element of G there corresponds a unique element of H, but distinct elements of G may have one and the same image in H. In a certain sense we can say that under an isomorphic mapping the group H is an accurate copy of G, but under a homomorphic mapping on transition from G to H distinct elements of G may coalesce; several elements, as it were, can be "merged" into a single element of H. However, this "coarse" nature of a homomorphic mapping is not a deficiency; on the contrary, it is a great advantage, because it enables us to use homomorphic mappings as a powerful tool in the investigation of group properties.

Homomorphic mappings make their appearance in many situations connected with transformations. For example, let us consider the symmetry group of the regular tetrahedron (figure 20). This group is isomorphic to the symmetric group of permutations of four elements, because there exists one and only one motion (of the first or second kind) that carries the vertices A_1 , A_2 , A_3 , A_4 into any other given arrangement.

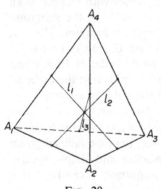

FIG. 20.

Now let us consider the lines l_1 , l_2 , l_3 that join the midpoints of opposite edges. Every motion that brings the tetrahedron into coincidence with itself generates a certain permutation of l_1 , l_2 , l_3 and every permutation of l_1 , l_2 , l_3 is generated by some symmetry of the tetrahedron. Clearly, the product of transformations of the tetrahedron corresponds to the product of the permutations of the lines l_1 , l_2 , l_3. From figure 20 it is

§6. FUNDAMENTAL CONCEPTS OF GENERAL THEORY 303

easy to read off how the homomorphic mapping of the symmetric group of permutations of the four elements A_1, A_2, A_3, A_4 onto the symmetric group of the permutations of the three elements l_1, l_2, l_3 can be realized in a natural way. It is not difficult to find the elements of the "larger" group that are "merged" in this homomorphism.

Let us discuss a few more examples. The set of all permutations of n elements is a noncommutative group for $n > 2$. On the other hand, the numbers $+1$ and -1 also form a group under multiplication. Now let us associate with every even permutation of arbitrary n elements the number $+1$ and with every odd permutation the number -1. This gives us a homomorphic mapping of the symmetric group of permutations of n elements onto the group $\{+1, -1\}$, because according to §3 the product of permutations of equal parity is an even permutation and the product of permutations of distinct parity is an odd permutation.

Another example: If we associate with every real number $x \neq 0$ its absolute value $|x|$, then the resulting mapping of the group of positive and negative real numbers under multiplication (zero excluded) onto the group of the positive real numbers only is a homomorphism under multiplication because $|xy| = |x||y|$.

We have already mentioned that in a plane every motion of the first kind A can be represented in the form of a product of a suitable rotation V_A around a fixed point O and a certain parallel shift D_A. Rotations around the point O form a group. Therefore the correspondence $A \to V_A$ uniquely maps the group of plane motions of the first kind onto the group of rotations of the plane around the point O. Let us show that this mapping is a homomorphism. From the decompositions $A = V_A D_A$, $B = V_B D_B$ it follows that

$$AB = V_A D_A V_B D_B = (V_A V_B)(V_B^{-1} D_A V_B D_B).$$

The first parenthesis is a rotation around O and the second is the product of the transformed translation $V_B^{-1} D_A V_B$ and the translation D_B and is consequently also a translation. This shows that the product of the motions AB is associated with the product of the corresponding rotation $V_A V_B$, i.e., that the mapping in question is a homomorphism.

Finally let us show that the factor group G/N of an arbitrary group G with respect to the normal subgroup N is a homomorphic image of G.

For by associating with every element g of G the coset gN containing g, we obtain the required homomorphic mapping of G onto G/N, since the product gh corresponds to the coset ghN, which is equal to the product of the cosets gN and hN corresponding to the elements g and h.

Turning now to general properties of homomorphic mappings, let us show that the neutral element goes over under any homomorphism into the neutral element and that inverse elements go over into inverse elements.

For if e is the neutral element of G, and e' its image in H, then it follows from $ee = e$ that $e'e' = e'$ so that, if we denote by ϵ the neutral element of H, we obtain: $e' = e'e'^{-1} = \epsilon$. This proves the first statement. Now let x and y be inverse elements in G and x' and y' their images in H. From $xy = e$ it follows that $x'y' = e' = \epsilon$, i.e., that x and y are inverse elements in H and hence

$$(x^{-1})' = x'^{-1}.$$

The facts we have proved make it easy to find the image of an arbitrary product of elements in G. For example,

$$(ab^{-1}c^{-1}dh^{-1})' = a'(b^{-1})'(c^{-1})'d'(h^{-1})' = a'b'^{-1}c'^{-1}d'h'^{-1}.$$

The following theorem is fundamental for the whole theory of homomorphic mappings.

Under homomorphic mapping of an arbitrary group G onto a group H, the set N of elements of G that are mapped into the neutral element e' of H is a normal subgroup of G; the set of elements of G that are mapped into an arbitrary fixed element of H is a coset of G with respect to N, and the one-to-one correspondence so established between the cosets of G with respect to N and the elements of H is an isomorphism between H and the factor group G/N.

Let us now prove the theorem. Let a, b be arbitrary elements of N. This means that $a' = b' = e'$ where, as before, the prime denotes the images of elements of G in H. But then

$$(ab)' = a'b' = e'e' = e',$$
$$(a^{-1})' = a'^{-1} = e'^{-1} = e';$$

i.e., ab and the inverse elements a^{-1}, b^{-1} belong to N so that N is a group. Furthermore, for an arbitrary element g of G we have

$$(g^{-1}ag)' = g'^{-1}a'g' = g'^{-1}e'g' = g'^{-1}g' = e';$$

i.e., $g^{-1}ag$ lies in N for every g of G and every a of N, and from this it follows obviously that N is a normal subgroup. This proves the first statement of the theorem.

To prove the second statement we choose in G an arbitrary element g and consider the set U of all those elements u of G whose image u' coincides with the image g' of g. Suppose that $u \in gN$, i.e., $u = gn$ where $n \in N$, so that $u' = g'n' = g'e' = g'$. Therefore $gN \subset U$. Conversely, if $u' = g'$, then $(g^{-1}u)' = g'^{-1}u' = g'^{-1}g' = e'$, i.e., $g^{-1}u = n$, where n is an element

§7. CONTINUOUS GROUPS

of N. Hence $u = gn$ and so $U \subset gN$. From $gN \subset U$ and $U \subset gN$ it follows that $U = gN$.

Finally, the third statement of the theorem is obvious: To arbitrary cosets gN, hN of the factor group G/N there correspond in H the elements g', h', and to the product of the cosets, by the formula

$$gN \cdot hN = ghN,$$

there corresponds $(gh)' = g'h'$, as required.

The theorem on homomorphisms shows that every homomorphic image H of a group G is isomorphic to the corresponding factor group G/H. Thus, to within an isomorphism all homomorphic images of a given group G are exhausted by its distinct factor groups.

§7. Continuous Groups

Lie groups; continuous groups of transformations. The progress that was made by means of the theory of groups in the solution of algebraic equations of higher degree induced mathematicians of the middle of the last century to attempt to use the theory of groups in the solution of equations of other forms, in the first instance the solution of differential equations, which play such an important role in the applications of mathematics. This attempt was crowned with success. Although the place occupied by groups in differential equations is entirely different from their place in the theory of algebraic equations, the investigations on the application of the theory of groups to the solution of differential equations led to a substantial extension of the very concept of a group and to the creation of a new theory of the so-called continuous groups and Lie groups which have proved to be extremely important for the development of the most diverse branches of mathematics.

Whereas the groups of algebraic equations consist only of a finite number of transformations, the groups of differential equations constructed in a similar way turn out to be infinite. However, the transformations belonging to a group of a differential equation can be given by means of a finite system of parameters, and by changing the numerical values of these all the transformations of a group can be obtained. Suppose, for example, that all the transformations of the group are determined by the values of the parameters $a_1, a_2, \cdots, a_r$. When we give these parameters the values $x_1, x_2, \cdots, x_r$, we obtain a certain transformation X; by giving to the parameters new values $y_1, y_2, \cdots, y_r$ we obtain another transformation Y. By hypothesis the product of these transformations $Z = XY$ also occurs in the group and hence is obtained by certain new values of the parameters

$z_1, z_2, \cdots, z_r$. The values z_i depend on $x_1, x_2, \cdots, x_r, y_1, y_2, \cdots, y_r$, i.e., they are certain functions of them

$$z_1 = \phi_1(x_1, x_2, \cdots, x_r; y_1, y_2, \cdots, y_r),$$
$$z_2 = \phi_2(x_1, x_2, \cdots, x_r; y_1, y_2, \cdots, y_r),$$
$$\cdots\cdots\cdots\cdots\cdots\cdots\cdots\cdots\cdots\cdots\cdots\cdots\cdots$$
$$z_r = \phi_r(x_1, x_2, \cdots, x_r; y_1, y_2, \cdots, y_r).$$

Groups whose elements depend continuously on the values of a finite system of parameters and whose multiplication law can be expressed by means of twice-differentiable functions $\phi_1, \cdots, \phi_r$ are called Lie groups in honor of the Norwegian mathematician Sophus Lie who first investigated these groups.

In the first half of the 19th century, N. I. Lobačevskiĭ developed a new geometric system which now bears his name. At approximately the same time projective geometry emerged as an independent geometric system; somewhat later the geometry of Riemann was created. As a result one could enumerate in the second half of the 19th century a number of independent geometric systems that investigated from different points of view the "spatial forms of the actual world" (Engels). To comprise all these geometric systems in a single point of view, but preserving their most important qualitative differences, proved possible by means of the theory of groups.

Let us consider a one-to-one transformation of the set of points of an arbitrary geometric space that does not change those basic relations between figures that are studied in this geometry. The collection of these transformations forms a group which is usually called the group of motions or of automorphisms of the given geometry. The group of motions completely characterizes the given geometry, in view of the fact that when the group of motions is known, the corresponding geometry can be regarded as the study of those properties of the collection of points that remain unchanged under the transformations of the group. The method of classifying the various geometric systems by their groups of motions was introduced in the second half of the last century by F. Klein. This method and the various geometric systems have been treated in Chapter XVII. Here let us only mention that the groups of motions of all geometric systems that were actually investigated in the last century turned out to be Lie groups. In view of this the task of studying Lie groups assumed particular importance.

Owing to its many connections with the most diverse domains of mathematics and mechanics the theory of Lie groups has been developed

§7. CONTINUOUS GROUPS

energetically from its foundation right to the present time. It so happens that certain problems that have not yet been solved for finite groups were solved comparatively rapidly for Lie groups. For example, little progress has been made so far in the problem of classifying the finite simple groups (i.e., the finite groups that have no nontrivial normal subgroups), but the corresponding classification of simple Lie groups was obtained by Killing and Cartan already at the end of the last century. By developing the theory of Lie groups the Soviet mathematicians V. V. Morozov, A. I. Mal'cev, and E. B. Dynkin have found a complete solution of the important and long outstanding problem of classifying the simple *subgroups* of Lie groups. In another direction the theory of Lie groups was developed by the Soviet mathematicians I. M. Gel'fand and M. A. Naĭmark who have found the so-called continuous representations of the simple Lie groups by unitary transformations of a Hilbert space; the latter task is of particular interest for analysis and physics.

The study of Lie groups proceeds by means of the peculiar apparatus of the so-called "infinitesimal groups" or Lie algebras. These will be discussed in more detail in §13.

Topological groups. Side by side with a wide extension of the classical theory of Lie groups, in the USSR exceptional advances were achieved in the more general theory of topological or continuous groups. In contrast to the concept of a Lie group, where it is required that the elements of a group be defined by a finite system of parameters and that the multiplication rule be expressible by means of differentiable functions, the concept of a topological group is simpler and wider. A group is called *topological* if apart from the ordinary group operation a concept of proximity is defined for its elements and if the proximity of group elements implies the proximity of their products and of their inverse elements.

Originally the concept of a topological group proved to be necessary to bring order into many of the fundamental concepts of the theory of Lie groups. But later the extreme importance of this concept for other branches of mathematics was recognized. The first papers on the theory of general topological groups fall into the early twenties of our century, but the fundamental results that make it possible to speak of the creation of a new discipline were not found until the end of the twenties and the beginning of the thirties. A considerable part of them was obtained by the Soviet mathematician L. S. Pontrjagin who deservedly is regarded as one of the founders of the modern theory of continuous groups. His book "Topological groups," which was the first in the world literature to contain a comprehensive account of the theory of continuous groups, still remains the basic textbook in this domain even after twenty years.

§8. Fundamental Groups

In all the concrete examples discussed in the preceding sections, groups have usually appeared as transformation groups of one set or another. The only exceptions were the groups of numbers with respect to addition and multiplication. We now wish to analyze an important example in which the group originally arises not as a group of transformations but as a certain algebraic system with one operation.

The fundamental group. Let us consider a certain surface S and on it a moving point M. By making M run on the surface along a continuous curve joining a point A to a point B, we obtain a definite path from A to B. This path may intersect itself any number of times and may even retrace part of itself in individual sections. In order to indicate the path it is not enough to give only the curve on which the point M runs. We also have to indicate the sections that the point traverses more than once and also the direction of its passage. For example, a point may range over one and the same circle a different number of times and in different directions, and all these circular paths are regarded as distinct. Two paths with the same beginning and the same end are called equivalent if one of them can be carried into the other by continuous change. In the plane or on a sphere any two paths joining a point A to a point B are equivalent (figure 21). However, on the surface of the torus, for example, the closed paths U and V (figure 22) that begin and end at the point A are not equivalent to each other.

If instead of a torus we consider an infinite circular cylinder extending in both directions and take on it the path X (figure 23), then it is easy to figure out that every closed path on the cylinder beginning at A is equivalent to a path of the form X^n ($n = 0, \pm 1, \pm 2, \cdots$), where we have to understand by X^n ($n > 0$) the path X repeated n times; by X^0 the zero path consisting only of the single point A; and by X^{-n} the path X^n traversed in the opposite direction; for example, $Z \sim X^{-1}$, $Y \sim X^2$, $U \sim X^0$ (figure 23). This example shows the significance of the concept of equivalence of paths: Whereas there exists an immense set of distinct closed paths on the cylinder, all these paths reduce, to within equivalence, to the circle X traversed in one or the other

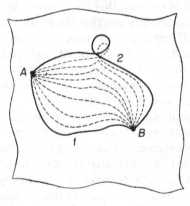

FIG. 21.

§8. FUNDAMENTAL GROUPS

direction a sufficient number of times. For $m \neq n$ the paths X^m and X^n are not equivalent.

Turning now to the discussion of an arbitrary surface, let us assume that two paths are given on it, namely a path U leading from a point A to a point B, and a path V leading from B to C. Then, by making a point run first through the path AB and then through BC we obtain a path AC which we naturally call the product of the paths $U = AB$ and $V = BC$

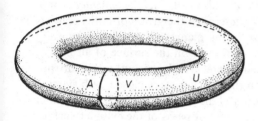
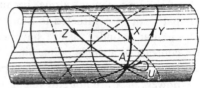

Fig. 22.   Fig. 23.

and denote by UV. If the paths U, V are equivalent to the paths U_1, V_1, respectively, then their products UV and U_1V_1 are also equivalent. The multiplication of paths is associative in the sense that if one of the products $U(VW)$ or $(UV)W$ is defined, then the other is also defined and the two products represent equivalent paths. If the moving point M is made to run through a path $U = AB$ but in the opposite direction, then we obtain the inverse path $U^{-1} = BA$ leading from B to A. The product of the path AB with its inverse path BA is a closed path equivalent to the zero path consisting only of the point A.

According to the definition we cannot multiply any two paths but only those in which the end point of the first coincides with the initial point of the second. This inadequacy disappears when we consider only closed paths starting from one and the same initial point A. Any two such paths can be multiplied and as a result we obtain again a closed path with the initial point A. Furthermore, for every closed path with initial point A its inverse path has the same properties.

Now let us agree to regard equivalent paths as distinct representations of one and the same "path," only drawn in distinct ways on the surface, and nonequivalent paths as representations of essentially distinct "paths." The remarks made previously show that then the set of all closed paths (we now omit the quotation marks) starting out from an arbitrary point A of the surface is a group under the operation of multiplication of paths. The unit (neutral) element of this group is the zero path, and the inverse element of a given path is the same path but traversed in the opposite direction.

The group of paths, in general, depends not only on the form of the surface but also on the choice of the initial point A. However, if the surface does not fall into separate pieces, i.e., if any two of its points can be joined by a continuous path lying on the surface, then the group of paths corresponding to distinct points are isomorphic and in that case we can talk simply of the group of paths of the surface S without indicating A. This group of paths of the surface is also called its *fundamental* group.

If the surface S is a plane or a sphere, then the group of paths consists of the unit element alone, because in the plane and on the sphere every path can be contracted to a point. However, on the surface of an infinite circular cylinder, as we have seen, there are closed paths that do not contract to a single point. Since on the cylinder every closed path starting from A is equivalent to a certain power of the path X (figure 23), and distinct powers of X are not equivalent, the group of paths of the cylinder surface is an infinite cyclic group. It can be shown that the group of paths on the torus (figure 22) consists of the paths of the form $U^m V^n$ ($m, n = 0, \pm 1, \pm 2, \cdots$) with $UV = VU$ and $U^m V^n = U^{m_1} V^{n_1}$ only for $m = m_1$, $n = n_1$ where we recall that in discussing the group of paths equality has to be understood in the sense of equivalence.

The importance of the group of paths is due to the following property. Let us assume that apart from the surface S another surface S_1 is given such that between the points of S and S_1 we can establish a one-to-one continuous correspondence. For example, such a correspondence is possible if the surface S_1 is obtained from S by means of a certain continuous deformation without tearing apart or fusing distinct points of the surface. To every path on the original surface S, there corresponds a path on S_1. Moreover, equivalent paths correspond to equivalent ones, the product of two paths to their product, so that the group of paths on the surface S_1 is isomorphic to the group of paths on S. In other words, the group of paths regarded from the abstract point of view, i.e., to within isomorphism, is an invariant under all possible one-to-one continuous transformations of the surface. If the group of paths of two surfaces are distinct, then the surfaces cannot be carried continuously into each another. For example, the plane cannot be deformed without fusions or tearings into the cylinder surface, because the group of paths of the plane consists of the unit element only and the group of paths of the cylinder is infinite.

Properties of figures that remain unchanged under one-to-one and bicontinuous transformations are studied in the fundamental mathematical discipline of topology, whose basic ideas have been explained in Chapter XVIII. Invariants of bicontinuous transformations are called *topological*

§8. FUNDAMENTAL GROUPS

invariants. The group of paths is one of the most remarkable examples of topological invariants. It is clear that the group of paths can be defined not only for surfaces but also for arbitrary sets of points, provided only that we can speak of paths in these sets and of their deformations.

Defining relations. In topology methods of computing the group of paths are studied in detail. As a rule it proves convenient to define these groups by a special method that is often applied in the theory of groups for the purpose of defining abstract groups in general and not only for fundamental groups in topology. It consists in the following.

Let G be a group. The elements $g_1, g_2, \cdots, g_n$ are called generators of G if every element g can be represented in the form

$$g = g_{i_1}^{\alpha_1} g_{i_2}^{\alpha_2} \cdots g_{i_k}^{\alpha_k},$$

where $i_1, i_2, \cdots, i_k$ are some of the numbers $1, 2, \cdots, n$; indices i that do not stand side by side may be identical; the number of factors k is arbitrary; the exponents $\alpha_1, \alpha_2, \cdots, \alpha_k$ are positive or negative integers.

To know the group G it is sufficient to know, apart from the generators, also which products represent one and the same element of the group and which represent distinct elements. Thus, in order to define the group we have to list all equations of the form

$$g_{i_1}^{\alpha_1} g_{i_2}^{\alpha_2} \cdots g_{i_k}^{\alpha_k} = g_{j_1}^{\beta_1} g_{j_2}^{\beta_2} \cdots g_{j_l}^{\beta_l}$$

that hold in G. Since the set of such equations is always infinite, we usually give, instead of describing them all, only such equations as imply all the remaining ones followed by the group axioms. These equations are called defining relations.

It is clear that there are various ways of giving one and the same group by defining relations.

Let us consider, for example, the group H with the generators a, b and the relations

$$a^2 = b^3, \quad ab = ba. \tag{10}$$

Setting $c = ab^{-1}$, we have

$$a = bc, \quad a^2 = b^2c^2, \quad b^3 = b^2c^2, \quad b = c^2, \quad a = c^3.$$

We see that all the elements of the group H can be expressed by the single element c, where

$$a = c^3, \quad b = c^2.$$

Since the relations (10) follow immediately from these equations, there are no nontrivial relations for c. Therefore H is an infinite cyclic group with the generating element c.

If we can choose in a group generators that are not connected by any nontrivial relations, then the group is called *free*, and these generators are free generators. For example, if a group has the free generators a, b, then every element of it can be *uniquely* written in the form

$$a^{\alpha_0}b^{\beta_1}a^{\alpha_1}b^{\beta_2}a^{\alpha_2} \cdots b^{\beta_k}a^{\alpha_k},$$

where $k = 0, 1, 2, \cdots, n$ and the exponents $\alpha_0, \beta_1, \alpha_1, \cdots, \beta_k, \alpha_k$ are positive or negative integers except that the "extremes" α_0 and α_k can also assume the value zero. A similar statement holds for free groups with a larger number of generators.

When we write out the generators and defining relations for two groups assuming that the groups have no elements in common, then by combining these relations we obtain a new group, the so-called free product of the given ones.

The theory of free groups, and also the more general theory of free products, has an important place in the theory of groups. From the geometrical point of view the free product of the groups H_1 and H_2 is a group of paths of that figure which can be represented in the form of a sum of two closed figures that are fused at only one point and have H_1 and H_2 as their groups of paths. We know already that the group of paths of a cylinder surface is a free group with one generator. From the remark just made it follows, for example, that the group of paths of the surface illustrated in figure 24 is a free group with two generators.

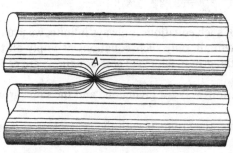

Fig. 24.

Similarly to the way in which the fundamental group of a surface was defined we can also introduce the fundamental group of spatial bodies, finite or infinite.

Knots and groups of knots. As we have already said, from the point of view of topology, two surfaces are regarded as identical if one of them can be carried into the other by a one-to-one and bicontinuous transformation. The problem of a topological classification of all closed surfaces was

§8. FUNDAMENTAL GROUPS

solved long ago. Every closed surface lying in our ordinary space is topologically equivalent either to a sphere, or to a sphere with a certain number of handles (figure 25). For example, the torus surface, illustrated in figure

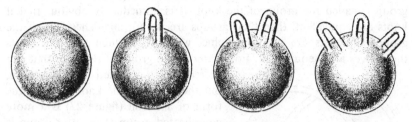

Fig. 25.

22, can be deformed continuously into a sphere with a single handle, the surface of a cube into the surface of a sphere, and so on. In view of this, the study of the fundamental groups of closed surfaces is not very interesting, since closed surfaces are completely classified even without these groups. However, there are very simple problems where so far almost nothing has been achieved without the fundamental groups. Among them is the famous problem of knots.

A knot is a closed curve lying in the ordinary three-dimensional space. As figure 26 shows, its position can be very varied. Two knots are called

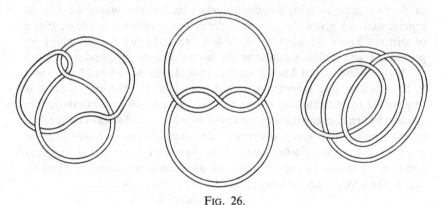

Fig. 26.

equivalent if one of them can be deformed into the other by a continuous process without breaking the curve and without self-penetration. Two problems arise at once: (1) how can we tell whether two knots given by their plane projections are equivalent or not; (2) how can we classify all nonequivalent knots?

Both problems remain as yet unsolved, but the substantial progress that has been made in a partial solution is connected with the theory of groups.

Let us remove from space the points that belong to the given knot and consider the fundamental group of the remaining set of points. This group is called the group of the knot. It is immediately obvious that if knots are equivalent, then their groups are isomorphic. Therefore, if the groups of knots are nonisomorphic, we can conclude that the knots themselves are inequivalent. For example, the group of the knot that can be reduced to a circle is a cyclic group, but the group of the knot that has the form of a trefoil (figure 27) is a more complicated group. The latter group is noncommutative and hence not isomorphic to the group of a circle. We can therefore state that it is impossible to deform the trefoil knot into a circle without breaking it, a fact that is completely obvious but requires a proof by precise mathematical arguments.

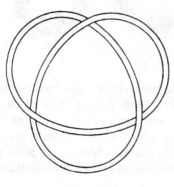

FIG. 27.

Unfortunately, in the discussion of the groups of knots there also arise difficult problems that have not so far been solved. The fact is that in topology very simple methods are known of finding generators and defining relations for the group of a knot represented in a given way. But in order to use groups for the comparison of distinct knots we have to be able to tell whether groups given by generators and defining relations are isomorphic or not, and a solution of this problem is not known so far. Indeed, the Soviet mathematician P. S. Novikov has recently proved the remarkable theorem that it is impossible to indicate any single regular process (more accurately, any so-called normal algorithm) by means of which it would always be possible to tell whether two given systems of defining relations for one and the same set of generators define one and the same group or not. This theorem compels us to doubt the existence of any uniform general method to decide the equivalence of knots given by their plane projections.

§9. Representations and Characters of Groups

The general theory of groups has a certain resemblance in its methods to elementary geometry: Both are founded on a definite system of axioms which is the starting point for the construction of the whole content of the theory. But the example of analytic geometry shows to what extent the

§9. REPRESENTATIONS AND CHARACTERS OF GROUPS

application of analytic, numerical methods can prove useful in the investigation of geometrical problems.

An application of the tools of analysis and classical algebra to the theory of groups is the so-called theory of group representations. Just as analytic geometry not only gives us methods of solving geometric problems by means of analysis but, conversely, throws a geometric light on many complicated problems of analysis, so to an even higher degree the representation theory not only serves as an auxiliary apparatus for investigating properties of groups, but by forging a link between deep concepts and problems of analysis and the theory of groups enables us to find expressions for group-theoretical facts in terms of numerical relations, and to find a group interpretation for analytic relationships. A large part of the present-day important applications of the theory of groups in physics is connected precisely with the theory of representations.

Representations of groups by matrices. In linear algebra (see Chapter XVI) we have discussed the operation of multiplication for matrices. This operation is associative, but in general noncommutative. The nonsingular square matrices of a given order form a group under multiplication, since the product of two nonsingular matrices is again nonsingular, the role of the neutral element is played by the unit matrix, and for every nonsingular matrix there exists its inverse which is also nonsingular.

Let us assume that a certain group G is given and that with every element g of it there is associated a definite nonsingular matrix of complex numbers A_g of order n such that when elements of the group are multiplied the matrices corresponding to them are also multiplied: $A_{gh} = A_g \cdot A_h$. Then we say that we have a representation of the group G by matrices of degree n. Usually the words "by matrices" are omitted and we simply speak of a representation of degree n of G. A representation of degree n of a given group G is simply a homomorphic mapping of G into the group of nonsingular matrices of degree n. From the general properties of homomorphic mappings it follows that in every representation the neutral element of G goes over into the unit matrix and inverse elements in G go over into inverse matrices.

Matrices of order 1 are simply individual complex numbers. Therefore a representation of degree 1 of G is a relation under which every element of G corresponds to a complex number and the product of the elements of the group corresponds to the product of these complex numbers. For example, the mapping under which we associate with an even permutation the number 1 and with an odd one the number -1 is a representation of degree 1.

By associating with every element of a group G the unit matrix E of degree n we obtain a representation of G which is called the unit representation of degree n. If G is a finite group containing more than one element, then G must necessarily also have infinitely many representations apart from the unit representation of the varying degrees. Methods of finding them will be indicated in the following.

When we know one representation of the group G, we can obtain an infinite set of others. For let $g \to A_g$ be the given representation of G by matrices of degree n. We choose an arbitrary nonsingular matrix P of the same degree n and set $B_g = P^{-1}A_gP$. The correspondence $g \to B_g$ is again a representation of G, since

$$B_{gh} = P^{-1}A_{gh}P = P^{-1}A_gA_hP = P^{-1}A_gPP^{-1}A_hP = B_gB_h.$$

The representations so obtained from a given representation by the choice of various matrices P are called *equivalent* to the given one. In the theory of representations equivalent representations are not regarded as essentially distinct, all representations being usually considered only to within equivalence.

Another method of finding new representations is the direct addition of representations, which consists in the following: Let $g \to A_g$, $g \to B_g$ be arbitrary representations of a group G by matrices of degree m and n, respectively. We consider the mapping

$$g \to \begin{bmatrix} A_g & 0 \\ 0 & B_g \end{bmatrix}.$$

By the rule for multiplication of matrices (see Chapter XVI) we have

$$gh \to \begin{bmatrix} A_{gh} & 0 \\ 0 & B_{gh} \end{bmatrix} = \begin{bmatrix} A_gA_h & 0 \\ 0 & B_gB_h \end{bmatrix} = \begin{bmatrix} A_g & 0 \\ 0 & B_g \end{bmatrix}\begin{bmatrix} A_h & 0 \\ 0 & B_h \end{bmatrix};$$

i.e., our mapping is again a representation of G. It is called the sum of the two given representations and is denoted by $A_g + B_g$. If the summands are rearranged, then we obtain another representation

$$g \to \begin{bmatrix} B_g & 0 \\ 0 & A_g \end{bmatrix},$$

which is, however, equivalent to the given one. Therefore, if we do not distinguish between equivalent representations, the addition of representations is a commutative operation. It is easy to see that under the same condition addition is also an associative operation. Having a certain stock of representations A_g, B_g, C_g, $\cdots$ of a group G we can obtain by means of addition representations of higher and higher degrees: $A_g + B_g + C_g$, $A_g + A_g + A_g + A_g$, and so forth.

§9. REPRESENTATIONS AND CHARACTERS OF GROUPS

For example, the numbers $1, -1, i, -i$ form a group under multiplication. By associating every number of this group with itself, we obtain a representation of degree 1. As a second representation we can take the mapping $1 \to 1, -1 \to -1, i \to -i, -i \to i$. The sum of these representations is the mapping

$$1 \to \begin{bmatrix} 1 & 0 \\ 0 & 1 \end{bmatrix}, -1 \to \begin{bmatrix} -1 & 0 \\ 0 & -1 \end{bmatrix}, i \to \begin{bmatrix} i & 0 \\ 0 & -i \end{bmatrix}, -i \to \begin{bmatrix} -i & 0 \\ 0 & i \end{bmatrix}.$$

Transforming this by means of the matrix

$$P = \begin{bmatrix} 1 & i \\ 1 & -i \end{bmatrix}$$

we obtain the equivalent representation

$$1 \to \begin{bmatrix} 1 & 0 \\ 0 & 1 \end{bmatrix}, -1 \to \begin{bmatrix} -1 & 0 \\ 0 & -1 \end{bmatrix}, i \to \begin{bmatrix} 0 & 1 \\ -1 & 0 \end{bmatrix}, -i \to \begin{bmatrix} 0 & -1 \\ 1 & 0 \end{bmatrix}.$$

It is interesting to note that all the matrices of this representation are real. Suppose now that all the matrices of a certain representation of degree n of a group G have the form

$$g \to A_g = \begin{bmatrix} B_g & C_g \\ 0 & D_g \end{bmatrix},$$

where B_g, D_g are square matrices and the left lower rectangle of A_g is entirely filled with zeros. By multiplying matrices A_g and A_h we obtain

$$A_{gh} = A_g A_h = \begin{bmatrix} B_g B_h & B_g C_h + C_g D_h \\ 0 & D_g D_h \end{bmatrix};$$

i.e., $B_{gh} = B_g B_h$, $D_{gh} = D_g D_h$. This shows that the mappings $g \to B_g$ and $g \to D_g$ are also representations of G, but of smaller degree. Here A_g is called a graduated representation of G and every representation equivalent to it is called *reducible*. A representation that is not equivalent to any graduated representation is called *irreducible*.

If in all the matrices of A_g not only the left lower but also the right upper rectangle C_g is filled with zeros, then A_g is said to *split* into the sum of the representations B_g, D_g. A representation that is equivalent to a sum of irreducible representations is called *completely reducible*.

In the theory of groups it is proved that every representation of a finite group is completely reducible.* Hence it follows that in order to find all the representations of a finite group it is sufficient to know its irreducible representations, because all the others are equivalent to various sums of irreducible ones.

The practical computation of irreducible representations of an arbitrary finite group is, as a rule, a fairly complicated task which is solved in an explicit form only for individual classes of finite groups; for example, for commutative groups, for the symmetric groups, and for some others, though from a theoretical point of view the properties of representations of finite groups have been studied in much detail.

Every finite group has a particular "regular" representation that is constructed as follows. Suppose that $g_1, g_2, \cdots, g_n$ are the elements of the given group G numbered in an arbitrary order and that

$$g_i g_k = g_{i_k} \quad (i, k = 1, 2, \cdots, n).$$

By choosing an arbitrary fixed value for k, we form the matrix of degree n which has a 1 at the i_kth place and zeros in the remaining places ($i = 1$, 2, $\cdots$, n), and we denote it by R_{g_k}. The correspondence $g_k \to R_{g_k}$ ($k = 1$, 2, $\cdots$, n) is called the *regular representation* of G. The fact that it is a representation can be shown by simple computations.

It can also be shown that by changing the numbering of the elements of the group we arrive at an equivalent representation and that consequently, to within equivalence, every finite group has only one regular representation.

Let us briefly formulate the fundamental theorems of the theory of representations of finite groups. The number of distinct (inequivalent) irreducible representations of a finite group is finite and is equal to the number of classes of conjugate elements (see §6) of the group. The degree of an irreducible representation is necessarily a divisor of the order of the group, and the regular representation is equivalent to a sum of all inequivalent irreducible representations in which every irreducible summand is repeated as often as its degree indicates.

This implies the following interesting relation between the order of a finite group and the degrees of its irreducible representations.

We denote the number of elements of the group G by n, the number of classes of conjugate elements by k, and the degrees of the irreducible representations of G by $n_1, n_2, \cdots, n_k$, respectively. From the construction of the regular representation it is clear that its degree is n. Furthermore, since the regular representation is equivalent to a sum of n_1 representations

---

* We recall that we are considering representations of groups by matrices whose elements may be arbitrary complex numbers.

equivalent to the first irreducible representation, plus n_2 representations equivalent to the second, and so forth, and since under addition of representations their degrees are added, we must have the following equation

$$n = n_1^2 + n_2^2 + \cdots + n_k^2. \tag{11}$$

By associating with every element of the group the number 1, we obtain the trivial irreducible representation of degree 1 which every group possesses. If in the formula (11) we take n_1 to be the degree of precisely this unit representation, then we can rewrite (11) in the equivalent form

$$n = 1 + n_2^2 + \cdots + n_k^2,$$

where $n_2, \cdots, n_k$ now denote the degrees of the nontrivial irreducible representations.

By using the fact that $n_2, \cdots, n_k$ must be divisors of n, we can occasionally, when k is known, find $n_2, \cdots, n_k$ from the equation (11) only. For example, the symmetric group S_3 of permutations of three elements has three classes of conjugate permutations: (1); (12), (13), (23); (123), (132). For $n = 6$, $k = 3$ the equation (11) admits only one system of solutions: $6 = 1^2 + 1^2 + 2^2$. Therefore S_3 has two distinct representations of degree 1 and one irreducible representation of degree 2.

Another example is finite Abelian groups. Here every element forms an individual class. Therefore $k = n$ and it follows from formula (11) that $n_1 = n_2 = \cdots = n_k = 1$, i.e.; all irreducible representations of these groups are of degree 1 and their number is equal to the order of the group.

The irreducible representations of Abelian groups are also called their characters, whereas for every representation of a non-Abelian group the name "character" is given to the set of the so-called traces (i.e., the sum of the diagonal elements) of the matrices forming the representation. The characters of finite groups have remarkable properties and relationships. The investigation of representations and characters of groups has enriched the theory of groups by interesting general results that have found extensive application in contemporary theoretical physics.

§10. The General Theory of Groups

We have already mentioned that almost throughout the last century the theory of groups was developed primarily as the theory of transformation groups. However, it gradually became clear that the study of groups as such was fundamental and the study of transformation groups can be reduced to that of abstract groups and their subgroups. The transition from the theory of transformation groups to the theory of abstract groups occurred first in the theory of finite groups, but the rapid development of the theory of Lie groups and the penetration of group theory into topology made it necessary to create the general theory of groups in which finite groups are regarded only as a certain special case.

The first textbook on the theory of groups in which this point of view was adopted in its full clarity was the book by O. Ju. Šmidt, which appeared in Kiev in 1916. Šmidt also obtained in the 1920's an important theorem on infinite groups which became the starting point of investigations of a number of other Soviet algebraists. Thanks to the activities of O. Ju. Šmidt and P. S. Aleksandrov, who did a great deal to popularize the ideas of contemporary algebra, a large school of group theory was formed in Moscow which later came under the leadership of their pupil, A. G. Kuroš. He became widely known, in particular, for his proof of the theorem that every subgroup of a free product is itself a free product of subgroups isomorphic to suitable subgroups of the factors and, possibly, a separate free subgroup. Later he published a monograph on the theory of groups which gave the first systematic account of the rich factual material obtained in the general theory of groups. This monograph is still the most complete textbook in the world literature on the general theory of groups and has become internationally famous.

Following the lead of the Moscow school, algebraists in Leningrad and other cities became interested in the general theory of groups and made their own contribution to its development. The researches on the theory of groups that are conducted at present in the Soviet Union comprise all its essential branches, and the results obtained by Soviet mathematicians have repeatedly exerted a decisive influence on the development of the subject.

§11. Hypercomplex Numbers

In solving practical problems by algebraic methods we usually arrive in the simplest cases at one or several equations from which the values of the unknown quantities have to be found. The unknown entities in this context are quantitative characteristics of the objects under investigation; the equations are formed by means of an analysis of the real relationships that hold between the objects.

This is the state of affairs in cases when we are dealing with the simplest quantities, such as mass, volume, or distance, that can be characterized quantitatively by a single number. However, in concrete problems we encounter objects unable to be characterized by a single number. Far

§11. HYPERCOMPLEX NUMBERS

from it, in the development of technology all the more important objects are of a much complicated nature and require for their characterization several numbers, even infinitely many. Even such important physical quantities as force, velocity, or acceleration are characterized by directed segments and require three numbers. Also it is well known that the position of a point in space is characterized by three numbers, the position of a plane also by three, the position of a line by four, and the position of a rigid body by six numbers. Therefore, when we wish to solve by algebraic means problems referring to more complicated objects, we obtain equations with a larger number of unknowns and often it turns out to be more tedious to analyze them than to solve the problem directly by making use of its geometric or physical peculiarities. Hence the idea naturally arose of trying to characterize more complicated objects not by systems of ordinary numbers, but by certain more complicated general numbers on which one might perform operations similar to the ordinary arithmetical operations. This statement of the problem was the more natural, since the history of science exhibited not the invariability of the concept of number but its flexibility, the gradual enrichment of the realm of numbers from the natural numbers to the fractional numbers, then to algebraic numbers, to real (rational and irrational) numbers, and finally to complex numbers.

Complex numbers. From Chapter IV the reader is already acquainted with the fundamental properties of complex numbers and their simplest applications. Here we shall only be interested in the foundation of the concept of a complex number. When we begin with a discussion of the ordinary real numbers, we notice that the square root of negative numbers has no meaning, because the square of every real number is positive or zero. One then shows that urgent needs of science compelled the mathematicians to regard expressions of the form $a + b\sqrt{-1}$ also as a special kind of number which became known as imaginary, as opposed to the ordinary real numbers. If it is assumed that these imaginary numbers are subject to the same laws of arithmetical operations as the ordinary numbers, then all square roots of negative numbers can be expressed in terms of the quantity $i = \sqrt{-1}$, and the result of arithmetical operations performed any finite number of times on real or imaginary numbers can always be represented in the form $a + bi$, where a and b are real numbers.

Clearly, this definition of imaginary numbers runs counter, in the highest degree, to common sense: First it was stated that the expressions $\sqrt{-1}$, $\sqrt{-2}$, and so forth, have no meaning, and then it was proposed that these expressions without a meaning should be called imaginary numbers. This circumstance caused many mathematicians of the 17th and 18th century to doubt the validity of the use of complex numbers. However, these doubts were dispelled at the beginning of the 19th century, when a geometrical interpretation was found for the complex numbers by points in a plane. Another purely arithmetical foundation of the theory of complex numbers was given somewhat later by the Hungarian mathematician Bolyai and the Irish mathematician Hamilton. This proceeds as follows.

Instead of the numbers $a + bi$ we shall simply speak of pairs of real numbers (a, b). Two pairs shall be regarded as equal if their first and second terms are equal, i.e., $(a, b) = (c, d)$ if and only if $a = c$ and $b = d$. Addition and multiplication of pairs are defined by the formulas

$$(a, b) + (c, d) = (a + c, b + d); (a, b) \cdot (c, d) = (ac - bd, ad + bc).$$

For example, we have

$$(2, 3) + (1, -2) = (3, 1), \quad (2, 3)(1, -2) = (8, -1),$$
$$(3, 0) + (2, 0) = (5, 0), \quad (3, 0)(2, 0) = (6, 0).$$

These examples show, in particular, that the arithmetical operations on pairs with a zero in the second place reduce to the same operations on their first terms, so that such pairs can be simply denoted by their first numbers. If we introduce the notation i for the pair $(0, 1)$ then we have

$$(a, b) = a(1, 0) + b(0, 1) = a + bi,$$
$$i^2 = (0, 1)(0, 1) = (-1, 0) = -1;$$

i.e., we have the usual notation for complex numbers.

Thus, from this point of view complex numbers are pairs of ordinary real numbers and operations on complex numbers are only a special kind of operations on pairs of real numbers.

Hypercomplex numbers. Various successful applications of the complex numbers induced mathematicians as early as the first decades of the 19th century to turn their attention to the problem whether one could not construct higher complex numbers to be represented by triplets, quadruplets, and so forth, of real numbers, similar to the way in which the complex numbers are constructed in the form of pairs of real numbers. From the middle of the last century onward many distinct special systems of such higher complex numbers or hypercomplex numbers were investigated, and at the end of the last and in the first half of the present century a general theory of hypercomplex numbers was developed that has found a number of important applications in neighboring domains of mathematics and physics.

§11. HYPERCOMPLEX NUMBERS

Hypercomplex number of rank n is the name for a number that can be represented by a collection of n real numbers $(a_1, a_2, \cdots, a_n)$ which for the time being we shall call its coordinates. The hypercomplex numbers $(a_1, a_2, \cdots, a_n)$ and $(b_1, b_2, \cdots, b_n)$ shall be called equal if their corresponding coordinates are equal, i.e., if $a_1 = b_1, a_2 = b_2, \cdots, a_n = b_n$. We define the operation of addition by the natural formula

$$(a_1, a_2, \cdots, a_n) + (b_1, b_2, \cdots, b_n) = (a_1 + b_1, a_2 + b_2, \cdots, a_n + b_n),$$

analogous to the formula of addition for complex numbers.

It is equally natural to introduce the operation of multiplication of a hypercomplex number by a real one: By definition we set

$$a(a_1, a_2, \cdots, a_n) = (aa_1, aa_2, \cdots, aa_n).$$

We now have to define the operation of multiplication of two hypercomplex numbers so that the result of this operation is again a hypercomplex number.

To extend the definition of multiplication of ordinary complex numbers to the general case is tedious. It can be done in various ways, and then we obtain various systems of hypercomplex numbers. Therefore, first of all we have to clarify what such a definition is to achieve. Undoubtedly it is desirable that the operations on hypercomplex numbers we are about to define should resemble in their properties the ordinary operations on real numbers. Now what are these properties of the ordinary operations?

In a careful discussion of the properties of numbers and the operations on them that are used most frequently in algebra, it is easy to observe that they reduce to the following:

1. For any two numbers, their sum is uniquely determined.
2. For any two numbers, their product is uniquely determined.
3. There exists a number zero with the property $a + 0 = a$ for every a.
4. For every number a, there exists the opposite number x satisfying the equation $a + x = 0$.
5. Addition is commutative

$$a + b = b + a.$$

6. Addition has the associative property

$$(a + b) + c = a + (b + c).$$

7. Multiplication is commutative

$$ab = ba.$$

8. Multiplication is associative

$$(ab) \cdot c = a \cdot (bc).$$

9. Multiplication is distributive

$$a(b + c) = ab + ac, \qquad (b + c)a = ba + ca.$$

10. For every a and every $b \neq 0$, there exists a unique number x satisfying the equation $bx = a$.

The properties 1 through 10 were selected as a result of a careful analysis; the development of mathematics in the last century proved their great importance. Nowadays every system of quantities satisfying the conditions 1 through 10 is called a *field*. Examples of fields are: the set of all rational numbers, the set of all real numbers, or the set of all complex numbers, because in each of these cases the numbers of the set can be added and multiplied and the result is a number of the same set, and the operations have the properties 1 through 10. Apart from these three very important fields we can determine infinitely many other fields formed from numbers. But beside the fields formed from numbers there is much interest in fields formed from quantities of another nature. For example, already at school we learn to operate with the so-called algebraic fractions, i.e., fractions in which the numerator and denominator are polynomials in certain letters. Algebraic fractions can be added, subtracted, multiplied, and divided, and these operations have the properties 1 through 10. Therefore, algebraic fractions form a system of objects that is a field. We could give many other examples of fields formed from quantities of a more complicated nature. In view of the importance of the properties 1 through 10 that define a field, the original formulation of the problem was to find such an operation of multiplication of hypercomplex numbers that they should form a field. In case of success one would then try to obtain new even more general complex numbers. However, already at the beginning of the last century it was discovered that this is only possible for hypercomplex numbers of rank 2 and that only the ordinary complex numbers could be obtained in this way. This result proved that the complex numbers have a very special position and that it is impossible to obtain an extension of the number system beyond the limits of the complex numbers, provided we insist that all the properties 1 through 10 are fulfilled.

Therefore, in further attempts to construct higher number systems it was necessary to omit one or several of the properties 1 through 10.

Quaternions. Historically, the first hypercomplex system that was discussed in mathematics is the system of quaternions, i.e., "fourfold

§11. HYPERCOMPLEX NUMBERS

numbers," which was introduced by the Irish mathematician Hamilton at the middle of the last century. This system satisfies all the requirements 1 through 10, except 7 (commutativity of multiplication).

Quaternions can be described as follows. For the quadruplets $(1, 0, 0, 0)$, $(0, 1, 0, 0)$, $(0, 0, 1, 0)$, $(0, 0, 0, 1)$ we introduce the abbreviations $1, i, j, k$. Then by the equation

$$(a, b, c, d) = a(1, 0, 0, 0) + b(0, 1, 0, 0) + c(0, 0, 1, 0) + d(0, 0, 0, 1)$$

every quaternion can be uniquely represented in the form

$$(a, b, c, d) = a \cdot 1 + b \cdot i + c \cdot j + d \cdot k.$$

The quaternion 1 will play the role of the unit of the system of quantities to be constructed; i.e., we shall assume that $1 \cdot \alpha = \alpha \cdot 1 = \alpha$ for every quaternion α. Further we set by definition: $i^2 = j^2 = k^2 = -1$;

$$ij = -ji = k,$$
$$ik = -ki = -j,$$
$$jk = -kj = i.$$

It is easy to memorize this "multiplication table" by means of figure 28 in which the points i, j, k on the circle represent the corresponding quaternions i, j, k. The product of two adjacent quaternions is equal to the third if the movement from the first sector to the second proceeds clockwise in the figure, and equal to the third with the minus sign if the motion is counterclockwise. Knowing the multiplication table for the quaternions i, j, k, we carry out the multiplication of arbitrary quaternions by using the distributive law 9. In fact:

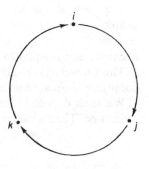

FIG. 28.

$$(a \cdot 1 + b \cdot i + c \cdot j + d \cdot k)(a_1 \cdot 1 + b_1 \cdot i + c_1 \cdot j + d_1 \cdot k)$$
$$= aa_1 \cdot 1 + ab_1 \cdot i + ac_1 \cdot j + ad_1 \cdot k$$
$$+ ba_1 \cdot i + bb_1 \cdot ii + bc_1 \cdot ij + bd_1 \cdot ik$$
$$+ ca_1 \cdot j + cb_1 \cdot ji + cc_1 \cdot jj + cd_1 \cdot jk$$
$$+ da_1 \cdot k + db_1 \cdot ki + dc_1 \cdot kj + dd_1 \cdot kk$$
$$= (aa_1 - bb_1 - cc_1 - dd_1) \cdot 1 + (ab_1 + ba_1 + cd_1 - dc_1) \cdot i$$
$$+ (ac_1 + ca_1 - bd_1 + db_1) \cdot j + (ad_1 + da_1 + bc_1 - cb_1) \cdot k.$$

The factor 1 in the first term of a quaternion is usually omitted and instead of $a \cdot 1$ we write a. The equations $ij = -ji$, $ik = -ki$, $jk = -kj$ show that the multiplication of quaternions is not commutative. The multiplicand and the multiplier are not of equal status here. Therefore, in computations with quaternions we have to adhere carefully to the order of the factors. Otherwise the operations with quaternions do not present any difficulties. In particular, the associative law 8 holds for the multiplication of quaternions. It is easily verified for the quaternions of the basis 1, i, j, k by means of the multiplication table; the transition to the general case is obvious.

The number a of the quaternion $a + bi + cj + dk$ is called its real or scalar part, and the sum $bi + cj + dk$ its vector part. The quaternions $a + bi + cj + dk$ and $a - bi - cj - dk$ that differ only in the sign of the vector part are called conjugate. Obviously, the sum of two conjugate quaternions is a real number. Furthermore, on multiplying conjugate quaternions by the previous formula we obtain

$$(a + bi + cj + dk)(a - bi - cj - dk) = a^2 + b^2 + c^2 + d^2; \quad (12)$$

i.e., the product of conjugate quaternions is also a real number.

The sum of the squares of the coefficients $a^2 + b^2 + c^2 + d^2$ of a quaternion $a + bi + cj + dk$ is called its norm. Since the square of every real number is nonnegative, the norm of every quaternion is also nonnegative, and is equal to zero only for a null quaternion.

The formula (12) shows that the product of any quaternion with its conjugate is equal to its norm.

We shall denote by an asterisk the quaternion that is conjugate to a given one. Then a direct multiplication verifies the following formula:

$$(\alpha\beta)^* = \beta^*\alpha^*.$$

This has an interesting consequence: The norm of the product of quaternions is equal to the product of the norms of the factors. For by the preceding we have

$$\text{norm } (\alpha\beta) = (\alpha\beta)(\alpha\beta)^* = \alpha\beta\beta^*\alpha^* = (\alpha\alpha^*)(\beta\beta^*) = \text{norm } \alpha \cdot \text{norm } \beta.$$

The properties of the norm enable us to give a very simple solution to the problem of division of quaternions. Let $\alpha = a + bi + cj + dk$ be an arbitrary nonzero quaternion. Then

$$(a + bi + cj + dk) \frac{1}{a^2 + b^2 + c^2 + d^2} (a - bi - cj - dk)$$

$$= \frac{1}{a^2 + b^2 + c^2 + d^2} (a^2 + b^2 + c^2 + d^2) = 1;$$

§11. HYPERCOMPLEX NUMBERS

i.e., the quaternion

$$\frac{1}{a^2 + b^2 + c^2 + d^2}(a - bi - cj - dk) = \alpha^{-1}$$

is the inverse of the given quaternion α.

Having found the inverse quaternion it is now easy to find the quotient of two quaternions. For suppose that two quaternions α, β are given, the first of them being different from zero. Then the quotients obtained by dividing β by α must be the solutions of the equations

$$\alpha x = \beta, \quad y\alpha = \beta.$$

Multiplying both sides of the first equation by the inverse quaternion α^{-1} on the left we obtain

$$x = \alpha^{-1}\beta.$$

Multiplying both sides of the second equation by α^{-1} on the right we have

$$y = \beta\alpha^{-1}.$$

Since the products $\alpha^{-1}\beta$ and $\beta\alpha^{-1}$ are in general distinct, we have to distinguish between two divisions for quaternions, on the right and on the left; both are always possible, except of course division by zero.

The algebra of vectors. Although the operations on quaternions are in many respects similar to those on complex numbers, the absence of the commutative law of multiplication makes the properties of quaternions very different from those of numbers. For example, from the algebra of complex numbers it is well known that a quadratic equation has two roots. But if we consider the quadratic equation

$$x^2 + 1 = 0$$

in the domain of quaternions, then we can easily find 6 roots: $\pm i, \pm j, \pm k$, and a more precise analysis shows that there is even an infinite number of other solutions. This circumstance strongly impedes the use of quaternions in mathematics, and notwithstanding the numerous attempts of Hamilton and other mathematicians to introduce quaternions into various branches of mathematics and physics, the role of the quaternions remains to the present day somewhat modest and can in no way be compared with the role of complex numbers.

However, quaternions have given a spur to the development of vector algebra which is an indispensable tool in modern technology and physics. The fact is that in mechanics and physics the concepts of velocity, acceleration, force, and so forth, which require three numbers for their characterization, play an essential role. Earlier we have seen that every quaternion can be regarded as an aggregate of a real number a and the vector part $bi + cj + dk$. Since the vector part of a quaternion is determined by three numbers, the most important physical quantities can be characterized by vector parts of quaternions.

Geometrically the vector part $bi + cj + dk$ of the quaternion $a + bi + cj + dk$ can be taken to represent the vector leading from the origin of a rectangular Cartesian system of coordinates to the point whose projections on the coordinate axes are equal to the numbers b, c, d, respectively. Therefore, every quaternion can be represented geometrically as an aggregate of a number and of a vector in space. Let us see how the operations on quaternions have to be interpreted.

We take two vector quaternions $xi + yj + zk$ and $x_1 i + y_1 j + z_1 k$ whose scalar parts are equal to zero. Geometrically, they are illustrated by vectors from the coordinate origin. The sum of these quaternions is again a vector quaternion $(x + x_1)i + (y + y_1)j + (z + z_1)k$. It is easy to see that the vector representing this sum is the diagonal of the parallelogram constructed on the first two vectors. Thus, the addition of vector quaternions corresponds to the well-known operation of addition of vectors by the parallelogram rule. Similarly, if we multiply a vector quaternion by an arbitrary real number, the representing quaternion vector is also multiplied by that number.

We come to a different situation when we multiply quaternions. Indeed,

$$(xi + yj + zk)(x_1 i + y_1 j + z_1 k)$$
$$= -xx_1 - yy_1 - zz_1 + (yz_1 - y_1 z)i + (zx_1 - z_1 x)j + (xy_1 - x_1 y)k;$$

i.e., on multiplying two vector quaternions we obtain a complete quaternion having a scalar part and a vector part.

The scalar part of the product of vector quaternions taken with the opposite sign is called the scalar product of the vectors representing the given quaternions, and the vector representing the vector part of the product is the vector product of the given quaternions. The scalar product of the vectors α and β is usually denoted by $(\alpha\beta)$ or simply by $\alpha\beta$, and the vector product of the same vectors by $[\alpha\beta]$. Let i, j, k be the vectors corresponding to the quaternions i, j, k, i.e., vectors of unit length lying along the coordinate axes. By definition, if $\alpha = xi + yj + zk$, and $\beta = x_1 i + y_1 j + z_1 k$, then

$$(\alpha\beta) = xx_1 + yy_1 + zz_1, \ [\alpha\beta]$$
$$= (yz_1 - y_1 z)i + (zx_1 - z_1 x)j + (xy_1 - x_1 y)k.$$

By means of the latter formulas it is easy to give also a geometrical

§11. HYPERCOMPLEX NUMBERS

interpretation of the scalar and the vector product of vectors. As it turns out, the scalar product of two vectors is equal to the product of their lengths and of the cosine of the angle between them, and the vector product of two vectors is the vector whose length is equal to the area of the parallelogram constructed on the given vectors and whose direction is perpendicular to the plane of this parallelogram on that side from which the rotation of the first given vector toward the second looks like the rotation of the x-axis toward the y-axis as seen from the z-axis.

Nowadays in mechanics and physics we do not, as a rule, use operations on quaternions, but instead we consider only the operations on vectors, and these operations are defined in a purely geometrical manner according to the results just stated.

In conclusion, we want to point out one problem in mechanics that can be solved by means of quaternions in a particularly elegant way. Its solution was actually one of the motives for the discovery of quaternions.

Suppose that a rigid body is first rotated by a certain angle ϕ in a given direction around the definite axis OA passing through a given point O, and that it is then rotated by an angle ϕ_1 around another axis OB passing through the same point. The question is: Around what axis and by what angle must the body be rotated in order to bring it from its first position at once to the third? This is the well-known problem of mechanics on the addition of finite rotations. True, it can be solved by means of the ordinary analytic geometry, as was done already by Euler in the 18th century. However, its solution assumes a far more lucid form by means of quaternions.

Let $\xi = xi + yj + zk$ and $\alpha = a + bi + cj + dk$ be two quaternions, the first of which will be regarded as variable and the second as fixed. The expression $\alpha^{-1}\xi\alpha$ is a vector quaternion, as can easily be verified by a computation. Now if the quaternions ξ, $\alpha^{-1}\xi\alpha$ and the vector part of α are represented by the vectors $\vec{\xi}, \vec{\xi_1}, \vec{\alpha}$, then it turns out that the vector $\vec{\xi_1}$ is obtained geometrically from $\vec{\xi}$ by a rotation around the axis passing through the vector $\vec{\alpha}$ by an angle ϕ defined by the formula

$$\cos\frac{\phi}{2} = \frac{a}{\sqrt{a^2 + b^2 + c^2 + d^2}}.$$

Therefore we can say that the quaternion $\alpha = a + bi + cj + dk$ represents the rotation of space by the angle ϕ around the axis $\vec{\alpha} = bi + cj + dk$.

Conversely, knowing an axis of rotation and an angle ϕ we can look for the quaternion that represents this rotation. There is an infinite set of quaternions but they all differ from one another only by a numerical factor.

Now let us consider another rotation by an angle ϕ_1 around a certain axis $\vec{\beta} = b_1 i + c_1 j + d_1 k$. Let this rotation be represented by the quaternion $\beta = a_1 + b_1 i + c_1 j + d_1 k$. Under the action of the first rotation an arbitrary vector $\vec{\xi} = xi + yj + zk$ goes over into the vector $\alpha^{-1}\vec{\xi}\alpha$ and under the action of the second rotation the latter vector goes over into $\beta^{-1}(\alpha^{-1}\vec{\xi}\alpha)\beta$. By the associative law the latter result can be represented in the form

$$\beta^{-1}(\alpha^{-1}\vec{\xi}\alpha)\beta = (\alpha\beta)^{-1}\vec{\xi}\alpha\beta.$$

Since the multiplication of a vector, namely the vector quaternion $\vec{\xi}$ by the quaternion $(\alpha\beta)^{-1}$ on the left and the quaternion $\alpha\beta$ on the right, is equivalent to a rotation of this vector by the corresponding angle around the corresponding axis, we come to the conclusion that the result of two successive rotations characterized by the quaternions α and β is the rotation characterized by the product $\alpha\beta$. In other words, to the addition of the rotations corresponds the multiplication of their representing quaternions.

Apart from geometric and physical applications quaternions have found remarkable applications in the theory of numbers. Of a succession of works in this domain we must mention, in particular, the papers of Ju. V. Linnik.

§12. Associative Algebras

General definition of algebras (hypercomplex systems). We have defined hypercomplex numbers as quantities for the description of which several real numbers are required, in fact for the sake of definiteness we have regarded hypercomplex numbers simply as systems of real numbers. However, this point of view is too narrow, and for theoretical investigations the following more general definition gradually became accepted.

A certain system of quantities S is called an algebra (or a hypercomplex system) over the field P if

a. For every element a of P and every quantity α of the system S, a certain element of the system is defined, which is called the product of a and α is denoted by $a\alpha$;

b. For any two quantities α, β of the system, a certain quantity of the same system is uniquely defined which is called the sum of the first two quantities and is denoted by $\alpha + \beta$;

c. For any two quantities of the system α, β, another quantity of the

§12. ASSOCIATIVE ALGEBRAS

same system is uniquely defined which is called the product of the first two and is denoted by $\alpha\beta$;

and if these three operations have the following properties:*

1'. $\alpha + \beta = \beta + \alpha$,
2'. $(\alpha + \beta) + \gamma = \alpha + (\beta + \gamma)$,
3'. The system S has a zero quantity θ with the property
$$\alpha + \theta = \alpha,$$
4'. $a(\alpha + \beta) = a\alpha + a\beta$,
5'. $(a + b)\alpha = a\alpha + b\alpha$,
6'. $(ab)\alpha = a(b\alpha)$,
7'. $\theta\alpha = \theta, 1 \cdot \alpha = \alpha$, where 1 is the unit element of the field P,
8'. Among the quantities of S there exist $\alpha_1, \alpha_2, \cdots, \alpha_n$, such that in terms of these every quantity of the system can be uniquely represented in the form $a_1\alpha_1 + a_2\alpha_2 + \cdots + a_n\alpha_n$,
9'. $(a\alpha)\beta = \alpha(a\beta) = a(\alpha\beta)$,
10'. $\alpha(\beta + \gamma) = \alpha\beta + \alpha\gamma$, $(\beta + \gamma)\alpha = \beta\alpha + \gamma\alpha$.

In this definition the elements of the arbitrary field P play the role that so far was played by the real numbers. From the condition 8' it is clear that every hypercomplex quantity is determined by a system of n elements $a_1, a_2, \cdots, a_n$ of P and that it can therefore, depending on the choice of P, be determined by n complex numbers, n rational numbers, n real numbers, and so forth.

The first eight postulates signify that S forms a linear finite-dimensional space (see Chapter XVI, §2) over the field P, which we shall call the ground field of the algebra.

The requirement 9' and 10' can be combined in the form of the equations

$$(a\beta + b\gamma)\alpha = a(\beta\alpha) + b(\gamma\alpha),$$
$$\alpha(a\beta + b\gamma) = a(\alpha\beta) + b(\alpha\gamma),$$

from which it follows that the operation of multiplication is linear with respect to each factor.

Of the two terms "hypercomplex systems" and "algebra" the second has been preferred in recent years, since the elements of very general "hypercomplex systems" may differ in their properties considerably from the ordinary numbers, so that it is inappropriate to call them "hypercomplex

---

* By letters of the Greek alphabet we denote arbitrary quantities of the system S, and by letters of the Latin alphabet elements of the field P.

numbers." The terms "hypercomplex systems," and "hypercomplex numbers" are now applied only to the simplest algebras, for example to the system of the ordinary quaternions.

From the requirements 1' through 10' it is clear that in algebras the commutativity and associativity of multiplication is not assumed, nor the existence of a unit element nor the possibility of "division."

Every algebra S has a basis, i.e., a system of elements $\alpha_1, \alpha_2, \cdots, \alpha_n$ in terms of which all the elements of the algebra can be uniquely represented in the form of linear combinations $a_1\alpha_1 + a_2\alpha_2 + \cdots + a_n\alpha_n$ with coefficients from the ground field P. Every algebra can have infinitely many bases, but the number of elements of each basis is one and the same and is called the rank of the algebra.

The system of complex numbers regarded as an algebra over the field of real numbers has a basis of the numbers 1 and i. But the pairs of numbers 2 and $3i$, 1 and $a + bi$ (a, b are real, $b \neq 0$) can also serve as bases.

Let $\epsilon_1, \epsilon_2, \cdots, \epsilon_n$ be the basis of an arbitrary algebra over a certain field P. By definition every element of the algebra can be written uniquely in the form

$$\alpha = a_1\epsilon_1 + a_2\epsilon_2 + \cdots + a_n\epsilon_n .$$

If $\beta = b_1\epsilon_1 + \cdots + b_n\epsilon_n$ is any other element of it, then by the properties 1' through 6' we have

$$\alpha + \beta = (a_1 + b_1)\epsilon_1 + (a_2 + b_2)\epsilon_2 + \cdots + (a_n + b_n)\epsilon_n .$$

Similarly, for every a of P we have

$$a\alpha = aa_1\epsilon_1 + aa_2\epsilon_2 + \cdots + aa_n\epsilon_n .$$

Therefore the operation of addition of quantities of the algebras and of their multiplication by elements of the field P are uniquely determined by the given formulas. The operation of multiplication of quantities of the algebra must be specially defined for each algebra; but we need not know how to multiply arbitrary quantities of the algebra, it is sufficient to know the law of multiplication of the basis quantities ϵ_i. Indeed, by the properties 9' and 10'

$$(a_1\epsilon_1 + a_2\epsilon_2 + \cdots + a_n\epsilon_n)(b_1\epsilon_1 + b_2\epsilon_2 + \cdots + b_n\epsilon_n) = \Sigma a_i b_j \cdot \epsilon_i\epsilon_j .$$

Each of the products $\epsilon_i\epsilon_j$ is a certain quantity of the algebra and can therefore be expressed in terms of the basis elements

$$\epsilon_i\epsilon_j = c_{ij1}\epsilon_1 + c_{ij2}\epsilon_2 + \cdots + c_{ijn}\epsilon_n .$$

Here c_{ijk} denote elements of the ground field P over which the algebra is

§12. ASSOCIATIVE ALGEBRAS

constructed. The first index denotes the number of the first factor, the second, the number of the second factor, and the third indicates the number of that element whose coefficient is c_{ijk}. The coefficients c_{ijk} are called the structure constants of the algebra, since a knowledge of these constants completely determines all the operations on the quantities of the algebra.

It is easy to count the number of structure constants of an algebra of rank n. Every constant has three indices i, j, k. Therefore the number of structure constants of an algebra of rank n is equal to the number of triplets formed from the natural numbers $1, 2, \cdots, n$, i.e., to n^3. For example, the system of complex numbers over the field of real numbers has a basis consisting of the numbers $1, i$. In virtue of the equations

$$1 \cdot 1 = 1 \cdot 1 + 0 \cdot i, \quad i \cdot 1 = 0 \cdot 1 + 1 \cdot i,$$
$$1 \cdot i = 0 \cdot 1 + 1 \cdot i, \quad i \cdot i = -1 \cdot 1 + 0 \cdot i,$$

the structure constants are equal, respectively, to

$$c_{111} = a, \quad c_{112} = 0, \quad c_{211} = 0, \quad c_{212} = 1,$$
$$c_{121} = 0, \quad c_{122} = 1, \quad c_{221} = -1, \quad c_{222} = 0.$$

Suppose, conversely, that n^3 elements c_{ijk} of an arbitrary field P are given, indexed by triplets of the natural numbers $(i, j, k = 1, 2, \cdots, n)$. Then they can be taken as the structure constants of an algebra over the field P using the equation $\epsilon_i \epsilon_j = \sum_{k=1}^n c_{ijk} \epsilon_k$ as the definition of multiplication in the algebra.

Previously we have seen that every algebra has, in general, infinitely many distinct bases. The structure constants depend on the choice of the basis and therefore one and the same algebra can be given by distinct systems of structure constants.

Which algebras should be regarded as distinct and which as equal? In the theory of algebras it is convenient to regard two algebras over one and the same field P as equal if they are isomorphic, i.e., if the quantities of one algebra can be put into one-to-one correspondence with the quantities of the other in such a way that the sum and the product of any two quantities of the first algebra are associated with the sum and the product of the corresponding quantities of the second algebra and that the product of any element of the field P by an element of the first algebra is associated with the product of the same element of P and the corresponding element of the second algebra.

This definition of identity of algebras shows that in the theory of algebras we only study those properties of the quantities and systems of quantities of the algebra that find their expression in the form of certain properties of the three basic operations. To put it briefly, the theory of algebra studies properties of the operations performed on the quantities of the algebra and has nothing to do with the nature of the quantities that form the algebra.

It is easy to show that if two algebras are isomorphic, then quantities that form a basis of one algebra correspond to quantities that form a basis of the other, and that the structure constants computed with respect to corresponding bases are equal. Conversely, if two algebras over one and the same field have equal structure constants in suitable bases, then such algebras are isomorphic.

Among all the algebras, the associative algebras have always played and are still playing a very important part, i.e., the algebras in which the operation of multiplication satisfies the associative law $\alpha(\beta\gamma) = (\alpha\beta)\gamma$. The present section will give an account of the properties of such algebras. Among the nonassociative algebras the most interesting are the Lie algebras, for which the following properties of multiplication are assumed to be satisfied:

$$\alpha\beta = -\beta\alpha, \quad \alpha(\beta\gamma) + \beta(\gamma\alpha) + \gamma(\alpha\beta) = 0.$$

They are of interest in view of the close connection that exists between Lie algebras and Lie groups, which were discussed in §7.

The algebra of matrices. We have pointed out earlier that in the first period of the development of the theory of hypercomplex systems the main attention was centered on the investigation of various systems which for one reason or another were of particular interest to the investigators. We have already examined some of these systems. The investigation of the algebra of matrices which plays a fundamental role in the general theory of algebras began approximately at the middle of the last century. Let us briefly recall here the definitions of the operations on matrices (see Chapter XVI, §1).

A matrix over a field P is a collection of elements of the field arranged in the form of a rectangular table. Two matrices are called equal if their elements in corresponding places are equal. Here we shall only consider square matrices for which the number of rows is equal to the number of columns. The number of rows or columns of a square matrix is called its order.

To add two matrices of equal order we add their corresponding elements. Multiplication of a matrix by a number is, by definition, multiplication of all the elements of the matrix by that number. The operation of multiplication of a matrix by a matrix is defined in a more complicated fashion: The product of two matrices of order n is the matrix of the same order in

§12. ASSOCIATIVE ALGEBRAS

which the element in the ith row and the jth column is equal to the sum of the products of the elements of the ith row of the first matrix into the corresponding elements of the jth column of the second. For example:

$$\begin{bmatrix} a & b \\ a_1 & b_1 \end{bmatrix} \begin{bmatrix} x & y \\ x_1 & y_1 \end{bmatrix} = \begin{bmatrix} ax + bx_1 & ay + by_1 \\ a_1 x + b_1 x_1 & a_1 y + b_1 y_1 \end{bmatrix}.$$

The motives for the choice of this definition of multiplication of matrices were explained in Chapter XVI.

In virtue of the definitions given, matrices of order n with elements from an arbitrary field P form a system of quantities which can be added, multiplied by elements of P, and multiplied among each other. Straightforward computations show that the properties 1' through 10' which define an algebra are satisfied. Furthermore, it is easy to show that the multiplication of matrices satisfies the associative law. Therefore the system of all matrices of a given order n with elements from a given field P form an associative algebra over this field.

The obvious equation

$$\begin{bmatrix} a & b \\ c & d \end{bmatrix} = a \begin{bmatrix} 1 & 0 \\ 0 & 0 \end{bmatrix} + b \begin{bmatrix} 0 & 1 \\ 0 & 0 \end{bmatrix} + c \begin{bmatrix} 0 & 0 \\ 1 & 0 \end{bmatrix} + d \begin{bmatrix} 0 & 0 \\ 0 & 1 \end{bmatrix}$$

shows that the four matrices on the right-hand side form a basis of the algebra of matrices of order 2. More generally, when we denote by ϵ_{ij} the matrix in which there is a 1 in the ith row and jth column and the remaining places are zeros, then we have the equation

$$\begin{bmatrix} a_{11} & \cdots & a_{1n} \\ \vdots & & \vdots \\ a_{n1} & \cdots & a_{nn} \end{bmatrix} = \sum_{i,j} a_{ij} \epsilon_{ij},$$

which shows that the matrices ϵ_{ij} form a basis of the algebra of matrices of order n. Since the number of matrices ϵ_{ij} is equal to n^2, the rank of the algebra of matrices is also equal to n^2. The multiplication table for the basis elements ϵ_{ij} has the form

$$\epsilon_{ij} \cdot \epsilon_{jl} = \epsilon_{il}, \qquad \epsilon_{ij} \cdot \epsilon_{kl} = 0, \qquad j \neq k, \qquad i,j,k,l = 1, 2, \cdots, n.$$

The algebra of matrices contains a unit element, namely the unit matrix.

Representations of associative algebras. Suppose that with every quantity of a certain algebra A over a field P a definite quantity of some other algebra B over the same field P is associated. If the sum and product of any two elements of A are associated with the sum and product of the corresponding elements of B and the product of every element of P by an arbitrary element of A with the product of the same element of P and the corresponding element of B, then we say that the algebra A is homomorphically mapped into the algebra B. A homomorphic mapping of an associative algebra into the algebra of matrices of order n is called a representation of A of degree n. If distinct elements of A correspond to distinct matrices the representation is called faithful or isomorphic. When an algebra A is isomorphically represented by matrices, we may assume that the operations on the quantities of the algebra reduce to the operations on the corresponding matrices. Therefore the task of finding representations of algebras is of considerable interest. Here we shall only consider some of the simplest methods of finding representations; however, these methods play an important role in the general theory.

Let us choose an arbitrary basis $\epsilon_1, \epsilon_2, \cdots, \epsilon_n$ in the given associative algebra A, and let α be an arbitrary quantity of A. The products $\epsilon_1\alpha$, $\epsilon_2\alpha, \cdots, \epsilon_n\alpha$ are again quantities of A and therefore must be expressible linearly in terms of $\epsilon_1, \epsilon_2, \cdots, \epsilon_n$. Suppose that

$$\epsilon_1\alpha = a_{11}\epsilon_1 + a_{12}\epsilon_2 + \cdots + a_{1n}\epsilon_n,$$
$$\epsilon_2\alpha = a_{21}\epsilon_1 + a_{22}\epsilon_2 + \cdots + a_{2n}\epsilon_n,$$
$$\cdots\cdots\cdots\cdots\cdots\cdots\cdots\cdots\cdots\cdots\cdots\cdots$$
$$\epsilon_n\alpha = a_{n1}\epsilon_1 + a_{n2}\epsilon_2 + \cdots + a_{nn}\epsilon_n.$$

As we can see, for a fixed basis we can associate with every element α a definite matrix $|a_{ij}|$. A very simple calculation shows that this correspondence is a representation of A. This representation is often called the regular representation of A. Its degree is obviously equal to the rank of the algebra.

The complex numbers can be regarded as an algebra of rank 2 over the field of real numbers with the basis 1, i. The equations

$$1 \cdot (a + bi) = a \cdot 1 + b \cdot i,$$
$$i \cdot (a + bi) = -b \cdot 1 + a \cdot i$$

show that in the corresponding regular representation the complex number $a + bi$ corresponds to the matrix

$$\begin{bmatrix} a & b \\ -b & a \end{bmatrix}.$$

§12. ASSOCIATIVE ALGEBRAS

The analogous representation of quaternions has the form

$$a + bi + cj + dk \to \begin{bmatrix} a & b & c & d \\ -b & a & -d & c \\ -c & d & a & -b \\ -d & -c & b & a \end{bmatrix}.$$

These representations of the complex numbers and the quaternions are faithful (i.e., isomorphic to the algebra). Examples show, however, that the regular representation is not always faithful. But if the algebra contains a unit element, then its regular representation is necessarily faithful.

It is easy to show that every associative algebra can be embedded in an algebra with a unit element. The regular representation of the containing algebra is faithful; therefore this representation of the given algebra is also faithful. Thus, every associative algebra has a faithful representation by matrices.

This method of finding representations is insufficient for constructing all the representations of an algebra. A more refined method is connected with the concept of an ideal of an algebra which plays an important role in modern mathematics.

A system I of elements of an algebra is called a right *ideal* if it is a linear subspace of the algebra and if the product of every element of I with an arbitrary element of the algebra again belongs to I. A left ideal is defined similarly (with an interchange of the order of the factors). An ideal that is simultaneously left and right is called two-sided. It is clear that the zero element of an algebra by itself forms a two-sided ideal, the so-called zero ideal of the algebra. Also the whole algebra can be regarded as a two-sided ideal. However, apart from these two trivial ideals, the algebra may contain other ideals, the existence of which is usually connected with interesting properties of the algebra.

Suppose that an associative algebra A contains a right ideal I. Let us choose a basis $\epsilon_1, \epsilon_2, \cdots, \epsilon_m$ in this ideal. Since in the general case I forms only part of A, the basis of I will, as a rule, have fewer elements than a basis of A. Let α be an arbitrary element of A. Since I is a right ideal and $\epsilon_1, \epsilon_2, \cdots, \epsilon_m$ are contained in I, the products $\epsilon_1 \alpha, \ldots, \epsilon_m \alpha$ are also contained in I and hence can be expressed linearly in terms of the basis $\epsilon_1, \cdots, \epsilon_m$; i.e.,

$$\epsilon_1 \alpha = a_{11} \epsilon_1 + \cdots + a_{1m} \epsilon_m,$$
$$\cdots\cdots\cdots\cdots\cdots\cdots\cdots\cdots\cdots$$
$$\epsilon_m \alpha = a_{m1} \epsilon_1 + \cdots + a_{mm} \epsilon_m.$$

By associating with the element α the matrix $\| a_{ij} \|$ we obtain, as before, a representation of the algebra A. The degree of this representation is equal to the number of elements of the basis of the ideal I and is, therefore, in general smaller than the degree of the regular representation. Obviously, the degree of the representation obtained by means of an ideal will be smallest if the ideal is minimal. Hence one can understand the fundamental role of minimal ideals in the theory of algebras.

The structure of algebras. By what has been said, every associative algebra A can be isomorphically represented by matrices of a certain order. The aggregate of matrices that correspond in this representation to the quantities of A is itself an algebra, but only part of the algebra of all matrices of the given order. If a certain part of the quantities of an algebra is itself an algebra, then it is called a subalgebra of the given algebra. We can therefore say that every associative algebra is isomorphic to a certain subalgebra of matrices.

Although this result is of interest in principle, since it reduces the problem of finding all algebras to that of finding all possible subalgebras of matrix algebras, it does not give a direct answer to the question of the structure of algebras. The first general answer to this question was given at the end of the last century in the works of F. E. Molin (1861–1941), Professor at the University of Dorpat (Tartu), who taught at the Polytechnic Institute at Tomsk around 1900.

An algebra is called *simple* if it does not contain any two-sided ideals other than the zero ideal and the whole algebra. Molin proved that every simple associative algebra of rank 2 or more over the field of complex numbers is isomorphic to the algebra of all the matrices of a suitable order over this field.

Continuing Molin's fundamental investigations, Wedderburn obtained at the beginning of the 20th century a number of results which give a very complete description of the structure of algebras over an arbitrary field.

An arbitrary system of elements of an algebra A (in particular the algebra A itself or an ideal or a subalgebra of it) is called *nilpotent* if there exists a natural number s such that the s-th power of any element of the system is equal to zero. Every associative algebra has the unique maximal two-sided nilpotent ideal which is called the radical of the algebra. An algebra whose radical is equal to zero is called *semisimple*. It can be shown that every semisimple algebra splits into a sum of a special kind of simple algebras; in virtue of this, the study of semisimple algebras reduces entirely to that of simple ones. Finally, an algebra A is called a *divison algebra* if every equation of the form $ax = b$ $(a \neq 0)$ has a solution in A.

The structure of simple algebras over the field of complex numbers is completely described by the theorem of Molin mentioned earlier. But if the ground field P is arbitrary, then the following more general theorem of Wedderburn holds: Every simple algebra of rank 2 or more over a field P is isomorphic to the algebra of all matrices of suitable order with elements from a certain division algebra over the same field P. Thus, Wedderburn's theorem reduces the problem of finding simple algebras over a given field P to that of finding division algebras over P. Over the field of complex numbers there exists only one division algebra, the field of complex numbers itself. Hence it follows by Wedderburn's theorem that all simple algebras over the field of complex numbers are isomorphic to an algebra of matrices over the same field, i.e., Molin's theorem.

Over the field of real numbers there exist only three associative division algebras: the field of real numbers itself, the field of complex numbers, and the algebra of quaternions. The proof of this statement is not very easy, and we shall not dwell on it here. By Wedderburn's theorem this implies that every simple algebra over the field of real numbers is isomorphic to the algebra of matrices of a suitable order either over the field of real numbers or over the field of complex numbers or over the quaternion algebra.

From these examples it is clear how the structure of semisimple algebras is described by the theorems of Molin and Wedderburn. In regard to algebras with a radical, for them the so-called fundamental theorem of Wedderburn is of great importance; according to this theorem, under certain restrictions to be imposed on the ground field, every algebra A with a radical R has a semisimple subalgebra L such that every element of the given algebra A can be uniquely represented in the form of a sum $\lambda + \rho$ ($\lambda \in L$, $\rho \in R$), where the subalgebra L is in a certain sense uniquely determined within A.

The fundamental theorems just formulated give an orderly idea of the possible types of associative algebras and reduce the question of their structure essentially to the analogous problem of the structure of nilpotent algebras. The theory of the latter is at present still in the process of development.

§13. Lie Algebras

In §12 we said that in addition to the theory of associative algebras at the present time the theory of Lie algebras has been worked out in great detail; for these, multiplication is subject to the rules

$$\alpha\beta = -\beta\alpha, \qquad \alpha(\beta\gamma) + \beta(\gamma\alpha) + \gamma(\alpha\beta) = 0.$$

The importance of these algebras can be explained by the fact that they are closely connected with Lie groups (see §7), i.e., with the most important class of continuous groups. As we have seen above, Lie groups play a remarkable role in contemporary geometry. Because of the origin of the theory of Lie groups and Lie algebras, the greatest interest lies in Lie algebras over the field of all real and of all complex numbers.

One of the simplest examples of a Lie algebra is the following. Let us consider the set of all square matrices of a given order n. We introduce an operation of commutation on them; by this we understand the formation of the so-called commutator $AB - BA$ of given matrices A and B, denoted by $[A, B]$.

It is easy to verify that

$$[A, B] = -[B, A],$$
$$[A, [B, C]] + [B, [C, A]] + [C, [A, B]] = 0.$$

Consequently, the set of all square matrices of a given order forms a Lie algebra with respect to the operation of commutation. It is clear that every subalgebra of the Lie algebra formed by matrices, i.e., every set of matrices that is closed with respect to the operations of addition, multiplication by a number of the ground field, and commutation, is in its turn a Lie algebra.

The question whether for every abstractly given Lie algebra there exists a matrix algebra isomorphic to it remained open for a long time. It was solved in the affirmative only in 1935 by *I. D. Ado*, a pupil of the famous algebraist H. G. Čebotarev.

Now let us sketch in general terms, without going into details and without giving rigorous statements, the connections between Lie groups and Lie algebras, restricting ourselves to the case when the Lie group and the Lie algebra are represented by matrices.

Let L be a certain Lie algebra of matrices. With every matrix A belonging to L we associate the matrix $U = e^A = E + A/1! + A^2/2! + \cdots$. Then the collection of all matrices obtained in this way forms a Lie group under the ordinary matrix multiplication. Conversely, for every Lie group we can find a unique Lie algebra (to within isomorphism) such that the group corresponding to it is isomorphic to the given one. For simplicity we have given not an accurate but a simplified formulation of the theorem on the connection between Lie groups and Lie algebras. Actually the relation $U = e^A$ exists only for U sufficiently close to the unit matrix and for A sufficiently close to the null matrix. A rigorous formulation would require the introduction of the rather complicated concepts of a local group and a local isomorphism.

Thus, the transition from the Lie algebra to the corresponding group proceeds by an operation similar to exponentiation and the inverse

§13. LIE ALGEBRAS

transition, from the group to the algebra, by an operation similar to taking logarithms.

If L coincides with the algebra of all matrices of order n, then the corresponding Lie group is the group of all nonsingular matrices, because every matrix U close to the unit matrix can be represented in the form $U = e^A$.

A matrix $A = \| a_{ij} \|$ is called skew-symmetric if its elements satisfy the relation $a_{ji} = -a_{ij}$. Skew-symmetric matrices form a Lie algebra, because if A and B are skew-symmetric, then the matrices $AB - BA = [A, B]$ and $\alpha A + \beta B$ are also skew-symmetric.

It is easy to verify that for every skew-symmetric matrix A the expression e^A is an orthogonal matrix and that every orthogonal matrix which is close to the unit matrix can be represented in this exponential form. Therefore the Lie algebra of the group of orthogonal matrices is the algebra of skew-symmetric matrices.

From analytical geometry it is known that every rotation of space around the coordinate origin is given by an orthogonal matrix and that the product of rotations corresponds to the product of the corresponding matrices. In other words, the group of rotations of space around a certain fixed point is isomorphic to the group of orthogonal matrices of order 3. Hence we deduce that the Lie algebra for the group of rotations of space is the algebra of all skew-symmetric matrices of order 3, i.e., the Lie algebra of matrices of the form

$$A = \begin{bmatrix} 0 & -a & -b \\ a & 0 & -c \\ b & c & 0 \end{bmatrix}.$$

Since each of these matrices is completely characterized by the three numbers a, b, c, it can be represented by the vector $\boldsymbol{a}$ having the projections a, b, c on the coordinate axes. Here a linear combination $\alpha A_1 + \beta A_2$ of matrices A_1 and A_2 of the given form obviously is associated with the linear combination of the corresponding vectors $\alpha \boldsymbol{a}_1 + \beta \boldsymbol{a}_2$, and the commutator of the matrices

$$[A_1, A_2] = A_1 A_2 - A_2 A_1$$

$$= \begin{bmatrix} 0 & -a_1 & -b_1 \\ a_1 & 0 & -c_1 \\ b_1 & c_1 & 0 \end{bmatrix} \begin{bmatrix} 0 & -a_2 & -b_2 \\ a_2 & 0 & -c_2 \\ b_2 & c_2 & 0 \end{bmatrix} - \begin{bmatrix} 0 & -a_2 & -b_2 \\ a_2 & 0 & -c_2 \\ b_2 & c_2 & 0 \end{bmatrix} \begin{bmatrix} 0 & -a_1 & -b_1 \\ a_1 & 0 & -c_1 \\ b_1 & c_1 & 0 \end{bmatrix}$$

$$= \begin{bmatrix} 0 & b_2 c_1 - b_1 c_2 & a_1 c_2 - a_2 c_1 \\ b_1 c_2 - b_2 c_1 & 0 & a_2 b_1 - a_1 b_2 \\ a_2 c_1 - a_1 c_2 & a_1 b_2 - a_2 b_1 & 0 \end{bmatrix}$$

is associated with the vector whose components are $b_1 c_2 - b_2 c_1$,

$a_2c_1 - a_1c_2$, $a_1b_2 - a_2b_1$, i.e., with the vector products of a_1 and a_2. So we have arrived at the remarkable result that the set of ordinary vectors under the operations of addition, multiplication by a scalar, and vector multiplication forms a Lie algebra which corresponds to the group of rotations of space around a fixed point. This shows at once how closely geometric concepts are connected with the group of rotations of space, in other words with the laws of motion of rigid bodies.

At the end of the last and the beginning of the present century, a number of results were obtained for Lie algebras that are similar to the fundamental results on associative algebras, although the proofs and statements are here more complicated. Thus, as a result of the efforts of Lie, Killing, and Cartan the concepts of a radical and of semisimplicity of a Lie algebra were successfully established at the beginning of the 20th century and all simple Lie algebras over the fields of real and complex numbers were found. In the early 1930's the theory of representations of Lie algebras by matrices was constructed, in principle, by Cartan and Weyl and proved to be a remarkable instrument for the solution of many problems. In the last 15 years the development of the theory of Lie algebras has occupied a number of Soviet mathematicians, who have obtained in this domain some significant results. In particular, they made important progress in the theory of representations of Lie algebras and gave definitive solutions to the problems of semisimple subalgebras of Lie algebras, of the structure of algebras with a given radical, and so forth.

§14. Rings

In §11 we have given the general definition of a field as an arbitrary set of elements on which the operations of addition and multiplication satisfying the postulates 1 through 10 are defined. By omitting in this definition the postulate 10, on the existence of a quotient, and the postulates 7 and 8, on commutativity and associativity of multiplication, we obtain a definition of the concept of a ring, one of the most important concepts of contemporary algebra.

Every field and also every algebra considered only with respect to the operations of addition and multiplication is a ring. An even simpler example of a ring is the set of all rational integers with the usual operations of addition and multiplication. Under the same operations the sets of numbers of the form $a + bi$, $a + b\sqrt{2}$, $a + b\sqrt[3]{2} + c\sqrt[3]{4}$ and so forth also form rings, where a, b, c are rational integers. The elements of these rings are numbers and the rings are therefore called number rings. Some important properties and applications of these rings were discussed in Chapters IV and X.

§14. RINGS

However, there exist important classes of rings whose elements are not numbers. For example, under the usual operations of addition and multiplication the sets of polynomials in given variables $x_1, x_2, \cdots, x_n$ with coefficients from any fixed ring or field form rings, also the set of all continuous functions defined on a certain domain, or the set of linear transformations of a linear space or a Hilbert space.

The arithmetic properties of number rings form the subject matter of the profound theory of algebraic numbers, which lies halfway between algebra proper and number theory proper. The investigation of properties of rings of polynomials is the object of the so-called theory of polynomial ideals, which is closely connected with the higher branches of analytical geometry. Finally, rings of functions and transformations play a fundamental role in functional analysis (see Chapter XIX).

On the basis of these and some other concrete theories, the general theory of rings and the theory of topological rings were rapidly developed in the present century.

For reasons of space we shall now give only some individual results relating to the rudiments of the theory of rings.

Ideals. A subset I of elements of a ring K (not necessarily associative) is called an *ideal* if the difference of any two elements of I is again contained in I and if the products ax, xa of an arbitrary element a of I and an arbitrary element x of K are contained in I.

Every ideal of a ring is itself a ring under the operations of addition and multiplication defined in the ring. Such parts are called subrings of the given ring, so that every ideal is at the same time a subring. The converse is not true, as a rule.

The intersection of an arbitrary system of ideals of a ring is again an ideal, in particular the intersection of all the ideals containing an arbitrary fixed element a of the ring is an ideal. This is called the principal ideal generated by the element a and is denoted by (a).

The concept of the ideal generated by two or several elements is defined in the same way. It is easy to show that if an associative commutative ring has a unit element, then the ideal generated by the elements $a_1, \cdots, a_n$ is simply the collection of all elements of the ring that admit a representation in the form of a sum $x_1 a_1 + \cdots + x_n a_n$, where $x_1, \cdots, x_n$ are arbitrary elements of the ring. In particular, the principal ideal (a) in a commutative associative ring with a unit element is simply the collection of all elements that are multiples of a, i.e., have the form xa.

In the ring of all rational integers every ideal is principal. The ring of polynomials in a single variable with coefficients from an arbitrary field has the same property, and so has the ring of complex numbers of the form $a + bi$, where a and b are rational integers, and also a number of other rings. However, the set of all polynomials in two variables x, y without a free term is an ideal, but not a principal ideal, in the ring of all polynomials in x and y with rational coefficients.

Just as we have done above for a normal subgroup in the theory of groups, so we can construct for every ideal I of a ring K a residue class ring (or factor ring) K/I. This is done as follows. Two elements a, b, of K are called congruent modulo the ideal I, in symbols $a \equiv b(I)$, if their difference $a - b$ is contained in I. It is easy to verify that the congruence relation is symmetric, reflexive, and transitive (see Chapter XV), so that all the elements of K are split into classes of congruent ones (modulo I). If we now consider these classes as elements of a new set, we can introduce for them the concept of a sum and of a product: The "sum" of two classes shall be that class which contains the sum of any two elements that occur in the given classes respectively, and the product is that class which contains the product of these representatives. From the definition of ideals it follows that the sum and product defined in this do not in fact depend on the choice of the representatives and that as a result the set of classes becomes a ring.

The role of the residue class ring in the theory of rings is entirely analogous to the role of the factor group in the theory of groups. In particular, the construction of residue class rings of known rings is a convenient method of forming new rings with various properties. Furthermore, it is easy to show, for example, that an arbitrary commutative ring K is isomorphic to the factor ring of a ring of polynomials with integer rational coefficients in a sufficiently large number of variables.

Arithmetic properties of rings. In number rings and in fields the product of several elements can only be equal to zero if at least one of the factors is equal to zero. In arbitrary rings this need not be true, for example the product of two nonnull matrices can be the null matrix. If in a certain ring $ab = 0$ and $a \neq 0$, $b \neq 0$, then a and b are called *divisors of zero*. If there are no such elements in a ring, then the ring is called a *ring without divisors of zero*.

For the investigation of the laws of divisibility in rings, we usually assume that the ring is commutative and has no divisors of zero. Such rings are often called integral domains. The number rings and polynomial rings mentioned previously are integral domains.

Let K be an integral domain. We say that an element a is divisible in K by the element b if $a = bq$, $q \in K$. From this it follows immediately that a sum of elements divisible by b is divisible by b and that the product of several elements of K is necessarily divisible by b if one of the factors

§14. RINGS

is divisible by b. When we try to introduce in the theory of rings the concept of a prime element similar to that of a prime number, we come across a complication that was already mentioned in Chapter X. Namely, to begin with we have to introduce the concept of associate elements of a ring, calling elements a, b associate if a is divisible by b and b divisible by a. Setting $a = bq_1$, $b = aq_2$, we have $ab = ab \cdot q_1 q_2$; i.e., $q_1 q_2 = e$, where e is the unit element of K. The quotients of associate elements are therefore called divisors of one or units. Every element of the domain is divisible by every unit. In the ring of rational integers the units are ± 1, in the ring of numbers of the form $a + bi$, where a, b are integers, the units are the numbers $\pm 1, \pm i$.

Every element of an integral domain K has decompositions of the form $a = a\epsilon \cdot \epsilon^{-1}$, where ϵ is an arbitrary unit. These decompositions are called trivial. If a has no other decompositions, then a is called a prime or indecomposable element of K. In connection with the very important theorem on the unique decomposition of integers into prime factors, it is of interest to find such classes of rings, and among them noncommutative ones, in which a similar theorem holds. For example this theorem holds in principal ideal rings, i.e., in integral domains in which all ideals are principal.

The very concept of ideal arose in connection with the problem of uniqueness of decomposition into prime factors. Approximately at the middle of the last century the German mathematician Kummer, trying to prove the famous proposition of Fermat that the equation $x^n + y^n = z^n$ has no integer solutions for $n \geqslant 3$, had the idea of considering numbers of the form $a_0 + a_1 \zeta + \cdots + a_n \zeta^{n-1}$, where $\zeta = \cos 2\pi/n + i \sin 2\pi/n$ is a solution of the equation $x^n = 1$ and $a_0, \cdots, a_n$ are ordinary integers. The numbers of this type form an integral domain and Kummer at first took it for granted as an obvious proposition that the theorem of unique decomposition into prime factors holds in this domain. On this basis he constructed a proof of Fermat's theorem. However, in checking his arguments he observed that this assumption of the uniqueness of decomposition is not true. Wishing to preserve the uniqueness of decomposition into prime factors, Kummer was compelled to consider decompositions of numbers of the domain into factors that do not occur in the domain itself. These numbers he called ideal. Subsequently, in the construction of the general theory, mathematicians introduced instead of the ideal numbers the sets of elements of the domain that are divisible by one ideal number or another, and they were called ideals. The discovery of the nonuniqueness of the decomposition into prime factors in number rings is one of the most interesting facts found in the last century and has led to the creation of the extensive theory of algebraic numbers.

One of the most striking applications of this theorem to the problem of the decomposition of ordinary integers into a sum of squares was mentioned at the end of Chapter X. The work of mathematicians of an older generation, E. I. Zolotarev, G. F. Voronoĭ, I. M. Vinogradov, and N. G. Čebotarev, has played a significant role in the development of the theory of number rings.

Algebraic varieties. Another source of the theory of ideals lies in algebraic geometry. When one first becomes acquainted with the theory of curves of the second order, one usually learns with astonishment that the single name hyperbola is given to the collection of two disconnected curves, namely the branches of the hyperbola, and also that a pair of straight lines is called a degenerate curve of the second order. This point of terminology is clarified in algebra: If equations of curves are considered in the form $f(x, y) = 0$, where $f(x, y)$ is a polynomial in x, y, then in the first case the left-hand side of this equation is an irreducible polynomial of the second degree, and in the second case a product of two factors of the first degree. A curve whose equation can be represented by means of an irreducible polynomial $f(x, y)$ is called irreducible, and otherwise reducible.

On transition to curves in space the matter becomes more complicated. A space curve can be represented by a system of two equations $f(x, y, z) = 0$, $g(x, y, z) = 0$, where the polynomials f and g are by no means uniquely determined by the curve. What shall we call here an irreducible curve?

The natural answer is given by the theory of ideals. Let $f_1, f_2, \cdots$ be an arbitrary set of polynomials in the variables x, y, z with complex coefficients. The set of points in the (complex) space whose coordinates make all these polynomials vanish is called the algebraic variety defined by the given polynomials. We denote this variety by M and consider all the polynomials in the variables x, y, z that vanish at every point of M. It is easy to see that the set I of all such polynomials is an ideal in the ring of polynomials in x, y, z. Moreover, this ideal has the property that if a power of some polynomial is contained in I, then the polynomial itself is contained in I. Now it turns out that, whereas distinct sets of polynomials may define one and the same algebraic variety, the correspondence between varieties and ideals with the aforementioned additional property is one-to-one.

Thus, in studying properties of varieties, it is natural to discuss not their more or less accidental "equations," but the corresponding ideals. If an ideal I can be represented in the form of the intersection of any two ideals I_1, I_2, then the variety M is the union of the varieties M_1, M_2, corresponding to the ideals I_1, I_2. Hence it is clear that a variety M must naturally be called irreducible when the corresponding ideal I cannot be represented in the form of an intersection of any two containing ideals. To the splitting of a curve into curves of lower orders, to the decomposition of a variety into irreducible ones, there now corresponds the representation of a corresponding ideal in the form of an intersection of indecomposable ones. The problem of uniqueness and possibility of such decompositions is one of the first in the theory of algebraic varieties and the general theory of ideals.

The structure of noncommutative rings. Every algebra is at the same time a ring with respect to the operations of addition and multiplication. Therefore, a considerable number of fundamental concepts and results in the theory of algebras remains valid for arbitrary rings. However, the transfer of more subtle results in the theory of algebras similar, in particular, to the theorems of Molin and Wedderburn (see §11) comes up against great difficulties which have only been partially overcome in the last 10 or 15 years. First of all there is the matter of finding such a definition of the radical of a ring that rings with a zero radical have some resemblance to semisimple algebras and that for all algebras results of the structure theory of algebras should be obtained as special cases from theorems in the theory of rings. There is at present in the theory of rings a number of definitions of a radical that enable us under some restriction or another to construct a satisfactory theory of the structure of semisimple rings. As we have mentioned earlier, the interest in the theory of noncommutative rings is stimulated to a certain extent by the very appreciable value of the theory of rings of operators in functional analysis.

§15. Lattices

As the reader is aware, a set of objects is called partially ordered if for certain pairs of its elements it can be determined which of these objects precedes the other or is subordinate to the other; here it is assumed that: (1) every object is subordinate to itself; (2) if a is subordinate to b and b subordinate to a, then it follows that a and b are identical; (3) if a is subordinate to b and b to c, then it follows that a is subordinate to c. The relation of subordination is usually denoted by the symbol $\leqslant$.

An important example of a partially ordered set is a system of all subsets of an arbitrary set where the relation of subordination means that one subset is part of another.

If the relation of subordination is defined for every pair of elements of a partially ordered set, then the set is called totally (or linearly) ordered. Ordered sets are, for example, the real numbers, where the relation $a \leqslant b$

means that a is not greater than b. By way of contrast, the partially ordered set of all parts of an arbitrary collection containing more than one element is not totally ordered, since subsets without common elements are not comparable with one another.

Suppose that the elements of a partially ordered set M have the property that every pair a, b have a unique nearest common larger element c, i.e., such that $a \leqslant c, b \leqslant c$ and that for every d of M satisfying the conditions $a \leqslant d, b \leqslant d$ we have $c \leqslant d$. Then M is called an *upper semilattice* and the element c is the "sum" of a and b. It is easy to verify that this "addition" has the following properties:

$$a + b = b + a, \quad (a + b) + c = a + (b + c), \quad a + a = a. \quad (13)$$

It is very remarkable that the converse can also be stated. If in a certain set an operation of addition is defined having the properties (13), then, by calling an element a subordinate to an element b if $a + b = b$, we obtain a partially ordered set in which $a + b$ is the unique nearest common larger element for a and b.

Similarly we can define lower semilattices by considering in place of the nearest larger elements the nearest smaller ones, which are here called "products"of the given elements. This operation has the same property as "addition," namely

$$ab = ba, \quad (ab)c = a(bc), \quad aa = a. \quad (14)$$

A partially ordered set which is at the same time an upper and a lower semilattice is called a *lattice*. By what has been explained, in every lattice we can define two operations subject to the conditions (13) and (14). However, these operations are connected with one another, since the relation $a \leqslant b$ in a lattice can be written in either of the forms $a + b = b$, $ab = a$. In other words, in lattices the equation $a + b = b$ and $ab = b$ must be equivalent. It turns out that the latter conditions can be written algebraically in the form of equations

$$a + ab = a, \quad a(a + b) = a, \quad (15)$$

and, by virtue of this, the study of lattices becomes a purely algebraic task of studying systems with two operations subject to the conditions (13), (14), and (15). The significance of the algebraic approach to the study of lattices consists, roughly speaking, in the fact that the peculiarities of one concrete lattice or another in individual cases can be conveniently expressed in the form of algebraic relationships of one kind or another between the elements; also we can take advantage of the rich apparatus of the classical theory of groups and rings.

§16. OTHER ALGEBRAIC SYSTEMS

As we have already mentioned, the set of all subsets of a set is a partially ordered set. It is not difficult to see that it is a lattice and that the lattice sum is here the union and the lattice product is the intersection of the corresponding subsets. If we consider not all but only some of the subsets, then we can obtain a variety of lattices. For example, lattices are the set of all subgroups and also the set of all normal subgroups of an arbitrary group, the set of all subrings and the set of all ideals of an arbitrary ring and so forth. In particular, in the lattices of all normal subgroups of a group and of all ideals of a ring apart from the fundamental identities (13), (14), and (15), the following so-called modular law holds also:

$$a(ab + c) = ab + ac.$$

The theory of lattices with the modular law (modular or Dedekind lattices) is an important chapter in the general theory of lattices.

A considerable number of theorems in the theory of groups and in the theory of rings are statements on the arrangement of subgroups, normal subgroups and ideals; consequently, these theorems can be reformulated as theorems on the lattices of subgroups or ideals. With some restrictions similar theorems hold for general lattices. In this way certain important theorems were transferred from the theory of groups, the theory of rings, and other disciplines to the theory of lattices. On the other hand, the application of the apparatus of the theory of lattices proved useful in finding properties of concrete lattices, for example in the theory of groups and the theory of rings.

The theory of lattices has grown up rather recently, in the twenties and thirties of our century, and has not yet found such important applications as, say, the theory of groups. However, at the present time the theory of lattices is a well-formed mathematical discipline with a rich content and a substantial range of problems.

§16. Other Algebraic Systems

In the preceding sections we have made an attempt to give an idea how the application of algebraic methods to an ever-expanding range of problems has led to an extension of the system of objects that are studied in algebra and to a generalization of the concept of algebraic operations. In this context an important part was played by the development of the axiomatic method which arose in the work of I. N. Lobačevskiĭ on the foundations of geometry and also the development of the general theory of sets.

One of the fundamental results was the gradual clarification of the general concepts of an algebraic operation, of an algebraic system, and the accumulation of the most important facts referring to the definition of algebraic systems. Instead of the concretely defined operations of school algebra, which concern mostly numbers, modern algebra starts from the general concept of an operation. Namely, suppose that a certain system of elements S is given and also a rule that associates with every system $a_1, a_2, \cdots, a_m$ of m elements of S, taken in a definite order, a well-defined element a of the same system. Then we say that on S an m-ary operation is given and that the element a is the result of this operation performed on the elements $a_1, a_2, \cdots, a_m$. A set of elements, together with one or several operations defined on it is called an algebraic system. One of the basic tasks of algebra is the study and classification of algebraic systems. However, in this form the problem has too general a character. In fact, only certain special algebraic systems have proved at present to be really important and capable of interesting theories. For example, of the systems with a single operation only the theory of groups, to which §§1 through 10 of this chapter were devoted, has grown to a deep mathematical science, and among the systems with two or more operations those of the greatest significance are fields, algebras, rings, and lattices. However, the number of algebraic systems that are actually considered for one reason or another increases continually. At the same time certain classical branches of algebra such as, for example, the study of homomorphisms, of free systems and free unions, of direct unions, and recently the study of radicals has been successfully extended to the general theory of algebraic systems. This enables us to speak of this theory as a new branch of algebra.

In discussing the character of algebra as a whole, it is often emphasized that the complete absence or the subordinate role of the concept of continuity is a distinguishing feature of it, so that algebra is regarded as a science with a preference for the discrete. This view undoubtedly reflects one of the important objective peculiarities of algebra. In the real world the discontinuous and the continuous are found in dialectic unity. But in order to know reality, it is sometimes necessary to dissect it into parts and to study these parts separately. Therefore the one-sided attention of algebra to discrete relationships must not be regarded as a deficiency.

From the example of the theory of groups it is clear that individual algebraic disciplines provide not only the tools for technical computation but also the language for the expression of deep laws of nature. However, apart from the direct practical value of a number of branches of algebra for physics, chemistry, crystallography, and other sciences, algebra occupies one of the most important places in mathematics itself. In the words of the well-known Soviet algebraist N. G. Čebotarev, algebra has been the cradle of many new ideas and concepts that arise in mathematics and has fertilized to a remarkable extent the development of branches of mathematics that serve as a direct basis for the physical and technological sciences.

Suggested Reading

W. Burnside, *Theory of groups of finite order*, Dover, New York, 1956.
M. Hall, Jr., *The theory of groups*, Macmillan, New York, 1959.
A. G. Kurosh, *The theory of groups*, 2 vols., Chelsea, New York, 1960.
D. Montgomery and L. Zippin, *Topological transformation groups*, Interscience, New York, 1955.
L. S. Pontryagin, *Topological groups*, Princeton University Press, Princeton, N. J., 1958.
H. Weyl, *Symmetry*, Princeton University Press, Princeton, N. J., 1952.

INDEX

This revised, enlarged index was prepared through the generous efforts of Stanley Gerr.

Abel, N.H., I/71, 263, 275, 276, 278; II/143; III/283, 293
Abelian group, III/283
Abel's theorem, I/275; II/143
Abscissa, I/213
Absolutely convergent series, I/170
Abstract algebra, I/264
Abstract geometry, III/178
Abstract group, III/320
Abstract space, III/133
Abstract topological space, III/133, 136
Action integral, II/133
Adams, I., I/4
Addition, III/12
Addition circuits, II/355, 356
Additive property, II/201
Address, II/338-339
Address code bus bars, II/352
Adherent point, III/221, 223
Adherent sets, III/160
Adjacent interval, III/23
Adjoining sets, III/222
Admissible functions, II/125
Admissible solutions, II/48, 50
Ado, I. D., III/340
Affine classification of curves, figures, surfaces, etc., I/233
Affine differential geometry, II/116
Affine geometry, III/131
Affine properties of figures, I/233
Affine space, III/56, 133
Affine transformation, I/231, III/131
Aggregate; *see* Set
Ahmes, I/4
Al-Biruni, I/4
Aleksandrov, A. D., I/250; II/114
Aleksandrov, P. S., III/25, 160, 200, 219, 221, 223, 224, 320
Alexander, J., III/220, 221
Algebra, I/41-42, 57, 261; III/263, 330

abstract, I/264
associative, III/330
axiomatic, I/264
basis of, III/332
division, III/338
fundamental theorem of, I/280-292; II/140
hypercomplex, systems, III/263, 330
ideal of, III/337
isomorphic, III/333
Lie, III/339
linear, II/37
matrix, III/334-335
nilpotent, III/338
rank of, III/332
semisimple, III/338, 342
simple, III/338
structure of, III/338
vector, I/214; III/327
Algebraic complement, III/65
Algebraic field, III/13
Algebraic geometry, I/258
Algebraic integer, II/203
Algebraic number, II/203
 of fixed points, III/220
 of singular points, III/216
Algebraic number theory, II/203
Algebraic operation, III/349-350
Algebraic solution, I/265
Algebraic system(s), III/263, 349-350
Algebraic variety, III/346
Algebraization, I/262
Algorithm, II/202
 Euclidean, II/202
Al-Kharizmi, I/39, 40, 41, 262
Almost everywhere convergent sequence, III/29-30
Alternating group, III/279
Altitude, I/213
Analogue machines, II/325

353

Analysis, I/45, 50, 58, 65
  functional, I/60, 258; III/223, 227
  indefinite integral in, I/137
Analytic continuation, II/187, 189
Analytic function(s), I/177; II/141
  uniqueness theorem for, II/187
Analytic geometry, I/45, 47, 183, 186
  infinite-dimensional, I/258
  in space, I/218-220
Analytic number theory, II/203
Analytic representation of functions, II/186
Analytic surface, II/101
Analyticity of complex function, II/185
AND element, II/358-361
Androhov, A. A., I/331, 355; III/214, 215
Angle, of parallelism, III/106-107, 108 111
  between vectors, III/56, 230
Apollonius of Perga, I/36, 37, 47
Approximate solutions, in calculus of variations, II/135-136
Approximating function, II/283
Approximation, best, II/267, 288
  best uniform, II/267, 282-283, 290
  by defect, I/302, 304
  of definite integrals, II/276
  by excess, I/302, 305
  formula, II/304
  of functions, II/265
  methods of, II/303
  by polynomials, II/268, 284
Arc length, II/63
Archimedes, I/15, 36, 37, 48, 65, 70, 131; III/16, 17
Archimedes' principle, III/16-17
Argument of complex numbers, I/282
Aristotle, I/65
Arithmetic, I/7
Arithmetic unit, II/334
Aryabhata, I/39
Associate elements, III/345
Associative algebra(s), III/330
Associativity, III/12-13
Asymptotic, convergence, III/108
Asymptotic law, II/208
"At a point," II/59
Automorphism, III/270
Automorphism group, III/295
Axiom(s), of addition, III/12
  of continuity, I/288; III/15, 124
  of distributivity, III/13
  of geometry, III/122
  of incidence, III/123
  of linear space, III/49
  of motion, III/123
  of multiplication, III/13
  of order, III/13
  of parallelism, III/124
  of probability theory, II/231
  triangle, III/222
Axiomatic algebra, I/264

Axis, of group, III/289
  of rotation, III/266
  of symmetry, III/265

Banach, S., I/60
Basic interval, II/219
Basis, of algebra, III/332
  of space, III/51-52
Bernoulli, D., I/51, 71, 307; II/246, 247, 262, 267, 268, 281, 287, 290, 291; III/239
Bernštein, S. N., II/264, 247, 262, 267, 268, 281, 287, 290, 291
Bernstein polynomial, II/290
Bertrand, J. L. F., II/207
Bessel, F. W., I/179
Betti, E., III/353
Betti number, III/211-212
Bézout, E., I/263
Bhaskara, I/39
Bicontinuous transformation, III/310
Bilateral symmetry, III/265
Binary system, II/347
Birkhoff, G. D., I/356; III/220
Blaschke, W., II/116
Bogoljubov, N. N., I/355, 356; II/257
Boltjanskiĭ, V. G., III/219/220
Bólyai, F., III/100, 104, 322
Bólyai, J., I/56; III/104, 105
Bolzano, B., I/58, 288, 289; III/3, 21
Bolzano-Weierstrass theorem, I/288; III/21
Bonnet, P. O., II/103
Bonnet's theorem; see Peterson's theorem
Borel, E., I/59; II/268
Boundary, of chain, III/210
  of oriented triangle, III/208, 209
  of set, III/160; see also Boundary point of set
Boundary conditions, II/17, 19
  homogeneous, II/21
Boundary point of set, III/20
Boundary value problem, II/15, 18, 128
Bounded infinite point set, III/21
Bounded set, III/19
  above, III/19
  below, III/19
Boyle, R., I/77, 242; II/14; III/142
Brachistochrome, II/119
Brahe, T., I/198
Brahmagupta, I/39
Branch point, II/152, 190
Briggs, H., I/42
Brouwer, L. E. J., III/218, 219
Brouwer's fixed point theorem, III/217, 218
Budan, I/297
Budan's theorem, I/297
Bunjakovskiĭ, V. Ja., III/57, 58, 235, 236

Calculus, differential, I/48, 131
  integral, I/48-49, 131
  matrix, III/91

INDEX

Calculus of variations, II/119
  admissible functions, II/125
  approximate solutions in, II/135–136
  variational problem in, II/130–131
Canonical box, III/83
Canonical form, of equation, I/208
  of matrix, III/86–87
Canonical Jordan matrix, III/84
Canonical quadratic form, III/88
Cantor, G., I/27, 55, 59; III/5, 15, 17–19, 21, 25, 219
Cantorian continuity, III/15
Cantor's perfect set, III/24
Čantor's principle of continuity, III/15
Čaplygin, S. A., I/160, 162
Cardan, G., I/263, 266, 267, 268, 270
Cardinal number of continuum, III/11
Cardinality of set, III/9
Card-punch, II/325
Cartan, E., III/177, 307, 342
Cauchy, A., I/51, 58, 71, 88, 89, 135; II/150, 153, 157, 162, 166, 171–173, 176–180; III/3, 16, 17, 57, 58, 235, 236
Cauchy-Bunjakovskiĭ inequality, III/57
Cauchy integral, I/177
Cauchy-Riemann equations, II/150
Cauchy's principle, III/16
Cayley square, III/300
Cauchy's theorem, I/176
Cavalieri, B., I/48
Cayley, A., III/138, 300
Čebotarev, H. G., I/279, 280; III/340, 346, 350
Čebyšev, P. L., I/59, 62, 71, 141; II/60, 203, 206, 207, 208, 211, 214, 223, 244–247, 275, 281–287, 291, 299; III/240
Čebyšev approximation, II/267, 282–283, 290
Čebyšev('s) inequality, I/245
Čebyšev polynomials, II/275, 285
Čech, E., II/116
Center, I/347
  of symmetry, III/266
Central limit theorem, II/247
Central symmetry, III/266
Certain event, II/229
Chain, III/209
Change of variable, I/139
Characteristic equation, of homogeneous linear equation, I/324, 325
  of matrix, III/82
Characteristic frequency; see Eigenfrequency
Characteristic function, II/30; III/32; see also Eigenfunctions
  of flow, II/157
Characteristic numbers, III/82
Characteristic oscillations, I/327; II/30; III/249
Characteristic polynomial, III/82

Characteristic quadratic equation, I/241
Characters, of commutative groups, III/220
  of groups, III/314
Circle of convergence, II/145
Circular model of projective plane, III/201
Circulation of flow, II/160
Circulatory motion, II/160
Clairaut, A. C., I/213; II/60
Classes, affine, of curves, I/233
Classical differential geometry, II/111
Closed class of curves or surfaces, II/113
Closed convex surface, II/113, 115
Closed interval, III/18
Closed orientable surface, III/211
Closed orthogonal system of functions, III/244
Closed set, III/22, 221
Closed surface, III/198, 205, 207
Codazzi, II/61, 103, 108
Codazzi's theorem, II/61, 103
Code bus bars, II/352
  address, II/352
Code of operations, II/338
Coding, II/336
  of instructions, II/347
  of numbers, II/347
Coefficient of dilation, II/164, 310
Coefficient matrix, III/43
  of transformation, III/77
Cohn-Vossen, S. E., III/293
Cohomology, III/220
Collinear set of points, I/246
Color space, I/152
Columbus, C., I/65
Column matrix, III/40
Combinatorial method of topology, III/204–205
Combinatorial topology, III/219–220
Commutation, III/340
Commutative group, III/283
Commutativity, III/12–13
Commutator, III/340
  of group elements, III/284
Commutator subgroup, III/284
Comparison instruction, II/339, 352–353
Comparison instruction code, II/353
Compatible orientations in topology, III/206
Complement, algebraic, in determinants, III/65
  of set, III/7, 23
Complemented plane, I/246
Complete analytic function, I/189
Completely reducible representation, III/317
Complete orthogonal system of functions, III/242
Complete system of events, II/233
Completeness of numerical line, III/18
Complemented plane, I/246
Complex coordinates, I/259

356   INDEX

Complex function, branch point of, II/152, 190
Complex integers, II/225, 227
Complex linear space, III/55
Complex numbers, II/139–153; III/321
 argument of, I/282
Complex plane, I/292
Complex root, I/271
Complex space, I/259
Complex variable(s), Cauchy-Riemann equations for, II/150
 coefficient of dilation in, II/164
 function of, II/139–195
 trigonometric function of, II/145
Composite function, II/104
Composite number, II/205
Composition of groups, III/297
Computation techniques, II/303
Computational mathematics, I/61
Concavity, I/113–114
Configuration space, III/176
Conformal geometry, II/116
Conformal transformation, II/166, 168
Congruences, II/110
Congruent surfaces, II/103
Conic sections, I/206
 focus of, I/202
Conjugate element in groups, III/298
Conjugate quaternions, III/326
Conjugate transformation, III/276, 284
Connectedness, III/160
Connectivity, of manifold, III/211
 of surface, III/197
Conservation of heat energy, equation of, II/9
Constant curvature, III/174
Constructing solutions, II/27
Continuation, analytic, II/187–189
 of function into complex domain, II/141–142
Continuity, Cantorian, III/15
 equation of, II/8
 of figure, III/160
 principles of, III/15
 of real number sequences, I/29, 89
Continuous deformation, II/101
Continuous field of directions, III/213
Continuous function, I/88
Continuous functional, III/253
Continuous group(s), III/305, 307
Continuous group of transformations, III/305
Continuously differentiable function, II/51; III/37; *see also* Smooth function
Continuous in interval, I/89
Continuous mapping, III/160
 of manifold, III/216
Continuous nonsmooth function, II/276
Continuous from one side, I/91
Continuous at a point, I/89
Continuous representation, III/307
Continuous spectrum of eigenvalues), III/258–259

Continuous transformation, II/101; III/160, 194, 221
 of continuous media, I/234–235
Continuous variable, I/33
Continuous vector field, III/213
Continuum
 cardinal number of, III/11
 uncountability of, III/17
Contour integral, I/162
Contour lines; *see* Level lines
Contraction of plane toward line, I/227, 235
Contraction theorems, II/311
Control circuit, II/358
Control memory block, II/352
Control unit(s), II/334
Convergence, asymptotic, III/108
 circle of, II/145
 domain of, II/143
 "in the mean," III/240
 of series, I/168
Convergent methods, II/308
Convergent series, I/167
 absolutely, I/170
Converging spherical wave, II/27
Convex body, II/113–115
Convex downward, I/114
Convexity, I/113–114
Coordinate frame, I/231
Coordinate-free definition, III/76
Coordinate method, I/186
Coordinate(s), of point, I/184
 vector, I/216
 of vector, III/53
Coordinate transformation, I/209; III/78–79
Copernicus, N., I/65, 183
Correlation coefficient, II/245
Corresponding homogeneous equation, II/28
Coset, III/301
 left, III/301
 right, III/301
Countability, III/13
Countable set, III/10
Covering of a set, III/26–27
Cross ratio, III/130
Crystallographic groups, III/285, 290, 292
Cubic equation, I/266
Čudakov, N. G., II/209
Curvature, of curve, II/68–69
 of space, III/172
 on a surface, II/80–86
Curvature tensor, III/172
Curve(s), affine classes of, I/233
 closed class of, II/133
 curvature of, II/68–69
 equation of, I/218–219
 family of, II/108
 integral, I/333; III/213
 irreducible, III/346
 length, II/63–64
 mathematical, II/58

Curve(s) (continued)
  n-dimensional, III/151
  space, arc length of, II/63
Curvilinear coordinates, I/91; II/103
Curvilinear integral, I/162–163
Cut, in complex functions, II/152
  Dedekind, III/14–15
Cycle(s), fundamental, III/212
  length of, III/281
  limit, I/353; III/213
  permutations into, III/280
  in topology, III/209
Cyclic permutation, III/280
Cyclic subgroup, III/282
Cyclotomic equations, I/276
Cylinder functions, III/239

d'Alembert, J., I/170; II/36; III/100, 239
d'Alembert, criterion of convergent or divergent series, I/170
  lemma of, I/290
Dandelin, G. P., I/307
Darboux, G., II/61, 101
Darwin, C. R., III/104
da Vinci, L., I/65; III/130
Decoding unit, II/336
Decomposition, of integers, II/225
  of permutation into cycles, III/280
  of a variety, III/347
de Coulomb, C. A., I/52
Dedekind, R., I/27, 59; II/203; III/15–17, 349
Dedekind, continuity of, I/288; III/15, 124
  cut, III/14–15
  lattice, III/349
  principle of, III/15
Defining relations of group, III/311
Definite integral, I/130
Deformability of surface, II/115
Deformation of surface, II/91, 100, 115
Degenerate transformation, III/77
Degree, of representation by matrices, III/315–336
  of symmetry, III/272
de Jonquières, III/206
de la Vallée-Poussin, C. J., II/268
del Ferro, Scipio, I/266, 272
Democritus, I/21, 25, 33
de Moivre, A., II/242
Density of set, III/13
Denumerability, III/13
Denumerable set, III/10
  of eigenfrequencies, II/31
Dependent magnitude, I/75
Derivative, of function, I/92
  of higher order, I/113
  of product, I/102
  of quotient, I/103
  of sum, I/101
Derived group, III/284
Desargues, G., III/130

Descartes, R., I/27, 38, 41, 46, 47, 48, 60, 69, 184–188, 190, 193, 213, 262, 294–298; II/104; III/204
Descartes' law of signs, I/295
Descriptive theory of sets, III/25
Determinant(s), algebraic complement in, III/65
  of coefficient matrix, III/67
  development of, III/65
  expansion of, III/65
  of matrix, III/61
  of nth order, III/64
Deterministic process, II/255
Developable surfaces, II/97
Deviation, II/244
  of frequency, II/244
Diagonal matrix, III/83–84
Diagonal process, III/11
Diameter, Newton's theory of, I/193
  of region, I/160n
Difference quotient of function, II/42
Difference of two sets, III/7
Differentiability of function, II/148
Differentiable function, I/99; II/163
Differential calculus, I/48, 131
Differential equation(s), I/311ff
  center in, I/347
  of Euler, II/124, 127
  field of directions, I/333; III/213–214
  focus in, I/347; III/217
  general solution of, I/322
  homogeneous linear, I/323
  "intensive," I/331
  linear, I/323
  order of, I/320
  ordinary, I/311, 313, 320
  stable, I/349
  stochastic, II/261–264
Differential, of function, I/117
  of several variables, I/148
Differential geometry, II/58–59, 61–63
  affine, II/116
  classical, II/111
  projective, II/116
Differential operator, III/255
Differentiation, rules of, I/101
  termwise, I/171
Digital machines, II/323
Dilation, coefficient of, II/164, 310
Dimension(s), of manifold, III/203–204
  of space, III/51, 218
Dimension theory, III/219
Diophantus, I/37, 38
Direct methods, I/336
Directed circular symmetry, III/287
Direction numbers, I/340
Directrix, of parabola, I/202
Dirichlet, P. G. L., I/80; II/53, 54; III/30, 33
Dirichlet, function of, III/30
Discontinuity, weak, II/49
Discontinuous function, I/89
Discrete group of motions, III/289, 292

Discrete symmetry group, III/289
Disjoint classes, in groups, III/298
Disjoint sets, III/7
Distance, between functions, III/222
  in n-dimensional space, III/221
  between points, III/18
Distribution of prime numbers, II/205–207
Distributive law, III/7
Distributivity, III/13
Divergence theorem of Gauss; see Ostrogradskiĭ, formula of,
Divergent series, I/167
Diverging spherical wave, II/27
Divisibility, II/200
Division algebra, III/338
Divisor(s), of one, III/345
  of zero, III/344
Domain, of convergence, II/143
  of definition, II/122; III/31, 254
  of function, I/75
  of rationality, III/294
  simply connected, II/168
  of values, III/157
  of variation of variables, III/157
Double-entry table, II/320
Double integral, I/161
Duality theorem, III/220
Dürer, A., III/130
Dynamic storage, of code, II/362

Eccentricity of conic section, I/205
Efimov, N. V., II/102
Egorov, D. F., III/30
Eigenfrequency, III/249
  denumerable set of, II/31
Eigenfunction, II/30; III/250
Eigenvalue, II/82; III/250
  continuous spectrum of, III/258–259
  of linear transformation, III/80–81
  of operators, III/256
Eigenvector(s), of linear transformation, III/80–81
  of operators, III/256–257
Eigenvibrations, I/327; II/30; III/249
Einstein, A., I/249; III/104, 165, 179, 185–188
Electronic commutator, II/353
Electronic computers, II/331
Element(s), associate, of ring, III/345
  of length, III/169
  of matrix, III/40
  of set, III/6
  unit, III/298
  zero, III/298
Elementary functions, I/107; II/141
Elementary number theory, II/203
Elementary symmetric polynomial, I/272
Ellipse, I/195–197
  of inertia, I/198–200
Ellipsoid, I/221
Elliptic cylinder, I/222
Elliptic paraboloid, I/221

Empty set, III/6
Ensemble, III/5
Entire functions, II/182
Envelope of family of curves, II/109
Equal sets, III/6
Equation(s), algebraic solution of, I/265
  canonical form of, I/208
  center in, I/347
  characteristic, homogeneous linear, I/324–325
  of continuity, II/8
  of curve, I/218–219
  cyclotomic, I/276
  differential, I/311–356
  first-degree, I/188, 267
  fourth-degree, I/268
  functional, I/311
  ground field of, III/294, 331
  integral, III/245–252
  Laplace's, II/15, 163
  of motion, II/9
  of oscillation, II/21
  parametric, II/62–63
  of plane, I/220
  Poisson, II/15
  quadratic, I/266
  Ricatti, I/330
  of straight line, I/220
  of surfaces, I/218
  third-degree, I/266
  of vibrating string, II/12
  wave, II/14
Equidistant curve, III/107
Equivalent knots, III/313
Equivalent paths, III/308
Equivalent quadratic forms, III/86
Equivalent representations, III/315
Eratosthenes, I/37; II/204
  sieve of, II/204
Error estimate, II/308, 369
Euclid, I/5, 16, 21, 22, 26, 27, 36, 37, 38, 41, 54; II/204, 207, 210; III/97–103, 112, 113, 115, 121, 122, 124, 130, 133, 174, 178, 179, 182
Euclidean algorithm, II/202
Euclidean geometry, III/113
Euclidean space, III/58
Euclid's parallel postulate, III/98, 124
Eudemus of Rhodes, I/20
Eudoxus, I/26
Euler, L., I/51, 54, 56, 71, 96, 161, 207, 213, 263, 264, 336, 337, 340; II/36, 60, 81, 83, 85, 91, 100, 108, 124, 127–129, 131, 133, 146, 147, 203, 205, 207, 208, 211, 220, 269, 293, 305, 316, 336–338; III/204–208, 212, 216, 239
Euler, angles of, I/221
  characteristic of triangulation, III/205
  differential equation of, II/124, 127
  formula(s) of, II/146; III/205
  functions of, II/220

INDEX

Euler *(continued)*
  method of broken lines, I/337
  theorem of, II/81, 83
Euler-Poincaré formula, III/212
Even function, II/200
Even permutation, III/279
Event(s), II/232
  certain, II/229
  impossible, II/229
  independent, II/234
  intersection of, II/233
  mutually exclusive, II/232
  necessary, II/229
  random, II/231, 252, 259
  stochastic, II/231
  union of, II/231
Everywhere dense covering, I/247
Evolute, II/114
Existence of solution, I/335, 336
Expansion, of determinant, III/65
  of function in power series, II/141–147; II/180
  of function in trigonometric series, II/293; III/242
  in orthogonal system of functions, III/237, 240
Exponential function of complex variable, II/145
Exponential matrix function, III/93
Exterior point of set, III/20
Extremal conditions, II/124
Extreme values of functionals, II/122
Extrinsic property of surface, II/102

Faber, G., II/276
Factor group, III/301
Factor ring, III/344
Factorization of polynomial, I/271
Faithful representation, III/336
False position, I/304, 306
Family, of curves, II/108
  of surfaces, II/108–109
Fedorov, E. S., I/23, 58, 62; III/147, 264, 290, 292, 293
Fedorov groups, III/285, 290, 292
Fejér, L., II/298, 301
Fejér sum, II/298
Fermat, P., I/184, 262, 263; II/225; III/345
Ferrari, L., I/41, 268
Field(s), III/324
  algebraic, III/13
  of directions, I/333; III/213–214
  vector, II/3; III/212
  continuous, III/213
Figures, affine classification of, I/233
  affine properties of, I/233
Finikov, S. P., II/116
Finite group, III/320
Finite set, III/8
Finite simple group, III/307
Finite-difference formula, I/120–121

Finite-dimensional linear space, III/51
Finsler, R., I/57; III/177
First fundamental quadratic form, II/105
First-degree equation, I/188, 267
Fixed point, III/216
Fixed-point machine, II/349
Fixed-point theorem, III/163
Flat point, II/82
Flip-flop, II/353
Floating-point system, II/349
Flux, II/7
Flux vector, II/7
Focus, of conic section, I/202
  in differential equations, I/347; III/217
Fok, V. A., III/188
Fokker, II/261
Forced oscillations, I/329
Four-dimensional cube, III/145
Fourier, J. B. J., I/297; II/28, 36, 40, 268, 269, 289, 291, 294, 296–298; III/35, 161, 259
Fourier, coefficient(s) of, II/294
  integral of, III/259
  series of, II/291–298
Fourth-degree equation, I/268
Fractional function, II/184
Fractions, origin of, I/24
Free generators, III/312
Free group, III/312
Free oscillations, III/248
Free product, III/312
Frenet, J. F., II/61
Frequency spectrum, III/249
Fresnel, A. J., II/333
Fridman, A. A., III/188
Fubini, G., II/116
Full additivity property of measure, III/27
Function(s), admissible, II/125
  analytic, I/177; II/141
    on interval, I/177
    at point, II/148
  representation of, II/186
  approximating, II/283
  approximation of, II/265
  characteristic, II/30; III/32
  closed orthogonal system of, III/244
  complex, analyticity of, II/185
  of complex variable, II/139–195
  composite, II/104
  continuation of, into complex domain, II/141–142
  continuous, I/88
    nonsmooth, II/276
    from one side, I/91
  derivative of, I/92
  difference quotient of, II/42
  differentiable, I/99; II/163
  differential of, I/117
  Dirichlet, III/30
  discontinuous, I/89
  domain of, I/75
  elementary, I/107; II/141
  entire, II/182

Function(s) (*continued*)
of Euler, II/220
even, II/200
expansion of, III/65
exponential, II/145
fractional, II/184
of function, I/105
generating, II/273
geometric theory of, II/163
Green's, III/245–248
graph of, I/76, 77, 108–116
harmonic, II/16
homogeneous linear, III/37–38
implicit, differentiation of, I/149
implicit definition of, I/144
infinite orthogonal system of, III/342
integrable, I/136, 158
with integrable square, III/254
integral, II/182
inverse, I/103
Lagrange, II/50–51
limit, III/4
linear, I/75; III/37
of matrix, III/91
argument, III/92
measurable, III/29
multiple-valued, II/152, 189
nonregular, II/113
number theoretic, II/220
odd, II/200
orthogonal, III/236
of real variables, III/3
of several variables, I/142
smooth, II/275; III/37, 75
spherical, III/239
stream, II/155
transcendental, II/182
trigonometric, of complex variable, II/145
vector, II/104–105
Functional, I/122; III/253
Functional analysis, I/60, 258; III/223, 227
Functional dependence, I/122
Functional equations, I/311
Functional space, III/157, 223, 254
infinite-dimensional, III/254
Fundamental coefficients, II/105
Fundamental cycle, III/212
Fundamental formula of integral calculus, I/165
Fundamental frequency, II/34
Fundamental group, III/308, 310
Fundamental quadratic form(s), II/105–108; III/168
Fundamental sequence, III/16
Fundamental theorem of algebra, I/280–292; II/140
Fundamental theorem of plane perspectivity, I/243–244

Galerkin, B. G., II/40, 41, 53, 305, 306
Galerkin's method, II/40

Galileo, I/43, 46, 47, 65, 66, 183, 249; II/255, 256
Galois, E., I/58, 263, 264, 273–279; III/293–297
Galois, group, I/278; III/293–295
theory, I/275–278
Gauss, K. F., I/71, 263, 276, 279; II/89, 90, 91, 103, 104, 108, 203, 206, 207, 281, 282; III/99, 100, 104, 164, 173, 178
Gauss, divergence theorem of; see Ostrogradskiĭ, formula of,
Gaussian curvature, II/86, 89; III/173
Gel'fand, I. M., III/307
Gel'fond, A. O., II/203
General affine transformation, I/231
General definition of group, III/297
General law of duality, III/220
General linear space, III/48
General solution of differential equation, I/322
General topological space, III/159
Generalized solutions of partial differential equations, II/48, 50
Generalized space, III/156
Generating function, II/273
Generators of group, III/311
free, III/312
Generators of subgroup, III/282
Genus of surface, II/194; III/198
Geodesic(s), II/97; III/114, 170
Geodesic curvature, II/95
Geodesic line, II/97
Geodesic segment, II/95
Geodesic "in the small," II/111
Geodesic surface, III/173
Geometric figure, I/19
Geometric method, III/162
Geometric number theory, II/203
Geometric theory of functions, II/163
Geometrical interpretation, I/332–335
Geometry, I/19, 56
abstract, III/178
affine, III/131
differential, II/146
algebraic, I/258
analytic, I/45, 47, 183, 186
in space, I/218–220
axioms of, III/122
classical differential, II/211
conformal, II/116
differential, II/58–59, 61
Euclidean, III/113
"in the large," II/111, 112
intrinsic, II/91–108; III/166
Lobačevskiĭ, III/105
n-dimensional, III/137–138
non-Euclidean, III/97
Riemannian, III/164
Gibbs, J. W., III/142
Gjunter, N. M., II/52
Goldbach, C., II/211, 217, 219, 224
Goldbach conjecture, II/211, 217

INDEX

Graduated representation, III/317
Graeffe, K., I/307
Graeffe's method, I/307n
Graph, of function, I/76, 77, 108–116
   of polynomial, I/303
Grassmann, H., II/104; III/138
Greatest lower bound, III/19
Greatest value, I/109
Green, G., III/245–248, 251
Green's function, III/245–248
Green's theorem, II/134; *see also* Ostrogradskiĭ, formula of
Ground field, III/294, 331
Group(s), Abelian, III/283
   abstract, III/320
   addition of, III/297
   of algebraic equations, III/294
   alternating, III/279
   automorphism, III/295
   axioms of, III/297
   axis of, III/289
   character of, III/314
   commutative, III/283
   characters of, III/220
   composition of, III/297
   conjugate element in, III/298
   continuous, III/305, 307
   crystallographic, III/285, 290, 292
   defining relations of, III/311
   definition, general, III/297
   derived, III/284
   disjoint classes in, III/298
   factor, III/301
   Federov, III/285, 290, 292
   finite, III/320
   finite simple, III/307
   free, III/312
   fundamental, III/308, 310
   Galois, I/278; III/293–295
   general definition of, III/297
   generators of, III/311
   graduated representation in, III/317
   infinite, III/320
   isomorphic, III/299
   of knots, III/312, 314
   Lie, III/305, 306
   local, III/340
   matrix representation of, III/315
   of motions, III/289
   multiplication of, III/297
   operation of, III/298
   order of, III/278
   of paths, III/310
   permutation, III/279
   representation of, III/315
   of rotations, III/278
   simple, III/301
   symmetric, III/279, 296
   topological, III/307
   of transformations, III/273–285, 297
      in differential geometry, II/116
Gyaseddin Jamschid, I/39, 40

Hadamard, J., I/208
Half-open interval; *see* Semi-interval
Hamilton, W. R., I/50, 51, 52, 104, 132, 133
Hamilton's principle; *see* Ostrogradskiĭ-Hamilton principle
Hardy, G. H., I/203
Harmonic function, II/16
   absolute maximum or minimum of, II/17, 18
Harmonic oscillations, III/248
Harmonic series, I/169
Heat-conduction vector, II/8
Hermitian vector space; *see* Unitary space
Hertz, H., II/5
Hilbert, D., II/60, 72, 279, 280
Hilbert space, III/161, 232–237
Hinčin, A. Ja., I/246; II/356
Hipparchus, II/36
Hippocrates, II/35
Holonomic system, II/132
Homeomorphic figures, III/197
Homogeneous boundary conditions, II/21
Homogeneous integral equation, III/249
Homogeneous linear differential equation, I/323
Homogeneous linear function, III/37–38
Homogeneous system of linear equations, III/73
Homogeneous transformation, III/76
Homological independence of cycles, III/210
Homological theory of dimension, III/219
Homologous to zero, III/211
Homology, III/210
Homomorphic image, III/303, 305
Homomorphic mapping, III/302
Homomorphism, III/302
Homotopic topology, III/220
Hooke, R., II/315; III/246
Hopf, H., III/216, 220
Hopf's theorem, III/216
Hurwitz, A., II/300, 302
Hyperbola, I/202
Hyperbolic cylinder, I/222
Hyperbolic paraboloid, I/219–220, 222
Hyperboloid, I/221, 225
   of revolution, I/225
Hypercomplex number(s), III/320–330
Hypercomplex system(s), III/263, 330
Hyperplane, III/139

Ideal, III/343
   of algebra, III/337
   left, III/337
   minimal, III/338
   principal, II/81
   right, III/337
   two-sided, 337
   zero, III/337
Ideal line, I/246
Ideal points, I/246

Identity theorem for analytic functions; *see* Uniqueness theorem for analytic functions
Identity transformation, III/282
Imaginary ellipse, I/208
Imaginary ellipsoid, I/221
Implicit definition of a function, I/144
Implicit function, differentiation of, I/149
Impossible event, II/229
Improper interval, III/19
Improper line, III/200
Improper motion, III/271, 292
Improper point, III/200
"In the large," I/352; II/59, 111–112
"In the small," II/59, 95
Incommensurable magnitudes, I/25
Increment, I/89, 117
Indecomposable element, III/345
Indecomposable ideal, III/347
Indefinite integral, in analysis, I/137
 of complex function, II/177
Independence, III/38
Independent events, II/234
Independent magnitude, I/75
Index, of isolated singular point, III/216
 of singular point, III/216
Inductive definition of dimension of sets, III/219
Inertia of quadratic form, III/88
Infinite-dimensional analytic geometry, I/258
Infinite group, III/320
Infinite orthogonal system of functions, III/342
Infinite plane figure(s), III/288
Infinite point set, III/21
 bounded, III/21
Infinite set, III/8
Infinite-dimensional functional space, III/254
Infinite-dimensional linear space, III/51
Infinite-dimensional space, III/157, 161, 232–237
Infinitely distant line, III/200
Infinitely distant point, III/200
Infinitely large magnitude, I/83
Infinitesimal, I/82
 of higher order, I/118
Inflection, point of, I/115
Inhomogeneous integral equation, III/250
Inhomogeneous linear integral equation, III/250
Initial condition, I/336; II/19
Initial value(s) of problem, I/336
Initial state of system, II/15
Initial-value problem, II/15
Input unit, II/334
Instruction, II/335
Instruction address, II/339
Instruction code, II/339
Instruction memory block, II/351
Integer, algebraic, II/203
 decomposition of, II/225

Integrable function, I/136, 158
Integral, I/128
 action, II/133
 Cauchy, I/177
 contour, I/162
 along curve, II/174
 over closed contour, II/177
 curvilinear, I/162–163
 definite (Riemann), I/130
 double, I/161
 indefinite, I/137
 Lebesque, III/30–36
 line, II/174
 multiple, I/158
 Riemannian, I/130; III/31
 surface, I/162
 two-fold, I/161
Integral calculus, I/48–49, 131
Integral curve, I/333; III/213
Integral domain, III/344
Integral equation(s), III/245–252
 homogeneous, III/249
 inhomogeneous, III/250
 of second kind, II/40
Integral functions, II/182
Integral line, I/333; III/213
Integral operator, III/255
Integrating machine, II/326
Integration, by parts, I/41
 path of, II/176
 termwise, I/171
"Intensive" differential equations, I/331
Interior point of set, III/20, 160
Interpolation polynomial(s), II/269, 286
Intersection, of events, II/233
 of sets, III/7
 of subsets, III/57
Interval, II/18
 adjacent, III/23
 basic, II/219
 closed, III/18
 improper, III/19
 nested, III/15
 open, III/18
Intrinsic geometry, II/91–108; III/166
Intrinsic metric, III/161
Intrinsic property, II/102
Invariant, I/238
 topological, III/310–311
Invariant subgroup, III/301
Inverse, of matrix, III/70
 of quaternion, III/327
Inverse element, III/298
Inverse function, I/103
Inverse transformation, III/76
Inverse vector, I/215
Irreducibility of polynomial, I/277
Irreducible curve, III/346
Irreducible factor, I/277
Irreducible representation, III/317
Irreducible variety, III/347
Irrotational motion in fluid, II/156
Isolated point, III/20

Isolated singular point, I/343; III/216
Isomorphic algebras, III/333
Isomorphic groups, III/299
Isomorphic mapping, III/299, 302
Isomorphic representation, III/336
Isomorphism, III/299
  local, III/340
Iteration method, II/310

Jackson, D., I/268, 290
Jordan, C., III/84, 95
Jukovski, N. E., I/5

Keldyš, L. V., III/25
Keldyš, M. V., II/290; III/260
Kepler, J., I/47, 48, 65, 69, 183, 198, 313
Kernel, of integral equation, III/250
  symmetric, III/251
Killing, III/307, 342
Kirchhoff, G. R., II/328
Kiselev, A. P., I/21
Klein, F., I/72, 280; III/115, 116, 120–125, 132, 133, 164, 177, 187, 199, 216, 218, 306
Klein bottle, III/199
Knot(s), III/312
  equivalent, III/313
  nonequivalent, III/313
Kolmogorov, A. N., II/262, 268, 269, 297; III/12, 25, 220
Kolosov, G. V., II/163
Korkin, A. N., II/267
Krashosel'skiĭ, M. A., III/224
Krylov, A. N., I/309, 339
Krylov, N. M., I/355, 356
Kuiper, N. H., II/115; III/115
Kummer, E. E., II/203; III/345
Kurnakov, N. S., III/142
Kuros, A. G., I/265; III/320

Lagrange, J. L., I/51, 54, 121, 154, 156, 157, 213, 214, 263, 272–275, 277, 278, 280; II/50, 203, 273, 281; III/3, 100, 137, 185, 293, 301
Lagrange function, II/50–51
Lagrange multipliers, I/54
Lagrange resolvent, I/273
Lagrange's formula, I/120–121
Lagrange's interpolation formula, II/273
Lagrange's theorem, III/301
Laguerre, E., I/213
Lambert, J. A., III/99, 100, 137, 185, 293, 301
Laminar flow; see Plane-parallel motion
Laplace, P.–S., II/15, 16, 17, 24, 28, 35, 36, 44, 49, 163, 244, 258, 307, 327, 328
Laplace equation II/15, 163
Laplacian, II/15
Lappo-Danilevskiĭ, I. A., III/95
Laptev, G. F., II/116

Lattice, III/347–348
  modular, III/349
Law(s), of inertia, III/88
  of large numbers, II/238, 246
  of probability, II/229
Lavrent'ev, M. A., I/337; II/174, 268, 290; III/25
Least upper bound, III/16–19
Least value, I/109
Lebesgue, H., I/59, 133, 158; II/52, 268, 297, 302; III/5, 26, 30–36, 237, 244
Lebesgue integrable, III/35
Lebesgue integral, III/30–36
Lebesgue integral sum, III/34
Lefschetz, S., III/220
Left coset, III/301
Left ideal, III/337
Left quotient, III/284
Legendre, A. M., II/206, 282; III/99, 100, 240
Legendre polynomials, II/282; III/240
Leibnitz, G. W., I/46, 48, 69, 70, 96, 132, 133, 134, 135, 136, 137, 262; II/59, 110, 332; III/104
Length, of curve, II/63–64
  of cycle, III/281
  of n-dimensional vector, III/229
  of vector, III/58
Leray, J., II/220, 224
Level curves, I/219
Level lines, I/145
Leverrier, U. J. J., I/4, 313
l'Hôpital, I/59
Lie, S., I/265; II/220, 305–307, 334, 340, 341, 342
Lie algebra, III/339
Lie group, III/305, 306
  continuous representation of, III/307
Limit, I/49, 80–82
  of sequence of matrices, III/92
  uniform, II/49
Limit cycles, I/353; III/213
Limit function, III/4
Limiting circle, III/111
Limit point; see Adherent point
Limit point of set, III/20
Line at infinity, III/200
Line integral, II/174
Linear algebra, III/37
Linear combination of vectors, III/49
Linear density, III/248
Linear dependence, III/49
Linear differential equation, I/323
Linear equations, homogeneous system of, III/73; see also First-degree equations
Linear finite-dimensional space, III/331
Linear form, III/39
Linear function, I/75; III/37
Linear independence, III/49
Linear interpolation, I/304, 306
Linear operators, III/252, 254
Linear point set(s), III/18

Linear space, III/48
  infinite-dimensional, III/51
Linear transformation, III/74
Linear vector space, III/48–49
Linearly ordered set, III/347
Linnik, Ju. V., II/210; III/330
Liouville, J., I/276, 330
Littlewood, J. E., II/203
Ljapunov, (Lyapunov), A. M., I/51, 59, 72, 348–351, 355; II/40, 247; III/25
Ljapunov stable, I/349
Ljapunov unstable, I/349
Ljusternik, L. A., II/112; III/219, 221, 224
Lobačevskiĭ, N. I., I/6, 21, 56, 57, 71, 80, 257, 264, 307, 308, 309; III/97, 101–129, 133, 159–166, 174, 178, 179, 182, 183, 184, 188
Lobačevskiĭ geometry, III/105
  equidistant curve in, III/107
Lobačevskiĭ's method, I/307
Lobačevskiĭ's parallel postulate, III/124
Local group, III/340
Local isomorphism, III/340
Local maximum, I/108
Local minimum, I/109
Logical operations, II/358
Lomonosov, M. V., III/151
Lorentz, H. A., I/227, 249, 250, 252–257; III/187
Lorentz transformation, I/249, 250
Lorentz transformation formulas, I/252
Lower semilattice, III/348
Luzin, N. N., I/59; III/25, 30

Mach, E., III/184
Magnetic drum storage, II/363
Mal'cev, A.I., III/307
Mandel'stam, L. I., I/355; III/214
Manifolds, III/202–204
  connectivity of, III/211
  continuous mapping of, III/216
  many-dimensional, III/218
  smooth, III/215
Many-dimensional analytic geometry, I/258
Many-dimensional manifold, III/218
Many-dimensional space, III/136, 142
Mariotte, E., I/77, 242; II/14; III/142
Markov, A. A., I/72, 356; II/246, 247, 260, 261, 267, 287, 371
Markov, V. A., II/267
Markov process, II/260
Markuševič, A. I., III/218
m-ary operation, III/350
Mathematical curve, II/58
Mathematical expectation, II/244
Mathematical logic, I/61
Mathematical physics, II/5
Matrix (Matrices), III/40
  canonical form, III/86–87
  canonical Jordan, III/84
  characteristic equation of, III/82
  characteristic numbers of, III/82

characteristic polynomial of, III/82
coefficient, III/43
  determinant of, III/67
  of coefficients, III/62–63
  column, III/40
  of coordinate transformations, III/79
  determinant of, III/61
  of equal structure, III/43
  function, III/91
  exponential, III/93
  nondegenerate, III/71
  nonnull, III/344
  nonsingular, III/71
  null, III/44
  orthogonal, III/341
  rank of, III/72
  row, III/40
  similar, III/80
  square, III/40
  unit, III/70, 335
Matrix algebra, III/334–338
Matrix calculus, III/91
Matrix element, III/40
Matrix multiplication, III/41
Matrix notation, III/69
Matrix operation, III/40
Matrix representation, of associative algebra, III/37
  of group, III/315
Maximum, I/108
Maxwell, J. C., I/4, 5; III/153
Mean curvature, II/86
Mean value theorem, I/120–121
Measurable function, III/29
Measurable set, III/27
Measure, of open set, III/26
  of set, III/25–26, 27
Memory location, III/340
Memory unit, II/334, 339, 362
Mendeleev, D. I., II/287
Men'šov, D. E., II/269
Mergeljan, S. N., II/290
Meromorphic function, II/184
Method, of broken lines, I/337; II/136, 137
  of chords, I/304, 306
  of coordinates, I/186
  of finite differences, II/40, 42
  Monte-Carlo, II/371
  of nets, II/40, 42
  of Newton, I/306
  of potentials, II/36
  of substitution, I/39
  of successive approximations, I/339; II/310
  of tangents, I/304
Metric, III/165
Metric space, III/161, 221–223
Metrizability of topological space, III/224
Meusnier, J. B. M., II/83, 85, 86, 91, 108
Meusnier's theorem, II/83
Michelangelo, I/65
Michelson, A. A., I/249, 250
Minding, F., II/60, 100, 111; III/115

Minimal ideal, III/338
Minimal principles, II/124
Minimal problem, II/120
Minimal surface, II/88
Minimizing sequence, II/136
Minimum, I/108
Minimum problem for multiple integral, II/133–134
Minkowski, H., II/113, 204; III/185
Minor of determinant, III/65
Mirror image, III/265
Mirror symmetry, III/265
Mixed partial derivative, I/147
Möbius, A. F., I/247; III/199–201
Möbius band, III/198–199
Möbius net, I/247
Möbius strip, III/199
Möbius surface, III/198–199
Modern mathematics, I/62
Modular lattice, III/349
Modular law, III/349
Modulus, of complex number, I/282
  of polynomial, I/286
Modulus surface, I/286, 292
Molin, F. E., III/338, 339, 347
Monge, G., II/110, 111
Monotony, law of, III/13
Monte-Carlo method, II/371
Moreland, II/322
Morozov, V. V., III/307
Morse, M., II/112; III/224
Motion, equation(s) of, II/9
  of first kind, III/271
  perturbed, I/350
  proper, III/271
  of second kind, III/271
  spiral, III/271
  unperterbed, I/350
Multiple integral, I/158
Multiple roots, I/293
Multiple-valued function, II/152, 189
Multiplication, III/13
  of matrices, III/41
  of paths, III/309
  of transformations, III/273
Multiplication circuit, II/357
Multiplication table, III/300
Multiplicative property, II/201
Multiplicity of a root, I/293
Multiplying punch, II/325
Multivalued function, II/152, 189
Mushelišvili, N. I., II/163
Mutually exclusive events, II/232

Naĭmark, M. A., III/307
n-dimensional angle, III/139
n-dimensional body, III/148–149
n-dimensional cube, III/144
n-dimensional curve, III/151
n-dimensional Euclidean space, III/56, 58
n-dimensional geometry, III/137–138
n-dimensional plane, III/138

n-dimensional point space, III/47
n-dimensional polyhedron, III/143–148
n-dimensional segment, III/139
n-dimensional set, III/219
n-dimensional simplex, III/146, 207
n-dimensional space, III/228–229
n-dimensional topology, III/218
n-dimensional triangulation, III/207
n-dimensional vector, III/229
n-dimensional vector space, III/46–47
Napier, J., I/42, 46
Nash, J., II/115
Nasireddin Tusi, I/39, 40; III/99
Necessary condition for convergence, I/168
Necessary event, II/229
Negative numbers, I/34
Neighborhood, III/159
Nemyckii, V. V., I/356; III/214, 223
Nest of intervals, III/15
Nested intervals, III/15
Nets of curved lines, II/60
Neutral element, III/298
Newton, I., I/33, 39, 40, 42, 46, 48, 50, 51, 52, 54, 69, 70, 86, 96, 132–137, 141, 186, 193, 194, 198, 213, 262, 263, 297, 306, 307, 313, 315; II/10, 11, 15, 59, 147, 255, 267, 273; III/104
Newtonian potential, II/37
Newton-Leibnitz formula, I/133
Newton's method, I/306
Newton's theory of diameters, I/193
Nilpotent algebra, III/338
Node, I/346; III/213
  of oscillation, II/34
Nonclosed convex surface, II/115
Noncommutative operation, III/274
Noncommutative ring, III/347
Noncommutativity of matrix multiplication, III/44
Nondeformability, II/115
Nondegenerate matrix, III/71
Nondegenerate transformation, III/77
Nondenumerability of continuum, III/17
Nonequivalent knots, III/313
Non-Euclidean geometry, III/97
Nonhomogeneous linear differential equation, I/323, 327–329
Nonnull matrix, III/344
Nonorientability, III/206
Nonregular functions, II/113
Nonregular surfaces, II/211, 212
Nonsingular matrix, III/71
Nonsingular transformation, III/77
Nonsmooth surface, II/114
Nontrivial normal subgroup, III/307
Nonuniformly convergent series, I/172
Norm of quaternion, III/326
Normal, II/78
Normal divisor, III/301
Normal orthogonal system of functions, III/241
Normal plane, II/74
Normal section, II/80

INDEX

Normal subgroup, III/301
Normal surface, III/198
Normal system, II/300
Novikov, P. S., III/25, 314
n-parameter family, II/109
nth-order parabola, I/303
Null matrix, III/44
Number(s), I/10
  additive property of, II/201
  algebraic, II/203
  analytic, II/203
  composite, II/205
  direction, I/340
  real, axioms of continuity for, I/288; III/15, 124
  transcendental, II/203
  whole, I/7; II/199
  winding, III/216
Number code, II/339
Number ring, III/342
Number theory, analytic, II/203
  elementary, II/203
  geometric, II/203
Number-theoretic function, II/220
Numerical line, III/4, 15, 18
Numerical methods, II/303

Odd function, II/200
Odd permutation, III/279
Odner, V. T., II/322
Omar Khayyam, I/39, 40
One-address system, II/345
One-dimensional chain, III/209
One-dimensional cycle, III/209
One-dimensional hyperplane, III/140
One-place semiadder, II/361
One-sided surface, III/198–199
One-sided torus, III/199
One-to-one correspondence, III/8
One-to-one transformation, III/268
Open interval, III/18
Open set, III/22
Operation code, II/352
Operation(s), on determinants, III/64
  on sets, III/6
  vector, III/64
Operative location, II/339
Operator(s), III/254
  linear, III/252, 254
Order, axioms of, III/123
  of connectivity, III/197, 211
  of contact, I/338
  of differential equations, I/320
  of group, III/278
  of infinitesimal, I/18
  of point sets, III/18
  of square matrix, III/40
  of symmetric group, III/279
  of symmetry, III/265
  of tangency, I/338
  of transformation, III/283
Ordered field, III/13

Ordered set, III/18, 347
Ordinary differential equations, I/311, 313, 320
Ordinary point, I/343
Ordinate, I/213
OR elements, II/358, 359–360
Orientability, III/206
Oriented circumference, III/287
Oriented simplex, III/211
Origin, I/184
Orthogonal complements, III/60
Orthogonal functions, III/236
Orthogonal invariant, I/239
Orthogonal matrix, III/341
Orthogonal projection, III/60
Orthogonal system, II/300; III/237, 238
Orthogonal transformation, I/237; III/78
  of quadratic form, III/89
Orthogonality, II/31
  of vectors, III/59, 230
Orthogonality property, II/295
Orthonormal basis, III/59
Orthonormal system of functions, III/238
Oscillation, equation(s) of, II/21
  forced, I/329
  free, III/248
  harmonic, III/248
  node of, II/34
Osculating plane, II/73–74
Ostrogradskiĭ, M. V., I/51, 71, 133, 142, 163, 165; II/8, 40, 132, 133, 134
Ostrogradskiĭ, formula of, I/163; II/8, 134
Ostrogradskiĭ-Hamilton principle, II/132
Outer measure, III/27
Output unit, II/334
Overtone(s), II/34

Pairwise orthogonal functions, III/237
Pairwise orthogonal vectors, III/231
Papaleksi, N. D., I/355
Parabola, I/202
Parabolic cylinder, I/222
Parabolic segment(s), I/338
Paraboloid, I/225
Parallel lines, III/106–107
Parallel postulate, III/98, 124
Parallel shift, III/271
Parallelism, III/106
Parameter, I/258
Parametric equations, II/62–63
Parity, III/280
Parseval, M., III/36
Parseval's equality, III/35–36
Partial derivative, I/145
  mixed, I/145
Partial differential, I/145–148
Partial differential equation(s), I/314; II/3
  admissible or generalized solutions of, II/48, 50
  constructing solutions of, II/27
  generalized solutions of, II/48, 50
Partial sum, I/166

INDEX

Partially ordered set, III/347
Pascal, B., I/70, 183; II/322
Path of integration, II/176
Perfect set, III/24
Periodic transformation, III/282–283
Permutation group(s), III/279
Perspective projection, I/242–243
Perturbed motion, I/350
Peterson, K. M., II/61, 103, 108
Peterson's theorem, II/61, 103
Petrovskiĭ, I. G., I/259; II/262
Phase plane, I/342
Phase space, I/343; III/154, 204
Physical system, characteristic oscillations of, I/327; II/30; III/249
Picard, E., II/183
Picard's method; see Method of successive approximations
Picard's theorem, II/183
Piecewise constant function, II/49
Piecewise smooth function, II/49
Planck, M. K. E. L., II/261
Plane, complemented, I/246
  equation(s) of, I/220
  n-dimensional, III/138
  of symmetry, III/265
Plane Fedorov group(s), III/290
Plane-parallel motion, II/154
Plane perspectivity, fundamental theorem of; see Uniqueness theorem of theory of projective transformations
Plateau, J. A. F., II/88
Pogorelov, A. V., II/115
Poincaré, H., I/51, 59, 72, 249, 348, 355; III/104, 121, 122, 124, 184, 185, 204, 212, 214, 218, 219
Poisson, S. D., II/15, 16, 17, 28, 36, 40
Point, adherent, III/221–223
  coordinates of, I/184
  of discontinuity, I/344
  of inflection, I/115
  of interpolation, II/273
  of intersection, I/187
  symmetrical, III/265
  of tangency, II/78
Point set(s), III/18
  bounded infinite, III/21
  linear, III/18
  order of, III/18
Poisson equation, II/15
Polar coordinates, I/257
Pole, II/185
Polyhedra, II/114
  in n-dimensional space, III/143
Polynomial, of best approximation, II/368
  characteristic, of a matrix, III/82
  graph of, I/303
  interpolation, II/269, 286
  irreducibility of, I/277
  Legendre, II/282; III/240
  of matrix, III/92
  symmetric, I/272; III/273
  trigonometric, II/268, 293

Poncelet, J. V., III/130
Pontrjagin, L. S., I/355; III/219–221, 264, 307
Pontrjagin's law of duality, III/220
Popov, A. S., I/5
Positional system, I/14
Positive definite quadratic form, III/89
Postnikov, M. M., III/220
Potential, of force, II/38
  of simple layer, II/39
  velocity, II/13, 156
Potential flow, II/13
Power series, I/175–176; II/141
Power(s) of square matrix, III/92
Prime element, III/345
Prime number(s), II/199, 201
Primitive, I/132
Principal curvature(s), II/81
Principal direction(s), II/81
Principal ideal, III/343
Principal ideal ring, III/345
Principal normal, II/74
Principal part, II/185
Principle(s), Archimedes', III/16–17
  of argument, I/299
  Cantor's, of continuity, III/15
  Cauchy's, III/16
  of continuity, III/15
  Dedekind's, III/15
Probability, II/229, 232
  addition of, II/233
Probability density, II/261
Probability distribution, II/244
Probable deviation, II/235
Proclus, III/99
Product, of cycles, III/280
  derivative of, I/102
  free, III/312
Programming, II/336
Projective differential geometry, II/116
Projective geometry, I/242, 248; III/129–131
Projective plane, in geometry, I/246
  in topology, III/200–201
Projective mapping, I/246
Projective transformation, I/246; II/116; III/127–128
Propagation of waves, II/25
Proper motion, III/271
Proportionality, III/38
Proximity, III/307
Pseudosphere, III/114
Ptolemy, Claudius, I/36, 38, 41
Punched card machines, II/323
Pythagoras, I/21, 25, 31, 46, 187, 195; III/112, 167

Quadratic equation, I/266
  characteristic, I/241
Quadratic form(s), III/84
  canonical, III/88
  equivalent, III/86

Quadratic form(s) (*continued*)
first fundamental, II/105
fundamental, II/105–108; III/168
inertia of, III/88
Qualitative theory of differential equations, I/348; III/214
Quasi-conformal transformation, II/171
Quaternion(s), III/324–325, 326–327
vector part of, III/326, 328

Radical, III/338
of ring, III/347
Radius of convergence, I/177; II/145
Radius vector, II/104
Ramanujan, S., II/203
Random event, II/231, 252, 259
Random process, II/255, 260
Random sampling method, II/371
Random variable, II/244
Rank, of algebra, III/332
of hypercomplex number, III/323
of matrix, III/72
of system of vectors, II/54
Raphael, I/65
Rapidity of convergence, II/308
Raševskiĭ, P. K., III/178
Rational number(s), III/12
Real line, I/246; *see also* Numerical line
Real number, I/27; III/12, 14
Real part of quaternion, III/326, 328
Real point, I/246
Real root, I/271
Real space, III/178
Real variable, function(s) of, III/3
Receiving register, II/355, 359
Reciprocity law, III/246
Rectangular array, III/40
Rectangular coordinates, I/213
Rectangular formula, II/280
Rectangular hyperbola, I/206
Rectilinear generator, I/226
Reduced equation, I/241
Reducible representation, III/317
Reduction, of quadratic form, III/86
to canonical form, I/207
Reflection(s), III/292
in line, III/289
regula falsi (false position), I/304, 306
Regular representation, III/318
Regular spatial point system, III/292
Regular surface, II/76–77
Remainder term, I/125
Representation(s), of associative algebras, III/335, 336
of degree n, III/315
of group, III/314
Reproducing punch, II/325
Residue class ring, III/344
Riccati, J. F., I/330
Riccati equation, I/330

Riemann, B., I/56, 57, 71, 91, 259; II/150, 153, 157, 162, 166, 169–173, 184, 191–193, 203, 208, 209; III/4, 29, 31, 33, 115, 136, 162–177, 179, 183, 184, 187, 218, 306
Riemann hypothesis, II/209
Riemann integral, I/130; III/31
Riemann surface, II/191–194; III/218
Riemannian geometry, III/164
Riemannian metric, III/169
Riemannian space, III/164–166
Riesz, F., I/60
Right coset, III/301
Right ideal, III/337
Right quotient, III/284
Ring, III/342
factor, III/344
without divisors of zero, III/344
noncommutative, III/347
of operators, III/347
semisimple, III/347
Ritz, W., II/53, 54, 306
Rolle, M., I/294, 296
Rolle's theorem, I/294
Roots, calculation of, I/302
multiple, I/293
multiplicity of, I/293
real, I/271
Rotation, of space, III/271
of transformation, II/165
Rotation group, III/278
Rotational flow, II/13
Row matrix, III/40

Saccheri, G. G., III/99, 100, 101, 104
Saddle point, I/346
Šafarevič,. I. R., I/280; III/296
Scalar, I/215; II/4
Scalar field, II/4
Scalar multiplication, III/58
Scalar part of quaternion, III/326, 328
Scalar product, n-dimensional space, III/230
of vectors, I/216; III/57, 328
Schauder, J., III/224
Schlaefi, L., III/143
Schrödinger, E., II/53
Schwartz's inequality, *see* Cauchy-Bunjakovskiĭ inequality
Schweikart, III/100, 104
Second-degree equation, I/266
Second derivative, I/113
Second fundamental quadratic form, II/107, 108
Second-order cone, I/223
Secular equation, III/82
Selberg, A., II/208
Self-adjoint linear operator, III/258
Self-adjoint operator, III/253, 257
Self-intersection, III/199, 200
Semiadder, II/361
Semi-interval, III/18, 23

INDEX

Semi-invariant, I/240
Semilattice, III/348
Semisimple algebra, III/338, 342
Semisimple ring, III/347
Separation of variables, II/28
Sequence, almost everywhere convergent, III/29–30
  fundamental, III/16
  minimizing, II/136
  of operations, II/336
  of real numbers, III/16
Series, I/166
  divergent, I/167
  harmonic, I/169
  power, I/175–176; II/141
  summable, I/167
  Taylor's, I/127
  trigonometric, III/241
  uniformly convergent, I/171
Series system, II/348
Serre, J.–P., III/220
Serret, J. A., I/264
Set(s), III/5
  adherent, III/160
  adjoining, III/222
  boundary point of, III/20
  bounded, III/19
    above, III/19
    below, III/19
      infinite point set, III/21
  cardinality of, III/9
  closed, III/22, 221
  compliment of, III/23
  countable, III/10
  covering of, III/26–27
  density of, III/13
  denumerable, III/10
  descriptive theory of, III/25
  empty, III/6
  equal, III/6
  finite, III/8
  greatest lower bound of, III/19
  infinite, III/8
  intersection, III/7
  of linear functions, III/39
  measurable, III/27
  n-dimensional, III/219
  open, II/22
  ordered, III/18, 347
  perfect, III/24
  theory of, III/4–5
  unbounded, III/19
  union of, III/6
  zero-dimensional, III/219
Set-theoretical topology, III/218–219
Sieve of Eratosthenes, II/204
Similar matrices, III/80
Simple algebra, III/338
Simple group, III/301
Simple subgroup, III/307
Simplex, III/146
Simplicial divisions of manifold, III/207
Simply connected domain, II/168

Simpson, T., II/276–281, 307, 314, 315
Simpson's formula, II/278–279
Simulating machines, II/325
Single-entry table, II/320
Single-valued branch, II/152, 191
Single-valued mapping, III/302
Singular point, I/343; III/213, 216
Singular transformation, III/77
Sitnikov, K. A., III/219, 220
Skew-symmetric matrix, III/341
Šmidt, O. Ju., III/264, 320
Smirnov, Ju. M., III/224
Smirnov, V. I., I/156
Smooth function, II/275; III/37, 75
Smooth manifold, III/215
Smooth surface, II/78
Smoothness, II/115
Šnirel'man, L. G., II/112; III/219
Sobolev, S. L., II/50, 53
Solution(s), admissible, II/48–50
  algebraic, I/265
  of differential equations, I/320
  generalized, II/48–50
  uniqueness of, I/335–336; II/15
Solution space, III/74
Solvability, of group, III–285
  of problems by ruler and compass, I/278
  by radicals, III/296
Solvable group, III/285
Solvable by radicals, I/280
Sorter, II/325
Space, abstract, III/133
  abstract topological, III/133, 136
  affine, III/56, 133
    transformation of, I/231; III/131
  basis of, III/51–52
  color, I/152
  configuration, III/176
  constant curvature, III/174
  curvature of, III/172
  dimension of, III/51, 218
  Euclidean, III/58
  functional, III/157, 223, 254
  general linear, III/48
  Hilbert, III/161, 232–237
  infinite-dimensional, III/157, 161, 232–237
  linear, III/48, 49
  many-dimensional, III/136, 142
  n-dimensional, III/228–229
  phase, I/343; III/154, 204
  real, III/178
  Riemannian, III/164–166
  rotation of, III/271
  topological, III/159, 221, 224
  unitary, III/60–61
  of zero curvature, III/174
Space lattice, II/203
Space-time, III/184
Span, III/54
Spherical functions, III/239
Spiral motion, III/271
Splitting field, III/294

INDEX

Splitting of representation, III/317
Square matrix, III/40
Stability, of computational process, II/315
  of numerical integration, II/370
Stable differential equation, I/349
Stable limit cycle, I/354
Stagnation point, II/161
Standard deviation, II/244
Steady motion, II/153
Stepanov, V. V., I/356; II/36, 257; III/214
Stevin, S., I/42
Stochastic differential equation(s), II/261–264
Stochastic event, II/231
Stochastic variable; see Random variable
Storage, II/335
Straight line, II/99
  equation(s) of, I/220
Straight line congruences, II/110
Stream function, II/155
Streamline, II/154
Structure, of algebras, III/338
  topological, III/220
Structure constant(s), III/333
Sturm, S., I/264, 297, 298
Sturm's theorem, I/297
Subalgebra, III/338
Subgroup(s), III/280–285
  generators of, III/282
  simple, III/307
Šubnikov, A. V., III/293
Subring, III/343
Subroutine, II/345
Subspace, III/53
Subspace spanned by system of vectors, III/54
Substitution, method of, I/139
Subtraction, III/12
Successive approximations method, I/339
Sum, derivative of, I/101
  Fejér, II/298
  of series, I/167
  of sets, III/6
  trigonometric, II/209, 210
  vector, I/215
Summable series, I/167
Sup (*supremum*), I/172
Superposition, method of, II/28
Surface(s), affine classification of, I/233
  analytic, II/101
  closed, III/198, 205, 207
    class of, II/113
    convex, II/113, 115
    orientable, III/211
  congruent, II/103
  connectivity of, III/197
  deformability of, II/115
  deformation of, II/91, 100, 115
  developable, II/97
  of discontinuity, II/49
  equations of, I/218
  family of, II/108–109
  genus of, II/194; III/198

geodesic, III/173
  minimal, II/88
  Möbius, III/199
  nonregular, II/211, 212
  nonsmooth, II/114
  normal, III/198
  regular, II/76–77
  of revolution of least area, II/120
  smooth, II/78
  theory of, II/76
  two-sided, III/198
  with varying Gaussian curvature, II/96–97
  of zero mean curvature, II/88
Surface element, II/96, 106–107
Surface force, II/10
Surface integral, I/162
Suslin, M. Ja., III/25
Sylvester, J. J., I/297; II/208
Symbols, I/10–15
Symmetric group, of permutations, III/279, 296
Symmetric kernel, III/251
Symmetric polynomial, I/272; III/273
Symmetric transformation, III/267
Symmetrical point, III/265
Symmetry, III/264, 270
  axis of, III/265
  bilateral, III/265
  center of, III/266
  central, III/266
  order of, III/265
  mirror, III/265
Symmetry group, III/278, 285
Systems, algebraic, III/263, 349–350
  binary, II/347
  of linear equations, III/61

Tables, II/319
Tabulator, II/325
Tangent, I/92–93; II/66–67
Tangent field, I/333
Tangent hyperplane, III/151
Tangent plane, II/77–78
Tartaglia, N., I/41, 263, 266, 267
Taurinus, III/100, 104
Taylor, B., II/273, 274
Taylor's formula, I/123–124, 153
Taylor's series, I/127
Temperature-gradient vector, II/14
Tensor, II/4
Termwise differentiation, I/171
Termwise integration, I/171
Thales, I/21
Theorem(s), Abel's, I/275; II/143
  Bolzano-Wierstrass, I/288; III/21
  Bonnet's, II/61, 103
  Brouwer's fixed point, III/217
  Budan's, I/297
  Cauchy's, I/176
  central limit, II/247
  Codazzi's, II/61, 103

INDEX

Theorem(s) (*continued*)
  contraction, II/311
  duality, III/220
  Euler's, II/81, 83
  fundamental, of algebra, I/280–292; II/140
  of plane perspectivity, I/243–244
  Gauss; *see* Ostrogradskiĭ, formula of,
  Green's, II/134
  Hopf's, III/216
  Lagrange's, III/301
  mean value, I/120–121
  Meusnier's, II/83
  Peterson's, II/61, 103
  Picard's, II/183
  Rolle's, I/294
  Sturm's, I/297
  uniqueness, for analytic functions, II/187
    of theory of projective transformations, I/243–244
  Wedderburn's, III/339
  Weierstrass', II/288
  Wilson's, II/225
  Žukovskiĭ's, II/159, 162
Theory, algebraic number, II/203
  analytic number, II/203
  of curves, II/61
  of functions of real variable, III/4
  Galois, I/275–278
  Newton's, of diameters, I/193
  of numbers, II/200–204
  of operators, III/253
Third-degree equation, I/266
Three-address system, II/338
Three-dimensional triangulation, III/207
Tihonov, A. N., III/224
Time-pulse code, II/348
Topological groups, III/307
Topological image, III/218
Topological invariance, III/206, 208, 218
Topological invariant(s), III/310–311
Topological mapping, III/221
Topological product, III/203
Topological property, III/194, 197
Topological space, III/159, 221, 224
Topological structure, III/220
Topological transformation, III/194
Topology, III/193
  of closed surfaces, III/218
  combinatorial, III/219–220
    method of, III/204–205
  compatible orientations in, III/206
  cycles in, III/209
  n-dimensional, III/218
Torricelli, E., I/70
Torsion, II/75; III/210
Torus, III/203
Total curvature, II/86; *see also* Gaussian curvature
Total variation of the integral, II/127
Totally ordered set, III/347
Trace, III/319
Tractrix, III/114

Transcendental number, II/203
Transcendental function, II/182
Transformation(s), affine, I/231; III/131
  bicontinuous, III/310
  coefficient matrix of, III/77
  conformal, II/166, 168
  conjugate, III/276, 284
  continuous groups of, III/305
  of coordinates, I/209; III/78–79
  degenerate, III/77
  general affine, I/231
  group, III/300
  of group, III/132
  homogeneous, III/76
  identity, III/282
  inverse, III/76
  linear, III/74
  Lorentz, I/249, 250, 252
  multiplication of, III/273
  nondegenerate, III/77
  one-to-one, III/268
  order of, III/283
  orthogonal, I/237; III/78
  singular, III/77
  symmetric, III/267
  topological, III/194
  unit, III/282
  unitary, III/307
Transpose, III/45
Transposed matrix, III/78
Transposition, III/280
Trapezoidal formula, II/279
Trefftz, J., II/53
Triangle axiom, III/222
Triangle inequality, III/161
Triangulation, III/205
Trigger cell, II/353
Trigonometric function of complex variable, II/145
Trigonometric polynomial, II/268, 293
Trigonometric series, III/241
Trigonometric sums, II/209, 210
Tschirnhansen, E. W., I/263, 270
Twin primes, II/205
Two-dimensional Betti number, III/212
Two-dimensional chain, III/210
Two-fold integral, I/161
Two-sided ideal, III/337
Two-sided surface, III/198

Ulug Begh, I/40
Unbounded set, III/19
Uncountability of the continuum, III/17
Undershooting and overshooting, method of, I/304
Undetermined multipliers, I/154
Uniform deformation of membrane, II/121
Uniform limit, II/49
Uniformly convergent series, I/171
Union, of events, II/232
  of sets, III/6

Uniqueness, of decomposition, III/345
  of solution, I/335-336; II/15
Uniqueness properties, II/187
Uniqueness theorem, for analytic functions,
    II/187
  of theory of projective transformations,
    I/243-244
Unit(s), III/345
Unit element, III/298
Unit matrix, III/70, 335
Unit representation, III/316
Unit transformation, III/282
Unitary space, III/60-61
Unitary transformation, III/307
Unoriented circumference, III/288
Unperturbed motion, I/350
Unsymmetrical body, III/272
Upper bound, III/16
Upper semilattice, III/348
Uryson, P. S., III/219, 220, 222, 224

Vagner, V. V., III/178
van der Waerden, B. L., I/265
von Helmholtz, H. L. F., II/317, 341
Variable, I/43
  change of, I/139
  continuous, I/33
  separation of, II/28
Variable magnitude(s), I/43, 82
Variance, II/244
Variation, II/127
Variational problem, II/130-131
Varičak, III/122
Variety, III/346
Vavilov, S. I., I/66
Veblen, O., III/220
Vector(s), I/214; III/46
  algebra of, I/214; III/327
  coordinates of, III/53
  flux, II/7
  of infinite-dimensional space, III/234
  inverse of, I/215
  length of, III/58
  n-dimensional, III/229
  velocity, II/154
Vector algebra, I/214; III/327
Vector coordinates, I/216

Vector field, II/3; III/212
  continuous, III/213
Vector function, II/104-105
Vector operation(s), III/46
Vector part of quaternion, III/326, 328
Vector product of vectors, III/328
Vector sum, I/215
  of two subspaces, III/54
Velocity potential, II/13, 156
Velocity vector, II/154
Vibrating string, equation of, II/12
Viète, F., I/39, 41, 262, 263, 265, 271, 272;
    III/294
Vinogradov, I. M., I/62; II/203, 204, 209-
    211, 217, 219, 220, 224, 269; III/346
Volume force. II/10
Voronoĭ, G. F., II/204; 147, 346

Wachter, III/100
Wallace, A. R., III/104
Wallis, J., III/99
Wave equation, II/14
Weak discontinuity, II/49
Wedderburn, III/338, 339, 347
Wedderburn's theorem, III/339
Weierstrass, K. T., I/27, 59, 71, 99; II/268,
    288-290; III/16, 17, 21
Weierstrass' principle, III/16
Weierstrass' theorem, II/288
Weight(s), II/240
Weyl, H., III/342
Whole number, I/7; II/199
Wilczynski, E. J., II/116
Wilson, C. T. R., II/87
Wilson's theorem, II/225
Winding number, III/216

Zeno the Eleatic, I/26
Zero, I/14
Zero curvature, space of, III/174
Zero element, III/298
Zero ideal, III/337
Zero-dimensional set, III/219
Zolotarev, E. I., I/264; II/203, 267; III/346
Žukovskiĭ, N. E., I/286; II/159, 160, 162
Žukovskiĭ's theorem, II/159, 162

CONTENTS OF THE SERIES

VOLUME ONE

PART 1

CHAPTER I   A GENERAL VIEW OF MATHEMATICS   1
A. D. Aleksandrov

§1. The Characteristic Features of Mathematics   1
§2. Arithmetic   7
§3. Geometry   19
§4. Arithmetic and Geometry   24
§5. The Age of Elementary Mathematics   35
§6. Mathematics of Variable Magnitudes   45
§7. Contemporary Mathematics   55
Suggested Reading   64

CHAPTER II   ANALYSIS   65
M. A. Lavrent'ev and S. M. Nikol'skiĭ

§1. Introduction   65
§2. Function   73
§3. Limits   80
§4. Continuous Functions   88
§5. Derivative   92
§6. Rules for Differentiation   101
§7. Maximum and Minimum; Investigation of the Graphs of Functions   108
§8. Increment and Differential of a Function   117
§9. Taylor's Formula   123
§10. Integral   128
§11. Indefinite Integrals; the Technique of Integration   137
§12. Functions of Several Variables   142
§13. Generalizations of the Concept of Integral   158
§14. Series   166
Suggested Reading   180

CONTENTS

PART 2

CHAPTER III   ANALYTIC GEOMETRY   183
B. N. Delone

§1. Introduction   *183*
§2. Descartes' Two Fundamental Concepts   *184*
§3. Elementary Problems   *186*
§4. Discussion of Curves Represented by First- and Second-Degree Equations   *188*
§5. Descartes' Method of Solving Third- and Fourth-Degree Algebraic Equations   *190*
§6. Newton's General Theory of Diameters   *193*
§7. Ellipse, Hyperbola, and Parabola   *195*
§8. The Reduction of the General Second-Degree Equation to Canonical Form   *207*
§9. The Representation of Forces, Velocities, and Accelerations by Triples of Numbers; Theory of Vectors   *213*
§10. Analytic Geometry in Space; Equations of a Surface in Space and Equations of a Curve   *218*
§11. Affine and Orthogonal Transformations   *227*
§12. Theory of Invariants   *238*
§13. Projective Geometry   *242*
§14. Lorentz Transformations   *249*
Conclusion   *257*
Suggested Reading   *259*

CHAPTER IV   ALGEBRA: THEORY OF ALGEBRAIC EQUATIONS   261
B. N. Delone

§1. Introduction   *261*
§2. Algebraic Solution of an Equation   *265*
§3. The Fundamental Theorem of Algebra   *280*
§4. Investigation of the Distribution of the Roots of a Polynomial on the Complex Plane   *292*
§5. Approximate Calculation of Roots   *302*
Suggested Reading   *309*

CHAPTER V   ORDINARY DIFFERENTIAL EQUATIONS   311
I. G. Petrovskiĭ

§1. Introduction   *311*
§2. Linear Differential Equations with Constant Coefficients   *323*

CONTENTS

§3. Some General Remarks on the Formation and Solution of Differential Equations   330
§4. Geometric Interpretation of the Problem of Integrating Differential Equations; Generalization of the Problem   332
§5. Existence and Uniqueness of the Solution of a Differential Equation; Approximate Solution of Equations   335
§6. Singular Points   343
§7. Qualitative Theory of Ordinary Differential Equations   348
Suggested Reading   356

INDEX   357

CONTENTS OF THE SERIES   377

CONTENTS OF THE SERIES

VOLUME TWO

PART 3

CHAPTER VI PARTIAL DIFFERENTIAL EQUATIONS 3
S. L. Sobolev *and* O. A. Ladyzenskaja

§1. *Introduction* *3*
§2. *The Simplest Equations of Mathematical Physics* *5*
§3. *Initial-Value and Boundary-Value Problems; Uniqueness of a Solution* *15*
§4. *The Propagation of Waves* *25*
§5. *Methods of Constructing Solutions* *27*
§6. *Generalized Solutions* *48*
Suggested Reading *54*

CHAPTER VII CURVES AND SURFACES 57
A. D. Aleksandrov

§1. *Topics and Methods in the Theory of Curves and Surfaces* *57*
§2. *The Theory of Curves* *61*
§3. *Basic Concepts in the Theory of Surfaces* *76*
§4. *Intrinsic Geometry and Deformation of Surfaces* *91*
§5. *New Developments in the Theory of Curves and Surfaces* *108*
Suggested Reading *117*

CHAPTER VIII THE CALCULUS OF VARIATIONS 119
V. I. Krylov

§1. *Introduction* *119*
§2. *The Differential Equations of the Calculus of Variations* *124*

CONTENTS

§3. *Methods of Approximate Solution of Problems in the Calculus of Variations* 135
Suggested Reading 138

CHAPTER IX   FUNCTIONS OF A COMPLEX VARIABLE   139
M. V. Keldyš

§1. *Complex Numbers and Functions of a Complex Variable* 139
§2. *The Connection Between Functions of a Complex Variable and the Problems of Mathematical Physics* 153
§3. *The Connection of Functions of a Complex Variable with Geometry* 163
§4. *The Line Integral; Cauchy's Formula and Its Corollaries* 174
§5. *Uniqueness Properties and Analytic Continuation* 187
§6. *Conclusion* 194
Suggested Reading 195

PART 4

CHAPTER X   PRIME NUMBERS   199
K. K. Mardzanisvili *and* A. B. Postnikov

§1. *The Study of the Theory of Numbers* 199
§2. *The Investigation of Problems Concerning Prime Numbers* 204
§3. *Čebyšev's Method* 211
§4. *Vinogradov's Method* 217
§5. *Decomposition of Integers into the Sum of Two Squares; Complex Integers* 225
Suggested Reading 228

CHAPTER XI   THE THEORY OF PROBABILITY   229
A. N. Kolmogorov

§1. *The Laws of Probability* 229
§2. *The Axioms and Basic Formulas of the Elementary Theory of Probability* 231
§3. *The Law of Large Numbers and Limit Theorems* 238
§4. *Further Remarks on the Basic Concepts of the Theory of Probability* 247

CONTENTS

§5. Deterministic and Random Processes   255
§6. Random Processes of Markov Type   260
     Suggested Reading   264

CHAPTER XII   APPROXIMATIONS OF FUNCTIONS   265
S. M. Nikol'skiĭ

§1. Introduction   265
§2. Interpolation Polynomials   269
§3. Approximation of Definite Integrals   276
§4. The Čebyšev Concept of Best Uniform Approximation   282
§5. The Čebyšev Polynomials Deviating Least from Zero   285
§6. The Theorem of Weierstrass; the Best Approximation to a Function as Related to Its Properties of Differentiability   288
§7. Fourier Series   291
§8. Approximation in the Sense of the Mean Square   298
     Suggested Reading   302

CHAPTER XIII   APPROXIMATION METHODS AND COMPUTING TECHNIQUES   303
V. I. Krylov

§1. Approximation and Numerical Methods   303
§2. The Simplest Auxiliary Means of Computation   319
     Suggested Reading   329

CHAPTER XIV   ELECTRONIC COMPUTING MACHINES   331
S. A. Lebedev and L. V. Kantorovič

§1. Purposes and Basic Principles of the Operation of Electronic Computers   331
§2. Programming and Coding for High-Speed Electronic Machines   336
§3. Technical Principles of the Various Units of a High-Speed Computing Machine   350
§4. Prospects for the Development and Use of Electronic Computing Machines   365
     Suggested Reading   374

INDEX   375

CONTENTS OF THE SERIES   395

Printed in the United States
By Bookmasters